Bernd Ulmann
Analog Computing

Also of interest

Analog and Hybrid Computer Programming
Bernd Ulmann, 2020
ISBN 978-3-11-066207-8, e-ISBN 978-3-11-066220-7

Quantum Information Theory
Concepts and Methods
Joseph M. Renes, 2022
ISBN 978-3-11-057024-3, e-ISBN (PDF) 978-3-11-057025-0

Multi-level Mixed-Integer Optimization
Parametric Programming Approach
Styliani Avraamidou, Efstratios Pistikopoulos, 2022
ISBN 978-3-11-076030-9, e-ISBN (PDF) 978-3-11-076031-6

Automata Theory and Formal Languages
Wladyslaw Homenda und Witold Pedrycz, 2022
ISBN 978-3-11-075227-4, e-ISBN (PDF) 978-3-11-075230-4

High Performance Parallel Runtimes
Design and Implementation
Michael Klemm, Jim Cownie, 2021
ISBN 978-3-11-063268-2, e-ISBN (PDF) 978-3-11-063272-9

Algorithms
Design and Analysis
Sushil C. Dimri, Preeti Malik, Mangey Ram, 2021
ISBN 978-3-11-069341-6, e-ISBN (PDF) 978-3-11-069360-7

Bernd Ulmann
Analog Computing

2nd edition

Mathematics Subject Classification 2010
Primary: 34-04, 35-04; Secondary: 92C45, 92D25, 34C28, 37D45

Author
Prof. Dr. Bernd Ulmann
Schwalbacher Str. 31
65307 Bad Schwalbach
ulmann@analogparadigm.com

ISBN 978-3-11-078761-0
e-ISBN (PDF) 978-3-11-078774-0
e-ISBN (EPUB) 978-3-11-078787-0

Library of Congress Control Number: 2022946448

Bibliographic information published by the Deutsche Nationalbibliothek
The Deutsche Nationalbibliothek lists this publication in the Deutsche Nationalbibliografie; detailed bibliographic data are available on the Internet at http://dnb.dnb.de.

© 2023 Walter de Gruyter GmbH, Berlin/Boston
Cover image: Bernd Ulmann
Printing and binding: CPI books GmbH, Leck

www.degruyter.com

To my beloved wife Rikka

Acknowledgments

This book would not have been possible without the support and help of many people. First of all, I would like to thank my wife RIKKA MITSAM, who not only did a lot of proofreading and a terrific job in preparing many of the pen-and-ink drawings, but also never complained about being neglected although I lived the last months more or less in seclusion writing this book.

I am particularly grateful for the support and help of JENS BREITENBACH, who did a magnificent job at proofreading and provided many suggestions and improvements enhancing the text significantly. He also spotted many LaTeX-sins of mine, thus improving the overall appearance of this book considerably.

In addition to that, I would like to thank BENJAMIN BARNICKEL, ARNE CHARLET, DANIELA KOCH, THERESA SZCZEPANSKI, Dr. REINHARD STEFFENS and FRANCIS MASSEN for their invaluable help in proofreading.

Additionally, I would like to thank TORE SINDING BEKKEDAL, who took the picture shown in figure 8.35, TIM ROBINSON, who built the incredible Meccano-Differential Analyser shown in figure 2.22, TIBOR FLORESTAN PLUTO, who took the photo shown on the title page, ROBERT LIMES, who took the picture shown in figure 9.3, and BRUCE BAKER, who donated the pictures shown in figures 13.60, 13.62, and 13.63, for their permissions to use the aforementioned pictures.

Without the continuous encouragement of Dr. habil. KARL SCHLAGENHAUF and his invaluable suggestions this book might very well not exist at all.

Last but not least, I would like to thank Prof. Dr. WOLFGANG GILOI, Prof. Dr. RUDOLF LAUBER, Dr. ADOLF KLEY, Prof. Dr. GÜNTER MEYER-BRÖTZ and MANFRED KLITTICH for sharing their memories of the early days of analog computing at Telefunken, etc. Furthermore, JEROEN BRINKMANN introduced me to the "Hammer Computer" and BERT BROUWER offered many the insights into solving partial differential equations in conjunction with heat transfer problems.

This book was typeset with LaTeX. Most schematics were drawn using EAGLE and ARNO JACOB's wonderful analog computing symbol library, other vector graphics were created with xfig.

Registered names, trademarks, designations, etc., used in this book, even when not specifically marked as such, are not to be considered unprotected by law.

Preface to the 2nd edition

This second edition of "Analog Computing" would not exist if it wasn't for Dr. Damiano Sacco, acquisition editor at DeGruyter, who asked me to prepare a new edition of "Analog Computing", which has established itself as a standard textbook since it was first published in 2013.

A lot has happened since then in the world of analog computing. First of all many "new" historic sources, papers, and photographs have become available since 2013. Accordingly, the bibliography of this 2nd edition has been expanded by more then 200 additional sources. Second, and even more importantly, interest in analog computing has grown enormously. This computing paradigm is about to change the world of computing in the near future with a plethora of interesting and commercially important applications ranging from low power computing as required for medical implants to high performance computing, artificial intelligence, and many, many more. It seems even plausible that analog computers might be able to do things typically ascribed to quantum computers while being much simpler to implement, run, and program.

Bringing back analog computers in much more advanced forms than their historic ancestors will change the world of computing drastically and forever. They will not replace the ubiquitous stored-program digital computers but they will complement them, thus making it possible to solve problems that are currently out of reach for standalone digital computers.

The author is especially indebted to Dr. Chris Giles for many valuable discussions and his meticulous proofreading of this 2nd edition and to Nicole Matje for her valuable corrections and suggestions. The author would also like to thank Achim Dassow for information about the QK-329 beam-deflection tube and additional information on early multiplication devices. Rainer Glaschick provided much background information on early differential analysers and on the hyperbolic field tube. Oliver Bach also proofread the book and took special care of the bibliography making sure that it is consistent and correct. Prof. Dr. Dirk Killat contributed figure 12.9. The new cover picture was taken by Tibor Florestan Pluto and shows the author with a GTE-22 analog computer from the late 1960s.

Contents

Acknowledgments —— VII

Preface to the 2nd edition —— IX

1	**Introduction** —— **1**	
1.1	Outline —— 1	
1.2	The notion of analog computing —— 2	
1.3	Direct and indirect analogies —— 4	
2	**Mechanical analog computers** —— **9**	
2.1	Astrolabes —— 9	
2.2	The Antikythera mechanism —— 9	
2.3	Slide rules —— 11	
2.4	Planimeters —— 13	
2.5	Mechanical computing elements —— 17	
2.5.1	Function generation —— 18	
2.5.2	Differential gears —— 20	
2.5.3	Integrators —— 21	
2.5.4	Multipliers —— 24	
2.6	Harmonic synthesizers and analysers —— 25	
2.7	Mechanical fire control systems —— 29	
2.8	Differential analysers —— 32	
3	**The first electronic analog computers** —— **41**	
3.1	Helmut Hoelzer —— 41	
3.1.1	The "Mischgerät" —— 42	
3.1.2	Hoelzer's analog computer —— 48	
3.2	George A. Philbrick's Polyphemus —— 57	
3.3	Electronic fire control systems —— 61	
3.4	MIT —— 67	
3.5	The Caltech Computer —— 68	
4	**Basic computing elements** —— **73**	
4.1	Operational amplifiers —— 73	
4.1.1	Early operational amplifiers —— 77	
4.1.2	Drift stabilisation —— 80	
4.2	Summers —— 85	
4.3	Integrators —— 89	
4.4	Coefficient potentiometers —— 92

4.5	Function generators —— 97	
4.5.1	Servo function generators —— 97	
4.5.2	Curve followers —— 98	
4.5.3	Photoformers —— 99	
4.5.4	Varistor function generators —— 100	
4.5.5	Diode function generators —— 100	
4.5.6	Inverse functions —— 102	
4.5.7	Functions of two variables —— 103	
4.6	Multiplication —— 105	
4.6.1	Servo multipliers —— 105	
4.6.2	Crossed-fields electron-beam multiplier —— 106	
4.6.3	Hyperbolic field multiplier —— 107	
4.6.4	Other multiplication tubes —— 108	
4.6.5	Time division multipliers —— 109	
4.6.6	Logarithmic multipliers —— 110	
4.6.7	Quarter square multipliers —— 111	
4.6.8	Other multiplication schemes —— 114	
4.7	Division and square root —— 114	
4.8	Comparators —— 115	
4.9	Limiters —— 116	
4.10	Resolvers —— 117	
4.11	Time delay —— 117	
4.12	Random noise generators —— 120	
4.13	Output devices —— 121	
5	**The anatomy of a classic analog computer —— 123**	
5.1	Analog patch panel —— 123	
5.2	Function generators —— 124	
5.3	Digital patch panel and controls —— 125	
5.4	Readout —— 127	
5.5	Control —— 129	
5.6	Performing a computation —— 131	
6	**Some typical analog computers —— 133**	
6.1	Telefunken RA 1 —— 133	
6.2	GAP/R analog computers —— 136	
6.3	EAI 231R —— 138	
6.4	Early transistorised systems —— 141	
6.5	Later analog computers —— 147	
6.6	THE ANALOG THING —— 150	

7	**Programming** —— 151	
7.1	Basic approach —— 151	
7.2	Kelvin's feedback technique —— 153	
7.3	Substitution method —— 155	
7.4	Partial differential equations —— 157	
7.4.1	Quotient of differences —— 158	
7.4.2	Separation of variables —— 160	
7.5	Scaling —— 162	
8	**Programming examples** —— 165	
8.1	Solving $\ddot{y} + \omega^2 y = 0$ —— 165	
8.2	Sweep generator —— 166	
8.3	Mass-spring-damper system —— 168	
8.4	Predator and prey —— 172	
8.5	Simulation of an epidemic —— 174	
8.6	Bouncing ball —— 176	
8.7	Car suspension —— 181	
8.8	Lorenz attractor —— 187	
8.9	Mathieu's equation —— 190	
8.10	Projection of rotating bodies —— 194	
8.11	Conformal mapping —— 196	
9	**Hybrid computers** —— 201	
9.1	Systems —— 201	
9.2	Programming —— 205	
9.3	Example —— 206	
10	**Digital differential analysers** —— 209	
10.1	Basic computing elements —— 210	
10.1.1	Integrators —— 210	
10.1.2	Servos —— 213	
10.1.3	Summers —— 214	
10.1.4	Additional elements —— 214	
10.2	Programming examples —— 215	
10.3	Problems —— 216	
10.4	Systems —— 217	
10.4.1	MADDIDA —— 217	
10.4.2	Bendix D-12 —— 220	
10.4.3	CORSAIR —— 224	
10.4.4	TRICE —— 225	

11 Stochastic computing — 229

12 Simulation of analog computers — 233
12.1 Basics — 234
12.2 DDA programming system for the IBM 7074 — 234
12.3 CSMP — 237
12.4 Modern approaches — 240

13 Applications — 243
13.1 Mathematics — 243
13.1.1 Differential equations — 243
13.1.2 Integral equations — 244
13.1.3 Roots of polynomials — 247
13.1.4 Orthogonal functions — 247
13.1.5 Linear algebra — 248
13.1.6 Eigenvalues and -vectors — 249
13.1.7 Fourier synthesis and analysis — 250
13.1.8 Random processes and Monte-Carlo simulations — 251
13.1.9 Optimisation and operational research — 252
13.1.10 Display of complex shapes — 253
13.2 Physics — 253
13.2.1 Orbit calculations — 255
13.2.2 Particle trajectories and plasma physics — 255
13.2.3 Optics — 258
13.2.4 Heat-transfer — 258
13.2.5 Fallout prediction — 262
13.2.6 Semiconductor research — 262
13.2.7 Ferromagnetic films — 264
13.3 Chemistry — 264
13.3.1 Reaction kinetics — 264
13.3.2 Quantum chemistry — 265
13.4 Mechanics and engineering — 266
13.4.1 Vibrations — 267
13.4.2 Shock absorbers — 268
13.4.3 Earthquake simulation — 268
13.4.4 Rotating systems and gears — 269
13.4.5 Compressors — 270
13.4.6 Crank mechanisms and linkages — 271
13.4.7 Non-destructive testing — 272
13.4.8 Ductile deformation — 272
13.4.9 Pneumatic and hydraulic systems — 273
13.4.10 Control of machine tools — 276

13.4.11	Servo systems —— 277	
13.5	Colour matching —— 278	
13.6	Nuclear technology —— 278	
13.6.1	Research —— 279	
13.6.2	Reactor/neutron kinetics —— 280	
13.6.3	Training —— 281	
13.6.4	Control —— 282	
13.6.5	Enrichment —— 283	
13.7	Biology and medicine —— 284	
13.7.1	Ecosystems —— 284	
13.7.2	Metabolism research —— 285	
13.7.3	Cardiovascular systems —— 285	
13.7.4	Closed loop control studies —— 286	
13.7.5	Neurophysiology —— 286	
13.7.6	Epidemiology —— 288	
13.7.7	Aerospace medicine —— 289	
13.7.8	Locomotor systems —— 289	
13.7.9	Dosimetry —— 290	
13.8	Geology and marine science —— 290	
13.8.1	Oil and gas reservoirs —— 290	
13.8.2	Seismology —— 291	
13.8.3	Ray tracing —— 292	
13.9	Economics —— 292	
13.10	Power engineering —— 295	
13.10.1	Generators —— 295	
13.10.2	Transformers —— 296	
13.10.3	Power inverters and rectifiers —— 296	
13.10.4	Transmission lines —— 297	
13.10.5	Frequency control —— 297	
13.10.6	Power grid simulation —— 298	
13.10.7	Power station simulation —— 300	
13.10.8	Dispatch computers —— 301	
13.11	Electronics and telecommunications —— 301	
13.11.1	Circuit simulation —— 301	
13.11.2	Frequency response —— 303	
13.11.3	Filter design —— 304	
13.11.4	Modulators and demodulators —— 304	
13.11.5	Antenna and radar systems —— 305	
13.12	Automation —— 305	
13.12.1	Data processing —— 306	
13.12.2	Correlation analysis —— 306	
13.12.3	Closed loop control and servo systems —— 307	

13.12.4	Sampling systems —— 307
13.12.5	Embedded systems —— 308
13.13	Process engineering —— 308
13.13.1	Mixing tanks, heat exchangers, evaporators, and distillation columns —— 309
13.13.2	Adaptive control —— 311
13.13.3	Parameter determination and optimisation —— 311
13.13.4	Plant startup simulation —— 312
13.14	Transport systems —— 313
13.14.1	Automotive engineering —— 313
13.14.2	Railway vehicles —— 317
13.14.3	Hovercrafts and Maglevs —— 317
13.14.4	Nautics —— 318
13.15	Aeronautical engineering —— 320
13.15.1	Landing gears —— 321
13.15.2	Aircraft arresting gear systems —— 323
13.15.3	Jet engines —— 323
13.15.4	Helicopters —— 323
13.15.5	Flutter simulations —— 324
13.15.6	Flight simulation —— 325
13.15.7	Airborne simulators —— 334
13.15.8	Guidance and control —— 336
13.15.9	Miscellaneous —— 338
13.16	Rocketry —— 338
13.16.1	Rocket motor simulation —— 338
13.16.2	Rocket simulation —— 339
13.16.3	Real-time data analysis —— 343
13.16.4	Spacecraft manoeuvres —— 343
13.16.5	Mercury, Gemini, and Apollo —— 345
13.17	Military applications —— 347
13.18	Education —— 348
13.19	Arts, entertainment, and music —— 349
13.19.1	Arts —— 349
13.19.2	Entertainment —— 352
13.19.3	Music —— 354
13.20	Analog computer centers —— 354

14	**Future and opportunities —— 357**
14.1	Challenges —— 359
14.2	Applications —— 361
14.3	Recent work —— 362

Bibliography —— 365

Index —— 427

"An analog computer is a thing of beauty and a joy forever."[1]

---

[1] JOHN H. MCLEOD, SUZETTE MCLEOD, "The Simulation Council Newsletter", in *Instruments and Automation*, Vol. 31, March 1958, p. 488.

1 Introduction

1.1 Outline

A book on analog computing and analog computers? You might ask: Isn't that 60 years late? No, it isn't – although the beautiful analog computers of the past are long since history and only few have been preserved in museum collections, the idea of analog computing is still a marvel of elegance and the following chapters will show that it has a bright and fruitful future.

The intention of this book is twofold: It gives a comprehensive description of the history and technology of classic analog computing but also shows the particular strengths of the analog computation paradigm, which, combined with current state of the art digital circuitry, will find applications in areas such as low power computing, high performance computing and maybe most importantly in *artificial intelligence* (*AI*) and many other fields.

The following chapters first introduce the notion of analog computing before describing the early development of analog computers starting with mechanical analog computers like the Antikythera mechanism, which was built around 100 B.C., and ending with the first analog electronic[1] analog computers developed by HELMUT HOELZER in Germany and GEORGE A. PHILBRICK in the United States.

Next, the basic elements of a typical analog computer are described followed by two chapters showing the anatomy of typical analog computers, ranging from classic systems to more recent implementations, and showing examples of some systems.

The next chapter gives an introduction to analog computer programming[2] followed by a number of practical programming examples ranging from the solution of simple differential equations to the simulation of more complex and non-linear systems.

Hybrid computers (analog computers coupled with stored-program digital computers) are covered in the next chapter. This is followed by a treatment of *Digital Differential Analysers* (digital implementations of analog computers), a chapter on stochastic computing, and a chapter on the simulation of analog computers on classic stored-program digital computers.

The next chapter covers a plethora of classic and current applications of analog computing.

---

[1] The notion of an *analog electronic analog computer* may look like a pleonasm, which it is not, as the following section will show.
[2] A much more in-depth treatment of analog and hybrid computer programming can be found in [ULMANN 2020/1].

The last chapter of this book covers the decline of analog computing in the late 1970s/early 1980s and the potentially bright future of analog computing in the 21st century.

1.2 The notion of analog computing

First of all it should be noted that the common distinction between *digital* and *analog* computers, based on the way values are represented, is not correct. It is often said that digital computers differ from analog computers by their way of representing numbers as sequences of bits (binary digits), while electronic analog computers work with continuous voltages or currents to represent variables. This erroneous view has even found its way into some encyclopedias.

Apart from the fact that even voltages or currents are not really continuous – eventually an operation like integration boils down to charging a capacitor with discrete electron charges – some analog computers have used a bit-wise value representation and have been implemented using purely digital elements.

If the type of values used in a computation – discrete versus continuous – is not the distinguishing feature, what else could be used to differentiate between *digital* and *analog* computers? It turns out that the difference is to be found in the structure of these two classes of machines: In our modern sense of the word, a digital computer's constituent elements have a fixed structure and it solves problems by executing a sequence (or sequences) of instructions that implement an algorithm. These instructions are read from some kind of memory, thus, a better term for this kind of computing machine would be *stored-program digital computer* since this describes both features of such a machine: Its ability to execute instructions fetched from a memory subsystem and working with numbers that are represented as streams of digits.[3]

An analog computer on the other hand is based on a completely different paradigm: Its internal structure is not fixed – in fact, a problem is solved on such a machine by changing its structure in a suitable way to generate a *model*, an *analog* of the problem.[4] This analog is then used to *analyse* or *simulate* the problem to be solved.[5]

Thus, the structure of an analog computer that has been set up to tackle a specific problem represents the problem itself while a stored-program digital

---

[3] Today these numbers are normally represented by *binary digits*, *bits* for short.
[4] [TSE et al. 1964, p. 333] characterized these analogs or analogies as follows: "*The term 'analogy' is defined to mean similarity of relation without identity.*"
[5] The path from analogy-making to modelling of a problem is treated comprehensivly by [CARE 2008].

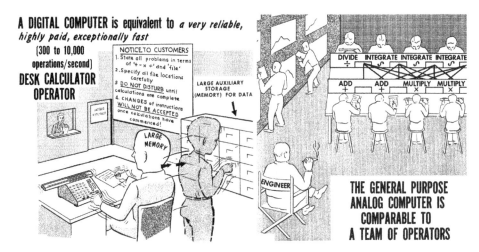

Fig. 1.1. Comparison of the basic structure of stored-program digital computers and analog computers (see [TRUITT et al. 1960, p. 1-40] and [TRUITT et al. 1960, p. 1-41])

computer keeps its structure and only its controlling program changes. This is summarized by CHARLESWORTH[6] as follows:

> "An analogue computer is a piece of equipment whose component parts can be arranged to satisfy a given set of equations, usually simultaneous ordinary differential equations."

Similarly [BERKELEY et al. 1956, p. 75] states that

> "[a]nalog computers, as the name is intended to imply, compute by means of setups that are analog of the problems to be solved."

Figure 1.1 shows this basic difference in architecture and operation between digital and analog computers.

Consequently, it is perfectly possible to build digital analog computers and this has been done in several ways.[7] In fact such machines may play a substantial role in the future when high precision is of the utmost importance and energy efficiency is a secondary consideration.

Employing the techniques of building models, analogs of problems to be solved or analysed, which have been developed through more than 50 years in the context of our current digital technology, can and will lead to systems with exceptional computational power as well as low power consumption.

---

[6] See [CHARLESWORTH et al. 1974, p. xi].
[7] Cf. sections 10 and 13.15.8.

	Basic technology	
	Analog electronic	Digital electronic
Stored-program control	N/A	stored-program (memory programmed) digital computer
Setting up an analog	traditional analog electronic analog computer	digital differential analyser

Table 1.1. Types of computing machines based on an analog/digital electronic implementation with control based on either a stored-program concept or the implementation of an analog.

Table 1.1 shows the four basic possible combinations of analog/digital implementation technology and stored-program control vs. setting up an analog. Of these combinations only three are of practical interest:

1. The modern stored-program digital computer,
2. the traditional analog electronic analog computer, which will be called an *analog computer* for simplicity in the following text, and, finally,
3. the *Digital Differential Analyser*,[8] which will be described in more detail in chapter 10.[9]

1.3 Direct and indirect analogies

When talking about analogies, it is necessary to distinguish between *direct* and *indirect* analogies. These two terms describe two extremes of abstraction levels in building analogies.

In the strict sense of the word direct analogies are models that are based on the same physical principles as the underlying problem to be solved just with a different scaling regarding size or time of a simulation. Well-known examples of such direct analogies are the determination of minimal surfaces using soap films,[10] the evaluation of tensile structures as they are used for roof structures

---

[8] *DDA* for short.
[9] It will be shown that DDAs are more capable machines than traditional analog computers since they can deal well with the highly important class of partial differential equations, which analog computers can do only with considerable difficulty and often only by means of discretisation.
[10] See [Bild der Wissenschaft 1970] for examples.

like the one built for the Olympic stadium in Munich,[11] wind tunnel models for the evaluation of aerodynamic properties of aircraft and rockets and many more. Other direct analogs employed *electrolytic tanks*. These are reservoirs filled with an electrolytic liquid with embedded electrodes to generate a desired potential distribution within the liquid. Using a two- or three-dimensional sensor carriage, much like an xy-plotter, the potential at any given coordinate within the tank can be determined. Such electrolytic tanks were widely used to solve problems in nuclear engineering, etc.

Over time the notion of direct analog analogies was also applied to computers consisting of networks of passive electronic components such as resistors, capacitors and inductors. [PASCHKIS et al. 1968, p. 5] defines a direct analog computer as being

> "based on the identity of the equations describing two or more systems and carrying out measurements on that system which appears most convenient for that purpose [...] In the direct analog, there is a one-to-one relationship between the (passive) components and the physical properties of the several parts of the prime system."

Due to their very nature such direct analogs are not very versatile and were often built and employed for a highly specific purpose.[12] Figure 1.2 shows a wonderful example for a direct analogy, a string-weight model used by ANTONI GAUDÍ[13] during the design and construction of the the Colónia Güell church. This model was built in 1908 at a scale of 1:10 with the weights scaled down by a factor of 10^{-4}. Such models are hanging down and simulate the pressure forces acting on pillars and columns by corresponding tensile forces through the maze of strings with their attached weights.[14]

In contrast, indirect analogies exhibit a much higher degree of abstraction and are thus much more versatile. The – mostly analog electronic – analog computers used for setting up indirect analogies are truly universal machines and cover a wide range of possible applications.[15] This higher level of abstraction makes the programming of this class of machines quite challenging since their setup does not

---

[11] The roof structure of the Olympic stadium in Munich was modelled to a large extent using curtain net lace as well as soap bubbles for determining the structure of single roof tiles.

[12] More detailed information about this basic class of analogs can be found in [JACKSON 1960, pp. 319 ff.], [PASCHKIS et al. 1968], [MASTER et al. 1955], [LARROWE 1955] and [KARPLUS 1958].

[13] 06/25/1852–06/10/1926

[14] See [KRÄMER 1989] for more details on this particular model. This approach was by no means new even in GAUDÍ's time. See [HAVIL 2019, pp. 173 ff.] for a mathematical and historic perspective.

[15] One of the earliest publications on the use of electronic analogs to simulate mechanical and acoustical systems was [OLSON 1943].

Fig. 1.2. Scale model for the Colonia Güell church

bear any direct resemblance of the problem to be solved. Therefore a thorough mathematical description of the basic problem is required as a precondition for programming an indirect analog computer,[16] as in the case of our modern stored-program digital computers.

Nevertheless, the level of abstraction required for the successful application of analog computers is still relatively small compared with the algorithmic approach of stored-program digital computers. Last but not least, analog computers, be they direct or indirect, are models.

Due to the fact that analog computers work by acting as a model for a given problem that is represented by direct or indirect means, the amount of circuitry necessary for a simulation is determined by the complexity of the underlying problem.

Accordingly, analog computers are not capable of the trade off between time to solution on the one hand and complexity of the underlying problem on the other that is characteristic of stored-program digital computers. This is both a

---

16 Direct analogs can also be employed in cases where no complete mathematical description of the problem to be solved exists – this may be caused by a principle lack of understanding or by the sheer complexity of the underlying problem. So in some cases direct analogs may even be employed today with success.

curse and a blessing: The curse being that an analog computer consisting of a given number of computing elements cannot solve a problem that requires more computing elements to be implemented. The blessing is that the time to solution on an analog computer is more or less constant and is not related to the size of the underlying problem.

Thus, large problems require large analog computers regardless of the acceptable time to solution – some classic problem areas, especially those found in aerospace and applications in the chemical industry, required well over 1 000 computing elements resulting in substantial, if not giant, analog computers.

In addition to this, a stored-program computer can always exchange compute time for precision – something an analog computer also cannot normally do.[17] The precision of an analog computer is given by its particular implementation and typically does not exceed about three to four decimal places for the variables involved in a computation.

---

[17] This does not hold true for digital differential analysers, cf. section 10.

2 Mechanical analog computers

The earliest analog computers were mechanical in their very nature but were far from being simple. In fact many mechanical analog computers were successfully employed to tackle complex problems ranging from peaceful tide computations to war-time applications like bomb trajectories, fire control, etc. The following sections give a short overview of the era of mechanical analog computers without going too much into detail since mechanical analogs will serve just as a prelude to this book's main theme of electronic analog computers.

2.1 Astrolabes

As early as about 150 B.C. the basics of *astrolabes* were developed. Such devices are basically inclinometers with some additional mechanics to model basic properties of spherical astronomy. Astrolabes are based on the apparent motion of celestial bodies, i.e., the observation that the paths described by stars in the sky are basically circles. Thus, the most common type of astrolabe is the *planispheric astrolabe*, developed in medieval times, which projects the firmament to the equatorial plane.

Using such an instrument it is possible to determine the position of some celestial bodies at a given time. As a navigational tool the planispheric astrolabe is far too imprecise. Nevertheless, it has been used to roughly located the stars for getting navigational fixes. Detailed information about astrolabes can be found in [DODD 1969] and [J. E. MORRISON 2007].

2.2 The Antikythera mechanism

More than 120 years ago, in 1900, sponge divers found a lump of corroded gears in a Roman ship wreck, which carried treasures from Greece dating back to about 100 B.C. It turned out that these were the remains of one of the most complicated mechanical and mathematical devices ever. Due to its location near the Greek island Antikythera (Αντικύθηρα) this impressive machine became known as the *Antikythera mechanism*. Figure 2.1 shows the main fragment of this early analog computer in its current state of preservation.[1]

---

[1] Picture taken by TILEMAHOS EFTHIMIADIS, protected by the *Creative Commons Attribution 2.0 Generic license*.

Fig. 2.1. Main fragment of the Antikythera mechanism as displayed in the National Archaeological Museum, Athens, Greece

Intrigued by this find, DEREK DE SOLLA PRICE[2] started investigating the inner workings of this device and summarized his astonishing discoveries as follows:[3]

> "*It is a bit frightening to know that just before the fall of their great civilization the ancient Greeks had come so close to our age, not only in their thought, but also in their scientific technology.*"

It turned out that the Antikythera mechanism was ahead of its time by at least 1 000 years. It is of such high complexity that recent research using modern X-ray tomography techniques, etc.[4] continues to deliver new insights. New capabilities and details were discovered as late as in 2021/2022.[5] The device modelled the movements of several celestial bodies, even taking into account various anomalies,

---

2 01/22/1922–09/03/1983
3 See [FREETH 2008, p. 7].
4 Cf. http://www.antikythera-mechanism.gr/.
5 See [FREETH et al. 2021] and [FREETH 2022].

which required differential gears[6] and much more complicated epicyclic gearing. This astonishing complexity of the Antikythera mechanism led MIKE EDMUNDS[7] to the following statement:

> "Nothing as sophisticated and complex is known for another thousand years. This machine rewrites the history of technology. It is a witness to a revolution in human thought."

This intricate mechanism allowed the calculation of sun and moon positions at given dates and phases of the moon as well as the prediction of solar and lunar eclipses.[8] The implementation of these functions required more than 30 gears, manufactured with extraordinary precision.[9]

2.3 Slide rules

One of the most common, simplest and well-known analog computers is the *slide rule*,[10] which comes in basically two configurations, *linear, circular*, and *helical*.[11] The basic idea of a slide rule is to reduce the problem of multiplication and division to that of addition and subtraction by employing logarithmically divided scales that may be displaced accordingly to each other in a lateral direction. The analog setup in this case is to mechanize the relation

$$\log(ab) = \log(a) + \log(b)$$

by using two logarithmic scales. Following the development of the logarithm by JOHN NAPIER[12] and HENRY BRIGGS,[13] who introduced the base 10 for logarithms, it was WILLIAM OUGHTRED,[14] who described the principle of the slide rule in his seminal two publications *The Circles of Proportion, and the Horizontall Instru-*

---

6 Prior to this discovery differential gears were thought to have been invented in medieval times.
7 See [FREETH 2008, p. 9].
8 There are still arguments whether the mechanism also featured indicators for the display of planet positions.
9 A wealth of information about this device may be found in [DE SOLLA PRICE 1974], [FREETH 2008] and [MCCARTHY 2009].
10 Also known as a *slipstick*.
11 Additional precision is achieved by this type of slide rule by wrapping extended scales along a helical path. The downside of this is that such slide rules typically feature only two scales.
12 1550–04/03/1617
13 February 1561–01/26/1630
14 03/05/1574–06/30/1660

Fig. 2.2. Typical scale (probably 18th century)

ment and *Two rulers of proportion*.[15] An early *scale* is shown in figure 2.2. Working with such a scale required a divider to transfer lengths from one of its engraved scales to another, a tedious and error-prone process that was greatly simplified by the introduction of a slider containing several scales, which led to the then ubiquitous slide rule.

Figure 2.3 shows one of the last, most complex and most versatile slide rules ever built, a Faber Castell 2/83N. Its three main parts are clearly visible:

- The *body*, which consists of the top and bottom *stator* or *stock*. The two stators are held together by two *end braces* or *end brackets*.
- The *center slide*, which can be moved laterally with respect to the body.
- The *cursor*, which slides in grooves of the body.

While simple slide rules only feature a couple of scales, complex ones as the 2/83N have up to 30 and more scales, which implement functions far beyond multiplication and division.[16] Using these scales, trigonometric functions, exponentiation, etc., can be evaluated. Until pocket calculators took over in the 1970s,[17] slide rules were as widely used as they were centuries ago. The following quotation from JOSEF VOJTĚCH SEDLÁČEK[18] shows this quite strikingly:[19]

---

15 A comprehensive history of the slide rule may be found in [CAJORI 1994] and [JEZIERSKI 2000]. A great introduction to the application and use of slide rules is given in [HUME et al. 2005].
16 If all scales are just on one side of the body, a slide rule is called *simplex*. If scales are found on both sides of the body, it is a *duplex* slide rule. In this case the cursor is also used to transfer partial results from one side of the body to the other.
17 Early pocket calculators such as the Hewlett Packard HP-35 or some models made by Texas Instruments like the SR-10, etc., were explicitly marketed as *electronic slide rules*. The fixed number format featured by early HP calculators that was often set to display only 2 or 4 decimal places also was a reverence for the slide rule.
18 02/24/1785–02/02/1836
19 Cf. [JEZIERSKI 2000, p. 16].

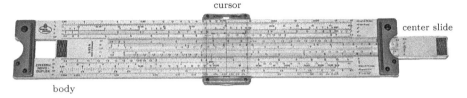

Fig. 2.3. Faber Castell slide rule model 2/83N

> "*It is said that the use of the slide rule in England is so widespread that no tailor makes a pair of trousers without including a pocket just for carrying a 'sliding rule'. During such a time, it is difficult to understand why the slide rule does not enjoy such well-deserved recognition in our own country.*"

Slide rules were essential tools for the scientific and technological progress of the last three centuries ranging from mathematics, civil engineering, commercial applications, electronics, chemistry, life sciences, etc., up to applications in aerospace technology.[20,21]

Although they were rendered more or less obsolete by pocket calculators 50 years ago, there are still areas of application where slide rules are employed regularly. For example, many aviators still use a flight computer like the E-6B, a special form of a circular slide rule, that allows the calculation of ground speed, i.e., the speed of an aircraft corrected for wind effects, and many other crucial parameters.[22]

Figure 2.4 shows a strange special purpose circular slide rule that was deployed in large amounts during the Cold War – a *Nuclear Weapon Effects Computer*, which allowed rough estimates of fatalities and damage should a nuclear air burst occur.

2.4 Planimeters

Planimeters are fascinating instruments. Their purpose, most aptly described by [HENRICI 1894, p. 497], is the following:

---

[20] In fact BUZZ ALDRIN (01/20/1930–) carried a Picket slide rule on the Apollo-11 mission, which was sold on September 20th, 2007 for $77,675.

[21] [KAUFMANN et al. 1955] describes some interesting electronic circuits, most of which requiring only passive components such as potentiometers and resistors, to implement an *electronic slide rule*.

[22] Apart from the ease of use, such specialized slide rules have the advantage of not requiring any electrical power or the like for their operation.

14 — 2 Mechanical analog computers

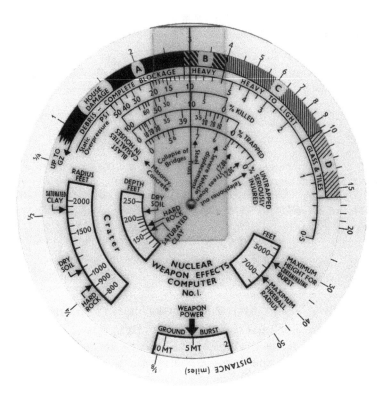

Fig. 2.4. Nuclear Weapon Effects Computer

> "*The object of a planimeter is to measure an area; it has, therefore, to solve a geometrical problem by mechanical means.*"

Measuring areas enclosed by some "good-natured" boundary curve is an important task in many branches of science as well as in commercial applications, registers of real estate and many more. A typical early application was to analyse pressure/volume indicator diagrams[23] as those written by recording steam engine indicators,[24] which requires the determination of the area enclosed by a curve, which is either plotted in a Cartesian or more often a polar coordinate system. A simple and direct method for performing this task is to cut out the area to be determined and weigh the resulting piece of paper yielding quite good results. Although this method is sometimes still used by chemistry students who regularly

---

23 See [Hütte 1926, pp. 380 f.].
24 The first of these devices was invented by JAMES WATT's[25] assistant JOHN SOUTHERN[26] around 1796 (cf. [MILLER 2011]).

have to determine integrals over curves generated by spectrometers and the like, it is not really suitable for everyday usage.

As early as 1814 J. M. HERMANN,[27] a Bavarian engineer, invented a planimeter that was built, after improvement by LÄMMLE, in about 1817. Unfortunately, this instrument seems to have gone unnoticed by his contemporaries and had no obvious influence on subsequent developments.[28] In 1824 an Italian professor for mathematics, TITO GONNELLA,[29] invented a *wheel-and-cone planimeter* that used a friction-wheel integrator (see section 2.5.3) to perform the necessary integration.[30] The first planimeter that was put into production was a device developed by a Swiss engineer named JOHANNES OPPIKOFER,[31] who developed two wheel-and-cone planimeters in 1827 and 1836 and a planimeter based on a friction-wheel rolling on a disk in 1849. In fact, there is a plethora of different planimeter principles and implementation variants.

The most successful type of planimeter is the *polar planimeter* that was developed in 1854 by the Swiss mathematician JACOB AMSLER-LAFFON.[32] Figure 2.5 shows a typical polar planimeter,[33] which is of a much simpler construction than most of the other instrument types.[34]

The basis of operation for planimeters in general is GREEN's theorem, which relates a double integral over a closed region, i.e., the area to be determined, to a line integral over the boundary of this region.[35] Thus, a planimeter is a mechanization of

$$\iint \left(\frac{\partial Q}{\partial x} - \frac{\partial P}{\partial y} \right) \mathrm{d}A = \oint F \mathrm{d}r.$$

Choosing P and Q in a way that the difference under the left integral equals one yields the area sought. Interestingly, it seems that the first explanation of the operation of planimeters using Green's formula wasn't given until [ASCOLI 1947].[36]

---

[27] 1785–1841
[28] Cf. [HENRICI 1894, p. 505].
[29] 1794–1867
[30] [HAEBERLIN et al. 2011] describes this instrument. See also [HENRICI 1894, p. 500].
[31] 09/15/1782–04/21/1864
[32] 11/11/1823–01/03/1912
[33] [FOOTE et al. 2007] shows how to build a simple polar planimeter.
[34] One notable exception is the *Prytz planimeter* (also known due to its physical shape as a *hatchet planimeter*) that was developed by the Danish mathematician and cavalry officer HOLGER PRYTZ (who published under the pseudonym "Z") around 1875. His instrument has no moving parts whatsoever but is of very limited precision. A good description of the principle of operation of this instrument can be found in [FOOTE et al. 2007, pp. 82 ff.].
[35] Cf. [ASMAR et al. 2018, pp. 177 ff].
[36] Explanations of the operation of planimeters with different chains of reasoning can be found in [HENRICI 1894, pp. 179 ff.], [MEYER ZUR CAPELLEN 1949], and [LEISE 2007]. Some patents describing interesting planimeter designs are [COFFIN 1882], [SNOW 1930], etc.

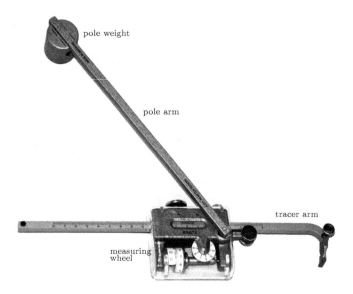

Fig. 2.5. Typical polar planimeter made by A. OTT, Bavaria, Germany

The main parts of a polar planimeter like the one shown in figure 2.5 are two arms connected with an elbow that is restricted to move on the circumference of a circle. The end of one arm, the *pole arm*, is fixed at a point called *pole* by means of a *pole weight* or a needle, while the end of the other arm, the *tracer arm*, can be moved freely by hand. This end is often formed as a needle or a magnifying lens to simplify tracing the boundary curve of the area to be measured. At the elbow is the integrating or measuring wheel.[37]

To measure the area of a circumscribed figure, the pole is fixed either outside or inside the figure, depending on its size. Then the measuring wheel with its vernier and the counting dial are reset with the needle or lens placed on the starting point of the boundary. Then the boundary line is followed manually in clockwise direction. After reaching the starting point again, the area can be read from the counting dial, the wheel and the vernier. If extended precision is necessary, the same procedure can be applied a second time while following the boundary line counterclockwise,[38] which, of course, yields a result with the opposite sign, the complement.

---

[37] This wheel is extremely delicate – its metal wheel tread should never be touched by hand since it is engraved with microscopic rills, which are vital for the overall precision of the instrument and are easily damaged or cluttered with dirt.
[38] Typical errors caused by a wheel with an axis not being perfectly horizontal can thus be compensated to a certain degree.

By setting the position of the elbow, which can normally be shifted along the tracer arm, various scaling values can be set, so areas can be measured in inches as well as in centimeters without the necessity to explicitly convert units.[39]

Determining large areas can be problematic since the counting dial can only represent one significant digit. Thus, as early as 1961 planimeters coupled with electronic counters were built. [Zuse Z80 1961] describes a linear planimeter[40] that was coupled to an electronic counter capable of processing up to 250 000 impulses per second. The pickup from the measuring wheel was done photoelectrically thus allowing even better precision than traditional mechanical instruments.

[LEWIN 1972] describes another development that was patented in 1972: Here the position of the tracer arm is sensed by two linear potentiometers generating voltages directly proportional to the current (x,y)-position of the tracer. These voltages in turn control an oscillator and some monostable devices. The integration process is then performed basically by a chain of decade counters. This scheme proved to be too complicated and costly for the market. Another quite recent development is described by [LIGHT 1975]. Here the position of a tracer needle or the like is determined by conductive foils, which are placed under the chart containing the curve to be integrated over.[41]

As old as planimeters are, some companies still manufacture polar and linear planimeters, which achieve accuracies of about $1^0\!/\!_{00}$. These instruments are still used regularly for surveying and mapping, for determining the area of furs and fabric, etc.

2.5 Mechanical computing elements

All of the instruments shown in the preceding sections are specialized analog computers, capable only of solving just one distinct problem each. This is obviously caused by their fixed structure – a slide rule can only add lengths, so the only variation possible is that of employing different scales to extend this basic operation to multiplication and division and many more.

In the same sense planimeters are specialized instruments for only determining areas. The following sections will now cover some basic mechanical computing

---

[39] [PALM 2014] gives a thorough treatment of the planimeters and other integrators made by A. OTT.

[40] These differ from polar planimeters in so far as the tracer arm is not free to rotate around the elbow connected to the pole arm but is guided in a strictly linear fashion, which is normally implemented by a two-wheel carriage that is dragged behind by the tracer arm while tracing the curve.

[41] Although this instrument was not a financial success due to its complexity, its pickup mechanism anticipated the basic techniques used for today's touch screens and the like.

elements that can be used to build true – in the sense of their reconfigurability – analog computers, *differential analysers*.[42]

Typically, mechanical analog computers represent values by rotations of shafts, which interconnect the various computing elements or by positions of linkages. As simple as most mechanical computing elements seem, they are quite powerful tools. Precisions up to 10^{-4} are possible given precise machining of the parts involved and clever scaling of the equations to be solved. In fact, mechanical analog computers even have one advantage over analog electronic analog computers: The former can integrate over every variable whilst the latter can only integrate over time, which then requires some ingenuity to solve partial differential equations[43] and other problems.

2.5.1 Function generation

A common task in simulations is the generation of functions, which are either defined analytically or by measured values. Functions of a single variable are easily implemented using cams driven by a shaft, the angular position of which represents the input variable. A follower measuring the position of the cam's surface yields the desired function value. The position of this pin can be translated into a rotational motion by a rack and pinion arrangement so this output value can be used as input for another element expecting an angular shaft position as its input value.

Another type of generator for a function of one variable is shown in figure 2.6. This is a *squaring cam*, which is used to generate a square function $f(x) = x^2$. If dx_1 denotes the element of rotation of the input shaft, the resulting movement of the string wound around the cone is $r_1 dx_1$ where r_1 denotes the average diameter of the cone at the current position of the string. With r_2 denoting the diameter of the drum, the drum rotation resulting from dx_1 is

$$dx_2 = -\frac{r_1 dx_1}{r_2}.$$

r_1 is proportional to the angle of rotation x_1 thus

$$dx_2 = -\frac{k x_1 dx_1}{r_2}$$

with k being a constant of proportionality. Integration finally yields

$$x_2 = -\frac{k}{2r_2} x_1^2,$$

---

[42] Information about mechanical computing elements can be found in [SVOBODA 1948], [MEYER ZUR CAPELLEN 1949], [WILLERS 1943], and [Bureau of Ordnance 1940].
[43] This particular advantage of mechanical analog computers is also exhibited by digital analog computers, see chapter 10, which make these devices quite interesting for future computer architectures.

2.5 Mechanical computing elements — 19

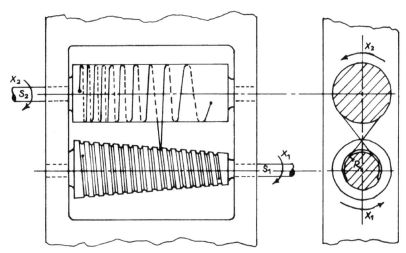

Fig. 2.6. Squaring cam yielding $x_2 = x^2$ (cf. [SVOBODA 1948, p. 22])

which is the desired square function. This mechanism does, of course, not work down to $x_1 = 0$ – if this is necessary, it may be combined with a differential gear as described in section 2.5.2.[44]

In many cases, functions of two variables are necessary – figures 2.7 and 2.8 show a *barrel cam*[45] used to implement a function $x_3 = f(x_1, x_2)$. The value of the input variable x_1 controls the lateral displacement of the three-dimensional barrel cam while the second input variable, x_2, controls the angular position of the cam. The output value is sensed by a follower that gauges the cam's surface. Other implementations feature a movable sensing pin carriage instead of a barrel cam that can be shifted laterally, which allows for a more compact design.

Barrel cams like this were often used in mechanical fire control systems but were also used in some cases in electronic analog computers since the generation of functions of two variables is a complicated task for such a machine. In this case the input shafts for x_1 and x_2 are controlled by electronic servo mechanisms while the output sensing pin is connected to a potentiometer, which delivers a voltage proportional to the desired function value.[46]

An important topic regarding cams and barrel cams is that their surface has to conform to several mechanical constraints to ensure that the sensing pin can follow the surface smoothly and that the pin can never block the cam in its movement due

---

44 Cf. [SVOBODA 1948, p. 22].
45 Three-dimensional cams like this have also been called *camoids*, cf. [SVOBODA 1948, p. 23].
46 See section 4.5.7.

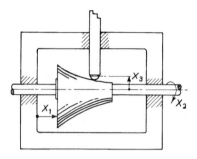 

Fig. 2.7. Three-dimensional cam as a function generator for a function of two variables $f(x,y)$ (cf. [SVOBODA 1948, p. 23])

Fig. 2.8. The barrel cam – a practical implementation of a function generator yielding $f(x_1, x_2)$ (cf. [Bureau of Ordnance 1940, p. 54])

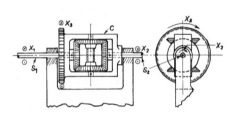 

Fig. 2.9. Structure of a bevel-gear differential (cf. [SVOBODA 1948, p. 6])

Fig. 2.10. A differential gear from the dead reckoning computer PHI-4A-2 used in Starfighter jets

to steep slopes or grooves. This might sometimes conflict with the mathematical requirements of the computer setup. In such cases, functions may be generated by solving suitable differential equations, which normally requires many more additional computing elements than a straightforward barrel cam implementation.

2.5.2 Differential gears

A barrel cam function generator like this could, in principle, be used to add or subtract two variables but a simpler, cheaper and more precise mechanism to implement this basic operation exists in form of the *bevel-gear differential*[47] shown in figure 2.9. This device adds two shaft rotations x_1 and x_2 with some scaling

---

[47] Such building blocks are also called *additive* or *linear* cells (cf. [SVOBODA 1948, p. 6]).

parameters s_1 and s_2, which result from the actual design of the differential, yielding $x_3 = s_1 x_1 + s_2 x_2$.

A practical example of such a differential gear is shown in figure 2.10. This was used in the dead reckoning computer PHI-4A-2 of a Starfighter jet.[48]

2.5.3 Integrators

In contrast to most other machines, integration is a fundamental as well as natural operation for an analog computer. Using mechanical computing elements, the operating principles of integration are remarkably simple.[49] The first mechanical integrators were developed in the early 19th century: In 1814 JOHANN MARTIN HERMANN developed an integrator consisting of a cone with a wheel rolling on its surface.[50] The position of the wheel on the envelope of the cone represents the values of the function to be integrated while the rotation of the cone corresponds to the variable of integration.[51]

Basically an integrator mechanizes the calculation of integrals like

$$x_3 = \int x_1 \, dx_2$$

where x_2 is represented by the rotation of the integrator disk (or cone) while the values of x_1 control the position of the friction-wheel rolling on the disk surface. Figure 2.11 shows the structure of such an integrator.

Obviously it is $dx_3 = k x_1 dx_1$ where $k = 1/r$ denotes the radius of the friction-wheel thus yielding

$$x_3 = \int_{x_{20}}^{x_2} \frac{1}{r} x_1 \, dx_2.$$

Figure 2.12 shows a friction-wheel integrator from the *Oslo analyser*, which was built from 1938–1942 – at its time one of the largest, most precise and most

---

[48] The central element carrying the bevel gears is called *spider block*.
[49] Far from being trivial is the implementation of integrators due to the necessary high precision in order to minimise the accumulation of errors within a simulation.
[50] See [PETZOLD 1992, p. 26].
[51] A similar mechanism was developed later by TITO GONNELLA and used as the basis for his planimeter.

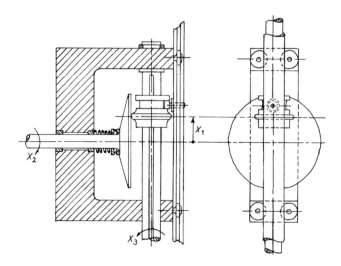

Fig. 2.11. A friction-wheel integrator (cf. [SVOBODA 1948, p. 24])

powerful differential analysers in the world.[52] On the right hand side[53] the rotating disk S and the friction-wheel h with its associated shaft are visible. The wheel position on the surface of the disk with respect to the disk center corresponds to $F(x)$.

In 1876, JAMES THOMSON,[54] the brother of WILLIAM THOMSON[55] – later Lord KELVIN – developed an integrator that replaced the friction-wheel with a steel sphere that runs in a movable cage (controlled by x_1).[56] This sphere provides a frictional connection between the rotating disk and a cylinder that acts as a pickup for the integration result x_3. Figure 2.13 shows a variation of this implementation – two stacked balls, instead of a single sphere, are guided in a movable cage. Such integrators are known as *double-ball integrators*.[57]

---

[52] This particular machine featured twelve integrators of this type and was used until 1954, see section 2.8.

[53] On the left side a *torque amplifier* and a *frontlash unit* can be seen. These are typical devices in a mechanical differential analyser but outside the scope of this book. Refer to [WILLERS 1943, pp. 236 ff.], [ROBINSON] or [FIFER 1961, p. 672] for more information about these units.

[54] 02/16/1822–05/08/1892

[55] 06/26/1824–12/17/1907

[56] See [THOMSON 1912, pp. 452 ff.].

[57] Such double-ball integrators were used well into the second half of the 20th century. This was mainly due to their high precision – [Librascope 1957] mentions a repeatability of 10^{-4} – quite remarkable for a mechanical device – as well as to their robustness. Accordingly, such integrators were used quite often in aerospace applications like dead reckoning computers, etc.

Fig. 2.12. Integrator from the Oslo differential analyser, see section 2.8 ([WILLERS 1943, p. 237])

Based on this integrator developed by his brother, Lord KELVIN devised the idea of a machine suitable for solving differential equations. The basic concept was to start with the highest derivative in the differential equation to be solved and to generate all of the necessary lower derivatives by repeated integration. Using these derivatives, the highest derivative – that was the starting point for this procedure – can now be synthesized by combining the lower derivatives in an appropriate way.[58] The discovery of this method is described in [THOMSON 1876] as follows:

> "But then came a pleasing surprise. Compel agreement between the function fed into the double machine and that given out by it [...] The motion of each will thus be necessarily a solution of [the equation to be solved]. Thus I was led to the conclusion, which was unexpected; and it seems to me very remarkable that the general differential equation of the second order with variable coefficients may be rigorously, continuously, and in a single process solved by a machine."

Interestingly, KELVIN did not build a practical machine based on this insight, which is all the more puzzling since a (double) ball integrator can transfer sufficient torque to allow a small number of such devices to be chained. Trying to implement this scheme using friction-wheel integrators would have required torque amplifiers, as these integrators can only transmit tiny amounts of torque. Consequently, this brilliant idea of setting up an analog to represent differential equations fell into oblivion. This happened again 45 years later when UDO KNORR[59] published a

---

[58] This *classical differential analyser technique*, as [KORN et al. 1964, p. 1-5] puts it, is described in more detail in section 7.2.
[59] 04/20/1887–07/10/1960

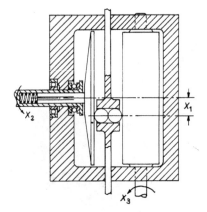 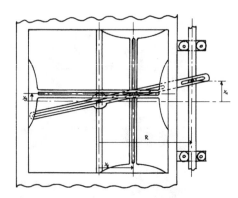

Fig. 2.13. Principle of operation of a double-ball integrator (see [SVOBODA 1948, p. 25])

Fig. 2.14. Basic structure of a slide multiplier (cf. [SVOBODA 1948, p. 12])

similar idea in 1921.[60] KNORR explicitly noted that coupled integrators can be used to solve a large group of differential equations of high degree.[61]

2.5.4 Multipliers

Multiplication is a quite difficult task for a mechanical analog computer. While multiplication with a constant can be easily performed with an appropriate gear mechanism or with a friction-wheel integrator where the position of the friction-wheel corresponds to the multiplier, the generalized operation where both input variables may vary is much more difficult to implement.

One common implementation is based on the product rule known from calculus. From $(uv)' = u'v + uv'$ it follows by integration that

$$uv = \int u\,dv + \int v\,du. \tag{2.1}$$

Thus, the multiplication of two variables can be implemented using two interconnected integrators.[62]

---

60 Cf. [PETZOLD 1992, p. 33].
61 See [WALTHER et al. 1949, p. 200].
62 This requires that the integrators are not restricted with respect to the variable of integration. Only mechanical analog computers and DDAs (cf. section 10) fulfill this requirement. Analog electronic analog computers can only use time as the variable of integration, so this class of machines requires different multiplication schemes.

Figure 2.14 shows the structure of a *slide multiplier*. It consists of two carriages sliding vertically and horizontally respectively and a lever that rotates around a center pin that is fixed to the enclosure. Both carriages and the lever are coupled with another pin that runs in grooves of these three elements. The input variables are x_1, represented by the displacement of the rotating lever's end, and x_2, which corresponds to the horizontal displacement of the second slider. The multiplication result $x_3 = kx_1x_2$ where k denotes a scaling factor that depends on the dimensions of the multiplier, is then available as the vertical displacement of the first slider.

2.6 Harmonic synthesizers and analysers

Although Lord KELVIN did not attempt to build a true general purpose mechanical analog computer, he did build some quite complex special purpose analog computers[63] to predict tides by means of harmonic synthesis.[64] Even the earliest sailors had a genuine interest in tide prediction since accurate predictions lead to fewer lay days in harbours. A good description of the term *tide* is given by KELVIN himself:[65]

> "*The tides have something to do with motion of the sea. Rise and fall of the sea is sometimes called a tide; but I see, in the Admiralty Chart of the Firth of Clyde, the whole space between Ailsa Craig and the Ayrshire coast marked 'very little tide here'. Now, we find there a good ten feet rise and fall, and yet we are authoritatively told there is very little tide. The truth is, the word 'tide' as used by sailors at sea means horizontal motion of water; but when used by landsmen or sailors in port, it means vertical motion of the water.*"

Tides are caused and influenced by the superposition of a number of effects that are, in fact, harmonic oscillations with different amplitudes, frequencies, and phases.[66] The most important of these effects are the earth's rotation around its own axis, the rotation of the earth around the sun, the rotation of the moon around the earth, the precession of the moon's perigee, the precession of the plane of the moon's orbit, etc.

In 1872 KELVIN developed the first *harmonic synthesizer*, a specialized analog computer for generating harmonics and adding these together to generate a

---

[63] For these machines the phrase "*substitute brass for brain*" was coined (see [ZACHARY 1999, p. 49] and [Everyday Science and Mechanics 1932]).
[64] Following the death of his first wife, MARGARET THOMSON nee CRUM on June 17, 1870, his interest for seafaring increased and he bought a 126 ton schooner, the LALLA ROOKH, which in turn sparked his interest in tide prediction.
[65] See [THOMSON 1882, Part I].
[66] These effects are called *partial tides*.

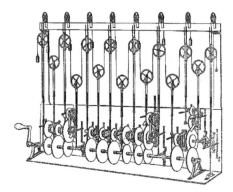
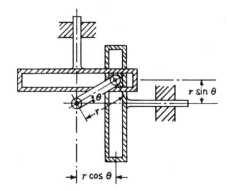

Fig. 2.15. Principle of operation of KELVIN's tide predictor (see [THOMSON 1911])

Fig. 2.16. A scotch yoke mechanism (cf. [KARPLUS et al. 1958, p. 242])

tide prediction.[67] This machine took ten partial tides into account. Its basic construction is shown in figure 2.15.[68] Using this machine it took only four hours to compute the 1 400 tides that occur during a year for a given harbour. Performing the same task manually took several months.[69]

Generally speaking, a harmonic synthesizer like a tide predictor generates a function $f(x)$ based on a given set of harmonics:[70]

$$f(x) = a_0 + a_1 \sin(x + b_1) + a_2 \sin(2x + b_2) + \cdots + a_n \sin(nx + b_n)$$

The basic harmonic functions are traditionally generated using a *scotch yoke mechanism* as shown in figure 2.16. It consists of a boom that is mounted on a shaft so that it can rotate around this mounting point. The angular position of this shaft represents the input variable θ. The free end of the boom guides two carriages that are restricted to perform horizontal respectively vertical movements only. The movements of these carriages then represent the values $r \sin(\theta)$ and $r \cos(\theta)$ where r denotes the radius of the circle described by the free end of the rotating boom.

Since the values $r \cos(\theta)$ and $r \sin(\theta)$ are represented by linear displacements, harmonics generated this way cannot be added together using a differential gear. Instead, a steel band of constant length is used, as can be seen in figure 2.15.

---

[67] Lord KELVIN's machine is on display at the Science Museum, London.
[68] Later machines generated even more partial tides. The *United States Coast and Geodetic tide-predicting machine No. 2* that was completed in 1910 generated 37 harmonic terms. [AUDE et al. 1936] describes a machine that took 62 harmonics into account (this machine was in operation in Hamburg, Germany, until 1968 and is now on display at the *Deutsches Museum* in Munich).
[69] Cf. [SAUER].
[70] Cf. [BERKELEY et al. 1956, pp. 135 ff.].

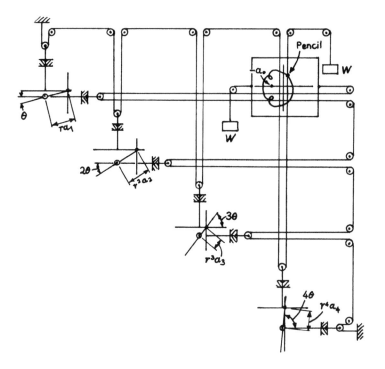

Fig. 2.17. Harmonic synthesizer for a fourth-degree equation (cf. [KARPLUS et al. 1958, p. 242])

While KELVIN's tide predictor only generated a single output function by the superposition of the various harmonics, later harmonic synthesizers often had a more generalized structure like that shown in figure 2.17.

This fourth degree harmonic synthesizer consists of four scotch yoke mechanisms as shown in figure 2.16, each of which generates a sine/cosine-pair and is parameterized by four crank length settings r, r^2, r^3 and r^4 (this rather unusual notation follows [KARPLUS et al. 1958, p. 242]). Two continuous tapes are threaded through the system adding all sine- and all cosine-values and driving the two coordinate inputs of a plotting mechanism.[71]

Later developments include a device described in [REDHEFFER 1953], which still used scotch yoke mechanisms but with an electronic pickup in form of potentiometers that sensed the positions of the sine/cosine carriages. The output

---

[71] A similar device, the high precision *Isograph*, which contained 10 sine/cosine-units was built at the Bell Telephone Laboratories in 1937: "*In the Isograph, gears were fitted to the bearings with an accuracy of one ten-thousandth of an inch for play and concentricity.*" (see [KARPLUS et al. 1958, p. 243]).

voltages of these potentiometers were then added by means of simple electronic networks and could directly be used to control an xy-plotter.

The inverse function of a harmonic synthesizer is performed by a *harmonic analyser*.[72] These devices perform a FOURIER analysis, i.e., they determine the amplitudes and phases of the *harmonics* that comprise a given input signal. The first such harmonic analyser was developed in 1876 by Lord KELVIN.[73] It consisted of five double-ball integrators and was demonstrated at the Royal Society, configured for a meteorological problem.

Its operation is based on the computation of integrals of the form

$$\int y \cos\left(\frac{2\pi\omega x}{c}\right) dy$$

$$\int y \sin\left(\frac{2\pi\omega x}{c}\right) dy$$

with $\omega \in \{1, 2\}$ (for meteorological applications) or $\omega \in \{\frac{39 \cdot 109}{40 \cdot 110}, 1\}$ for work on tides.[74]

One of the most complex harmonic analysers was devised by ALBERT ABRAHAM MICHELSON[75] of MICHELSON-MORLEY experiment fame and SAMUEL WESLEY STRATTON.[76] This machine was capable of performing harmonic synthesis as well as analysis with up to 20 harmonic terms – a remarkable feat. Due to its very clever principle of operation and implementation it weighed only 69 kg and could fit on a standard desk.[77] This particular machine came back into the limelight again in the early 2010s when BILL HAMMACK et al. published a video series detailing on the operation of this device, which they had recently restored.[78]

The determination of FOURIER coefficients is also possible using only a planimeter. This startling technique is described in detail in [WILLERS 1943, pp. 171 ff.]. Since this manual process is not only tedious but also time consuming and error-prone, semi-automatic harmonic analysers based on planimeters have been developed. Figure 2.18 shows a planimeter-based harmonic analyser that was developed and produced by Mader-Ott.[79]

---

72 See [BERKELEY et al. 1956, pp. 132 ff.], [WILLERS 1943, pp. 168 ff.] and [McDONAL 1956], which give a detailed description of a practical harmonic analyser.
73 See [THOMSON 1878] and [BERKELEY et al. 1956, p. 133].
74 See [THOMSON 1878].
75 12/19/1852–05/09/931
76 07/18/1861–10/18/1931
77 See [MICHELSON et al. 1898].
78 [HAMMACK et al. 2014] describes this remarkable machine in incredible detail.
79 See [WILLERS 1943, pp. 178 ff.]. TATJANA JOËLLE VAN VARK has built outstanding modern implementations of harmonic analysers and synthesizers (among other mathematical instruments) displaying an incredible degree of scientific craftsmanship, see http://www.tatjavanvark.nl.

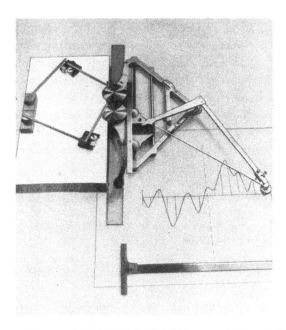

Fig. 2.18. Harmonic analyser made by Mader-Ott (cf. [WILLERS 1943, p. 180])

Apart from obvious applications in fields like mechanical engineering where harmonic analysers were used regularly to analyse complex vibrations and the like, they also played an important role in resource and oil exploration where seismograms must be analysed to reveal the structure of earth's interior.

2.7 Mechanical fire control systems

Another type of very complex and highly specialized analog computers are *fire control systems*. These military systems perform the following tasks:[80]

> "*Fire control equipment, that takes in indications of targets from optical or radar perception and using extensive calculating equipment puts out directions of bearing and elevation for aiming and time of firing for guns, according to a program that calculates motion of target, motion of the firing vehicle, properties of the air, etc.*"

Until the late 19th century fire control was a simple task given the relatively short ranges of typical gunnery. In previous centuries, firing *broadsides* was a common

---

[80] See [BERKELEY et al. 1956, p. 163].

mode of fighting in naval warfare.[81] During the 19th century shell ranges grew from a few hundred meters up to six miles, which greatly complicated the task of aiming. Long range guns with ranges up to 17 miles were common at the end of World War II[82] – the time of flight of such a shell was in the range of about one minute – resulting in an aiming task so complex that it would have been impossible to hit a target reliably without the support of complex fire control systems, which were implemented as large mechanical analog computers.[83] These machines were not only rugged but also extremely long-lived regarding their service life as [BROMLEY 1984, pp. 1 f.] puts it:

> "*Mechanical analog devices were first used for naval gunnery in World War I, were greatly developed for naval and anti-aircraft gunnery between the wars, were further developed and extended to aircraft systems during World War II, and continued in service in refined versions into the 1970s.*"

Typical effects that have to be taken into account by a fire control system include:

Wind and current effects: Flight times of up to one minute and more make shells susceptible for forces exerted by side winds and the like. Current effects acting on torpedoes are equally, if not even more, complicated to model and predict, mainly due to the long run time of such a weapon.

Heading: Both opponents are normally not at rest but moving (in the case of aircraft very fast).

Environmental conditions: Barometric pressure, temperature, humidity, etc., all influence the path of a shell or torpedo.

Aiming point vs. sensor position: Especially in the case of torpedoes there is normally a mismatch between the position of the sonar system with respect to the periscope and the launching tubes on the side of the attacker and, in the case of sonar contact, the sound source and the desired aiming point on the target vessel.[84]

Signal run time: In case of sonar bearing analysis, the run time of the sound waves can often not be neglected regarding the ships' positions.

---

[81] The typical range of fire during the historic encounter between the *USS Monitor* and the *CSS Virginia* on March 9, 1862 was only about 100 meters (see [CLYMER 1993, p. 21]).
[82] See [BROMLEY 1984, p. 2].
[83] At the battle of Jutland in 1916, the hit rate was less than 5% even though both navies were using mechanical fire control systems.
[84] Details of the German torpedo control system can be found in [RÖSSLER 2005, pp. 79 ff.], including a schematic diagram of a typical analog lead-lag computer.

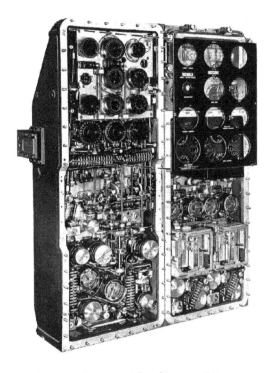

Fig. 2.19. TDC Mark-3 with front panels removed (see [Bureau of Ordnance 1944, p. 34])

Figure 2.19 shows the *Torpedo Data Computer Mark-3*[85] developed by the *Arma Corporation*[86] and gives an impression of the immense complexity of such mechanical analog fire control systems. This particular system is deservedly regarded as a masterpiece mechanical analog computer. The two interconnected racks have a width, height, and depth of 160 cm × 115 cm × 100 cm and weigh about 1.5 metric tonnes. The power consumption was quite impressive at 55.7 to 140 A at 115 V operating voltage.

A wealth of additional information about fire control systems can be found in [BERKELEY et al. 1956], [CLYMER 1993, pp. 286 ff.], [FRIEDMAN 2008, pp. 16 ff.], [GRAY et al. 1955], and [Admiralty 1943].

---

85 *TDC Mark-3* for short.
86 See [CLYMER 1993, p. 28].

2.8 Differential analysers

A central figure in the development of the mechanical and electromechanical *differential analyser* was VANNEVAR BUSH,[87] who developed a keen interest in analog computing techniques in the early 20th century.[88] He was not aware of the fundamental works of JAMES THOMSON and Lord KELVIN[89] and when he was informed about these previous developments in later years he still claimed the intellectual property on his inventions for himself, stating that *"[i]nventors are supposed to produce operative results"* based on the fact that none of his predecessors had successfully built a working general purpose mechanical analog computer.[90]

In 1925, triggered by the increasing computational requirements for solving power grid problems, BUSH pursued the idea of a large-scale mechanical general purpose analog computer. Six years later, in 1931, this machine, the first true general purpose mechanical analog computer, was completed[91] at the *Massachusetts Institute of Technology*.[92] It was, in fact, a collection of individual computing elements like those described in the preceding sections. In addition to these components, special units like torque amplifiers, frontlash units[93] and *helical gearboxes* had to be developed. The latter allowed the interconnection of the various computing elements on a central interconnect unit that housed all input and output shafts, which were called *bus rods* (18 in total).

Figure 2.20 shows this early differential analyser.[94] To set up the machine to solve a particular problem, its computing elements had to be connected using these bus rods, a lengthy process. Set up times of several days were not uncommon.[95] One eight hour day of operation was charged at US$ 400.[96] Much of the machine operational time was devoted to work regarding the stability of long distance lines

---

[87] 03/11/1890–06/30/1974

[88] As early as 1912 he submitted a patent describing a *profile tracer* that allowed to trace and plot a ground profile (see [BUSH 1912]). This device contained two mechanical integrators in series that took the displacement of a pendulum mass as input thus yielding a profile trace by integrating twice over the acceleration of the mass.

[89] Cf. [DODD 1969, p. 5] and [ZACHARY 1999, p. 49].

[90] See [ZACHARY 1999, p. 51]. [PUCHTA 1996] gives a thorough overview of his early analog computers.

[91] Its construction cost about US$ 25,000.

[92] *MIT* for short.

[93] See [WILLERS 1943, pp. 236 ff.], [ROBINSON], and [FIFER 1961, p. 672].

[94] More information about this particular machine can be found in [Meccano 1934/2], [ZACHARY 1999], and [GLEISER 1980]. A rather simple but still useful mechanical differential analyser is described in [KASPER 1955].

[95] See [ZACHARY 1999, p. 51].

[96] See [MACNEE 1948, Sec. I].

2.8 Differential analysers

Fig. 2.20. VANNEVAR BUSH's differential analyser (source: [Meccano 1934/2, p. 443])

based on [CARSON 1926].[97] As a computer it proved to be highly influential and successful.[98] It inspired other researchers to build their own differential analysers such as those:

- A Meccano based differential analyser built by DOUGLAS HARTREE[99] and ARTHUR PORTER[100] at Manchester University in 1934,[101]
- a similar Meccano based machine built by J. B. BRATT at Cambridge University in 1935,[102]
- the *Oslo Analyser*,[103] which was built from 1938 to 1942 at Oslo University's Institute of Theoretical Astrophysics under the auspices of SVEIN ROSSELAND,[104] who visited the MIT and knew about the MIT differential analyser,[105]

---

[97] See [OWENS 1986, p. 67].
[98] It was soon called *thinking device*, *mechanical brain* or even *man-made brain* by the awed press (see [ZACHARY 1999, p. 51]).
[99] 03/27/1897–02/12/1958
[100] See [ROBINSON 2008] for an oral history of ARTHUR PORTER.
[101] See [Meccano 1934] and [Meccano 1934/2].
[102] This machine is now part of the collection of the *Museum of Transport and Technology*, Auckland, New Zealand: http://www.motat.org.nz/explore/objects/differential-analyser (retrieved 11/30/2012).
[103] Cf. [HOLST 1982] and [HOLST 1996].
[104] 03/18/1894–01/19/1985
[105] A bit of trivia: When Germany occupied Norway during World War II, ROSSELAND removed the integrator wheels from the machine and buried them safely packaged behind the institute to make sure that his differential analyser would not support Germany's war efforts.

- a differential analyser with six integrators that was built and completed in 1938 in Russia,[106]
- the *Integrieranlage IPM-Ott* that was built in Germany[107] (start of development in 1938),
- the giant ROCKEFELLER[108] *Differential Analyser* devised by VANNEVAR BUSH and installed at MIT's *Center for Analysis* in 1942, weighing about 100 tons, and being by far the most complex on this list of machines,[109]
- a differential analyser built by General Electric at a cost of $125,000 between 1945 and 1947, which was in use at the University of California Los Angeles,[110]
- and finally a mechanical differential analyser with only three integrators, which was built as late as 1959 at the *Institut für Angewandte Mathematik und Mechanik*, Friedrich-Schiller University Jena (Germany), under the leadership of ERNST WEINEL.[111]

As an example a simple differential analyser setup is shown in figure 2.21. The differential analyser is used to integrate over a function $y = f(x)$, which is provided as a plot on the *input table* shown on the upper left.[112] The integral over $f(x)$ is to be determined between the limits x_1 and x_2. The motor drives three shafts of the differential analyser: Two shafts control the x-direction movements of the cross-hairs of the input and the pen of the *output table* (upper right) while the third shaft drives the integrator disk, whose rotation and thus time is the variable of integration.

In the simplest case a human operator would now track the curve on the input table by turning the crank accordingly to keep the curve in the cross-hairs of the input table's magnifier lens. This in turn changes the displacement of the friction-wheel of the integrator in a way that the angular displacement of w represents

$$x = \int_{x_1}^{x_2} f(x)\,dx.$$

The shaft coupled to the friction-wheel of the integrator finally drives the y-input of the output table effectively plotting the desired integral of $f(x)$.

---

106 See [ETERMAN 1960, pp. 39 ff.].
107 See [WALTHER et al. 1949].
108 The *Rockefeller Foundation* initially funded this machine in 1935.
109 See [OWENS 1986] for details on that particular machine.
110 See [N. N. 1978]. A noteworthy study performed on this machine is described in [GUIBERT et al. 1949].
111 This machine was primarily intended to be used in education, see [KRAUSE 2006].
112 The problem of determining the area under a curve is, in fact, so simple and could be solved by employing a planimeter (see section 2.4) so that such a setup would never have been found in a practical differential analyser installation.

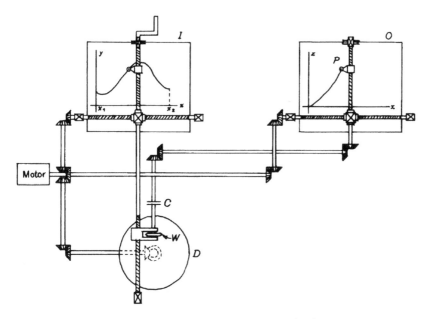

Fig. 2.21. A simple differential analyser setup for integration (cf. [KARPLUS et al. 1958, p. 190], [SOROKA 1962, p. 8-10])

Even today mechanical differential analysers are still fascinating devices. Figure 2.22 shows a modern implementation of such an analog computer using Meccano parts, which was developed and built by TIM ROBINSON in the early 2000s.

Apart from their slow overall operation due to the mechanical nature of the computing elements, the main problem of these differential analysers were the excessive setup times due to the direct mechanical interconnections between the various computing units. Figure 2.23 shows the schematic representation of BUSH's differential analyser configured for the VALLARTA-LEMAITRE cosmic ray problem,[113] which ran for a total of 30 days of computer time.[114] A setup like this could easily require several days to complete and check out – unacceptable for an agile environment and way too expensive for commercially viable applications of this technology.

Accordingly, *electromechanical differential analysers* were conceived, which used intricate rotational sensors and servo mechanisms to render the direct mechanical interconnects obsolete while still using mechanical computing elements.

---

[113] See [LEMAITRE et al. 1936] and [LEMAITRE et al. 1936/2].
[114] See [SONI et al. 2017, p. 30].

2 Mechanical analog computers

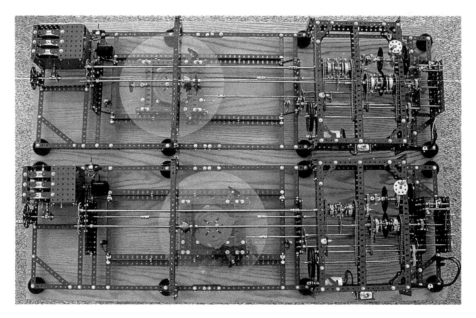

Fig. 2.22. TIM ROBINSON's Meccano differential analyser (reprinted with permission of TIM ROBINSON)

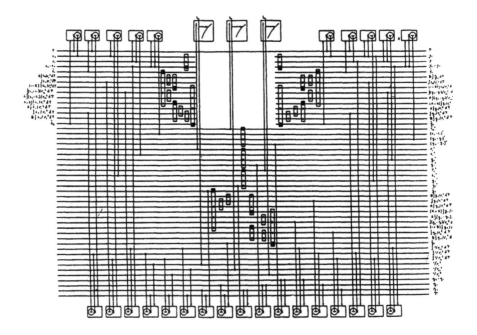

Fig. 2.23. Setup of the BUSH differential analyser for the VALLARTA-LEMAITRE cosmic ray problem (see [OWENS 1986, p. 77])

The first such machine was the ROCKEFELLER *differential analyser*, which VAN-NEVAR BUSH had planned as a follow up to his first differential analyser.[115]

Apart from more computing elements and increased precision, special emphasis was put on the capability to "*switch rapidly from one problem to another in the manner of the automatic telephone exchange*".[116] At its heart was a *crossbar switch* like those found in automatic telephone exchanges of that era. Instead of rearranging rotating rods or changing the wiring on a patch panel, the machine could be configured for a specific problem just by devising a "configuration bitstream" for the crossbar as it would be called today. This configuration data was read into the machine by a paper tape reader.[117]

Of special importance were the rotational sensors required for transforming the rotational outputs of the various computing elements into corresponding voltages. These are described by [BERKELEY et al. 1956, p. 116] as follows:

> "*[G]reat ingenuity and pains have been devoted to making it a precision mechanism. In particular, the take-off of wheel rotation is effected by an electrostatic angle indicator. This imposes virtually no load at all on the turning of the wheel [...]*"

The signals generated by these take-offs were then fed into servo circuits that drove motors, which were assigned to the various input shafts of the individual units. The interconnection between these inputs and the servo outputs was accomplished by the aforementioned crossbar switch, which allowed problems to be set up quickly and thus eliminated the main drawback of its predecessor machines. In addition to this, the summers had remotely controllable gear boxes for every single input shaft.[118] Thus, even the change of coefficients could be performed from a central control panel. Figure 2.24 shows this 100 ton behemoth differential analyser, which has been aptly described as "*the most important computer in existence in the United States at the end of the war*"[119] due to its work on war related computational problems.

In the following years a number of these electromechanical differential analysers were built around the world. The *Minden* system is probably the last example of an electromechanical differential analyser but was not as advanced as BUSH's

---

[115] Another such electromechanical differential analyser was built in Germany from 1939 to 1945. This particular machine contained four integrators, six function tables with optical sampling and two multipliers. This machine was used successfully and commercially well into the late 1950s. More information can be found in [WALTHER et al. 1949].

[116] See [OWENS 1986, p. 78]. This capability will be of central importance for highly integrated analog computers for the 21st century – see chapter 14.1.

[117] Details on this machine can be found in [BUSH et al. 1945, pp. 275 ff.].

[118] See [BERKELEY et al. 1956, p. 117].

[119] See https://www.computerhistory.org/revolution/analog-computers/3/143, retrieved 02/02/2022.

2 Mechanical analog computers

Fig. 2.24. The ROCKEFELLER differential analyser (see [BUSH et al. 1945, p. 312])

last machine. In 1948 the German company *Schoppe & Faeser* started the development of a system that was named Minden after the German city in Westphalia where development took place. Of this machine three machines were eventually built and sold. The last system contained twelve integrators, 20 summers, and ten function tables.[120] The computing elements had a precision of 0.01%. Figure 2.25 shows a dual integrator unit of a Minden system while figure 2.26 gives an overview of a typical Minden installation.[121]

Despite their advantages over purely mechanical differential analysers, these electromechanical systems could not compete with the emerging analog electronic analog computers, which offered much greater speed, flexibility and ease of construction and maintenance. The last attempt to build a competitive electromechanical differential analyser was undertaken by HANS F. BÜCKNER,[122] who de-

---

120 These function tables were based on input tables and allowed to trace a curve given an x coordinate as input yielding the y value of the curve at this particular point.
121 This last and largest Minden system was used by the German company *Siemens-Schuckert* until 1971 (see [PETZOLD 1992, p. 53]). Most parts of the machine are now part of the collection of the Deutsches Museum in Munich.
122 11/22/1912–?

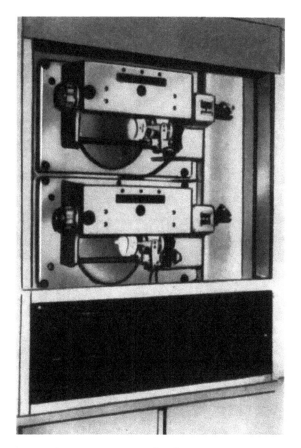

Fig. 2.25. Dual integrator of the *Minden* system (see [Rationalisierungskuratorium 1957, p. 51])

veloped the *Integromat* in 1949.[123] This system was built by Schoppe & Faeser and was still based on mechanical computing elements that were interconnected electromechanically. The rationale behind this development was that traditional electromechanical differential analysers were too complicated and thus too expensive to manufacture due to the high precision of their computing elements while for some applications this precision was not necessarily a requirement.

Accordingly, the Integromat used mechanical computing elements that were digital regarding the way in which values were represented. These integrators were called *step integrators*[124] and were based on sets of fine toothed gears, which were interconnected by differential gears. The functions to be integrated controlled the

---

123 See [BÜCKNER 1950] and [BÜCKNER 1953].
124 *Stufenintegrator* in German.

Fig. 2.26. Differential analyser *Minden* (see [Rationalisierungskuratorium 1957, p. 51])

gear ratios selected[125] while the variable of integration drove the input shaft of this multi-gear device.[126]

These integrators had the advantage of relatively simple construction and high reliability but their precision turned out to be inadequate. Even a simple problem, the solution of the differential equation $\ddot{y} = -y$, which is often called *circle test* since it yields a sine/cosine signal pair that can be used to draw a circle on a two-coordinate output device, showed an error of about 0.5% after only one period.[127] Research on damped oscillations showed errors in the range of 4.5% after only five periods – unacceptable for real applications.

Another problem was the inherently slow speed of these step integrators, which was worsened by the need to use very small step widths in order to minimise errors. Times of up to 45 minutes for the simulation of a single damped oscillator's oscillation were not uncommon.[128] Accordingly, this machine turned out to be a dead-end although it was used well into the 1950s.

---

125 Thus, only discrete values could be used.
126 All input shafts were driven by forerunners of today's stepper motors.
127 Here and in the following a dot over a variable denotes its derivative with respect to time. Accordingly, two dots represent the second derivative and so on.
128 See [EGGERS 1954].

3 The first electronic analog computers

It is difficult to write about the "first" electronic analog computers because there is no such thing as the single first machine. In fact, the 1930s and 1940s were ripe for the idea of building a differential analyser based on analog electronic components. Consequently, the following sections focus on five largely independent early developments, namely the works of HELMUT HOELZER, the early developments by GEORGE A. PHILBRICK, the birth of electronic fire control systems, MIT's first electronic differential analyser, and finally the Caltech analog computer. These distinct lines of development form the foundation for nearly all of the analog computer developments in subsequent decades.[1]

3.1 Helmut Hoelzer

In 1935 HELMUT HOELZER,[2] a student of electrical engineering at Darmstadt University in Germany, realized that mathematical operations such as integration or differentiation could be implemented quite easily using electronic circuits. Being an avid glider pilot, he knew that there was no device available in the aerospace industry to indicate the true ground speed of an aircraft and comprehended that by measuring the acceleration of an aircraft and integrating over the electric signal from the accelerometer a signal proportional to the velocity should be easily obtained.[3]

Unfortunately, his ideas were neglected for a rather long time – especially by mathematicians who were resistant to the general idea of "computing" with arcane electronic devices instead of solving complex problems by pure thought. It was not until the development of the *A4 rocket*[4] that he could eventually explore and implement his ideas, which resulted in the *Mischgerät* – the world's first fully electronic onboard computer for controlling a ballistic missile. Eventually, these

---

[1] In 1938 HANS KLEINWÄCHTER (1915–10/26/1997) proposed an electronic analog computer based on special cathode ray tubes, see [KLEINWÄCHTER 1938]. Unfortunately this early foray into electronic analog computing seems never to have been pursued further.
[2] 02/27/1912–08/19/1996
[3] See [HOELZER 1992, p. 4]: "*[Es gab] in der ganzen Fliegerei nicht ein einziges Gerät [...], welches die absolute Geschwindigkeit eines Flugzeuges [...] gegenüber der Erde messen kann. Aha, dachte ich, das ist ja ganz einfach, man nimmt die Beschleunigung, die man ja messen kann, integriert sie und – voilà! – hier ist die Geschwindigkeit.*"
[4] See [LANGE 2006], [REISIG 1999], and [DUNGAN 2005] for detailed information about the A4 rocket. The abbreviation "A4", short for "aggregate 4", denoted the rockets used during the extremely tedious and long development process, while the actual weapon system became known as the "V2", German for "Vergeltungswaffe 2", "vengeance weapon 2".

developments led to HOELZER's development of a general purpose electronic analog computer, which was brought to the United States after World War II and used well into the 1950s during the development of rockets like the Redstone.

3.1.1 The "Mischgerät"

When World War II broke out, HOELZER was working as an engineer for the renowned German company *Telefunken*,[5] developing radio transmission systems. One evening he was approached by WERNHER VON BRAUN[6] and two of his friends, HERMANN STEUDING and ERNST STEINHOFF,[7] who led the department of *Orientierung, Steuerung und Bordinstrumentierung*[8] in *Peenemünde*, where the A4 rocket was under development. They asked him to come to Peenemünde and help designing a radio guidance system for this new ballistic missile.

HOELZER started working on the proposed radio guidance system shortly thereafter.[9] A particular problem that aroused his interest was the need to compute derivatives and integrals in flight – something completely unheard of before. Based on his ideas from his time at university he decided to use a capacitor as the central element for a circuit capable of differentiation and integration.

The current $i(t)$ charging a capacitor is generally described by

$$i(t) = C\dot{v}(t)$$

where C denotes the capacity and $v(t)$ is the voltage across the capacitor plates at time t.[10] This can also be written as

$$v(t) = \frac{1}{C} \int i(t)\,\mathrm{d}t,$$

so both operations, differentiation and integration, can be implemented using a capacitor. A simple, yet in many cases sufficient approximation is implemented by a simple passive[11] RC circuit.[12] In the case of differentiation the current flowing through the capacitor can be measured as voltage drop across a sufficiently small

---

[5] After the war Telefunken played a major role in the development of analog computers in the European market.
[6] 03/23/1912–06/16/1977
[7] 02/11/1908–12/02/1987
[8] German for *guidance, steering, and onboard instrumentation*.
[9] See [TOMAYKO 1985, p. 230]. One of the resulting radio guidance systems was named *Hawaii I* and was built by Telefunken (see [LANGE 2006, pp. 163 ff.]).
[10] A dot over a variable denotes its derivative with respect to time.
[11] I. e., there is no active element such as an operational amplifier in the circuit.
[12] A circuit consisting of a resistor and a capacitor.

resistor, in the case of integration the voltage across the capacitor's plates can be used directly as an approximation for the time integral of some varying input voltage.

As simple as this idea looks from today's perspective, its realization was far from easy in the 1940s due to the fact that it requires DC coupled amplifiers because the signals to be amplified and processed further are mainly DC voltages. An ideal DC amplifier would yield an output voltage of 0 Volts if its input voltage is 0 Volts, regardless of its gain, which should be as high as possible. Due to unavoidable imbalances, aging of components, temperature effects, etc., this is not true for a real-world DC coupled amplifier in which the individual amplifier stages are coupled directly. Accordingly, a small error voltage occurring in an early amplifier stage will be propagated and amplified by the following stages, resulting in a non-zero output signal even for an input signal of 0 V.

This precludes the application of such an amplifier for an integrator circuit, since the capacitor would integrate this offset signal, which would not only invalidate the computation but eventually saturate the integrator.

To solve this inherent problem HOELZER decided to use an AC coupled amplifier instead. This idea solved the drift problem as AC coupled amplifiers employ amplification stages, which are not coupled galvanically but by means of capacitors, which effectively block any DC (drift) voltage. Although this characteristic of AC coupled amplifiers is very desirable, it also makes this kind of amplifier unsuitable for the use with DC (or very low frequency) voltage signals as they occur in the control circuitry of a missile.

So HELMUT HOELZER faced the problem of converting an input DC voltage to an AC voltage suitable for feeding an AC coupled amplifier and rectifying the output voltage again to get the desired DC signal. Transforming DC signals into AC voltage was done by means of a *ring modulator*,[13] whose structure is shown in figure 3.1. This circuit was described first in the early 1930s and was used widely in radio applications.[14] Remarkably, most of the earliest ring modulators used a very early form of solid state diodes, which were called *Sirutor*[15] in Germany. These devices were developed in 1934 and used a small stack of copper(I) oxide[16] pills within a little tube with electrodes attached to its ends. Figure 3.2 shows a typical Sirutor like those employed by HELMUT HOELZER for his ring modulators.

---

[13] The name stems from the fact that the central four diodes are often arranged in a ring-like structure.
[14] See [ASCHOFF 1938] and [CHANCE et al. 1949, p. 379].
[15] Short for *Siemens-Rundfunk-Detektor*.
[16] Cu_2O

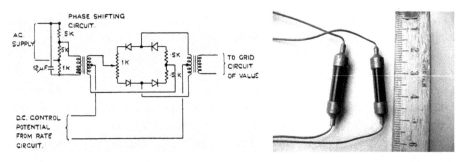

Fig. 3.1. Ring modulator circuit used in the Mischgerät (see [HUDSON et al. 1945, p. 13])   **Fig. 3.2.** Sirutor (scale in centimeters)

In 1940 the first usable radio guidance system for the A4 rocket was developed[17] and HOELZER turned his interest to the development of a gyro based stabilisation and guidance system. Like earlier developments by ROBERT GODDARD[18] the A4 used movable vanes in the rockets engine exhaust, *exhaust rudders*, as well as *air rudders* to control the flight path of the rocket. The exhaust rudders were the most effective means of control but could only work during the powered ascent phase. After re-entry into earth's atmosphere the air rudders became effective.[19]

This system turned out to be difficult to control due to the sheer size of the A4 and the resulting moment of inertia. The control system had to take this into account to avoid wild oscillations, which would result in the destruction of the rocket during flight. Figure 3.3 illustrates this problem: The upper half shows the buildup of oscillations in the case of a simple control system unable to take the moment of inertia into account.[20] The lower half shows the desired behaviour of the controller – the oscillation is quickly damped out, which requires more than just a linear controller, it requires the ability to take a derivative term into account.

At first, the company *Kreiselgeräte GmbH* attempted to develop a gyro based controller, which showed the undesirable behaviour shown in the top half of figure 3.3. This first implementation did not take any derivatives of the signals generated by the gyro platform into account. Another approach, using bank-and-turn indicators in addition to the main gyro, turned out to be too complex and too costly. After several unsuccessful attempts to build a working control system for the A4, VON BRAUN approached HELMUT HOELZER as he was aware of the fact

---

17 See [TOMAYKO 1985, p. 230].
18 10/05/1881–08/10/1945
19 [BATE et al. 1971] give a thorough description of guidance and control of ballistic missiles in general.
20 In today's parlance this would be a P-controller, using only a proportional feedback path.

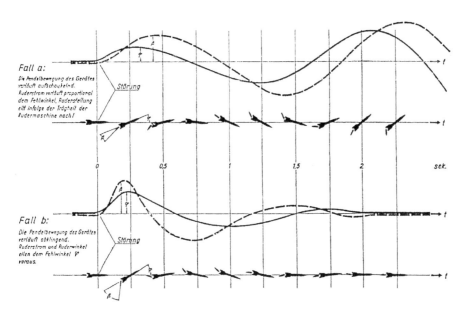

Fig. 3.3. Control problem of the A4-rocket ([N. N. 1945, fig. 82])

that HOELZER's radio guidance system also needed derivatives and time integrals of sensor output voltages, which were generated using RC combinations instead of costly additional instruments. This turned out to be a viable approach – while the initially proposed bank-and-turn indicators initially cost about US$ 7 000, the simple circuit developed by HOELZER only cost about US$ 2.50[21] and worked very well even under the severe stresses encountered during the flight of the rocket.

The idea of employing a differentiating circuit to derive rate of change signals from sources such as a gyro system was completely novel at that time and industry representatives were sceptical to say the least. HOELZER encountered substantial hostility from the companies involved in the development work for the A4. Only the obvious success of his control scheme eventually convinced the management. The resulting control system was dubbed *Mischgerät*, which can be translated as *mixing unit*, a camouflage term invented to disguise the real purpose of this first fully electronic onboard computer.

Figure 3.4 shows the main parts of the A4 guidance and control system. At the top of the rocket, directly below the war head, is a compartment containing the gyro system, an optional radio receiver/transmitter, and the all important Mischgerät, which controls the servo motors that operate the exhaust and air rudders at the lower end of the rocket.

---

[21] See [TOMAYKO 1985, p. 232].

46 — 3 The first electronic analog computers

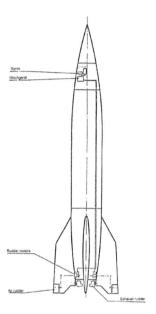

Fig. 3.4. Main elements of the A4 control system ([N. N. 1945, fig. 83])

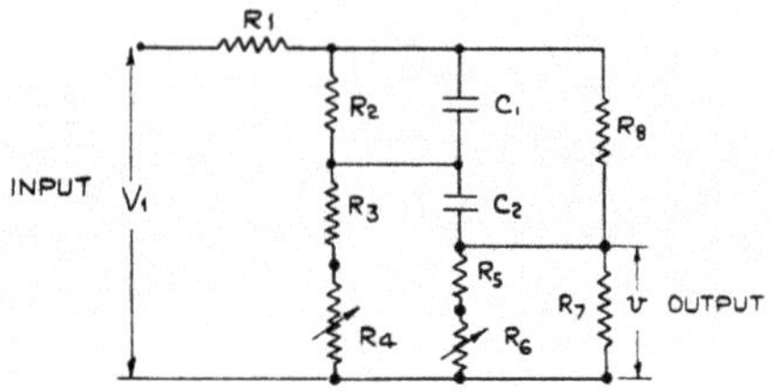

Fig. 3.5. Structure of the basic RC combination used in the Mischgerät (see [HUDSON et al. 1945, p. 13])

A more detailed schematic of the control system is shown in figure 3.6.[22] Shown on top of the picture is the main bus of the rocket supplying all subsystems with 27 Volts DC. Using two motor-generator units, denoted by *Umf. I* and *Umf. II*,[23] this supply voltage is converted into the voltages required for the gyro motors and the Mischgerät. The two gyros, *Richtgeber D* and *Richtgeber EA* are connected to the Mischgerät, which is shown in the middle of the figure. Using RC combinations denoted by *RC Glied* the input signals are differentiated, which effectively implements a PD controller.[24] The resulting output signals are used to drive the servo motors controlling the rudders. Figure 3.5 shows the structure of the basic RC combination used in the Mischgerät.

---

22 A much more in-depth treatment of the Mischgerät from the British perspective can be found in [HUDSON et al. 1945]. HOELZER, too, wrote in great detail about this device in the second (secret) part of this Ph. D. thesis (see [HOELZER 1946/2]).

23 Short for *Umformer*, German for a motor-generator unit.

24 Using simple RC circuits directly to generate derivatives and integrations yields quite imprecise results. Since the gyro based control scheme for the A4 rocket only required linear terms and first and second derivatives, this was not a problem as long as the controller itself was operating in a stable way. More precise computing circuits as used in general purpose analog computers would have been too costly, too large and too complicated for the intended use of the rocket.

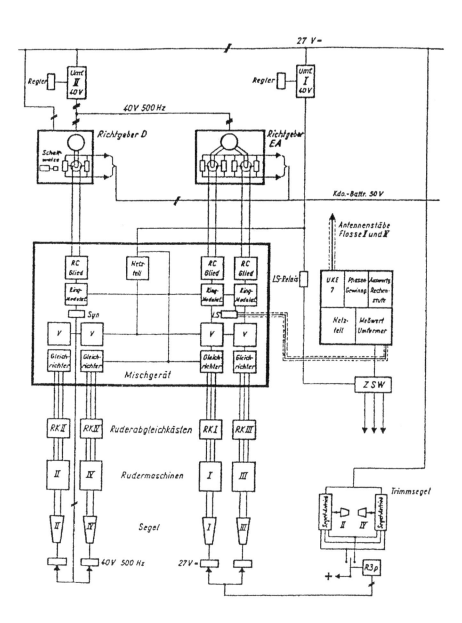

Fig. 3.6. Simplified schematic of the A4 control system (see [N. N. 1945, fig. 80])

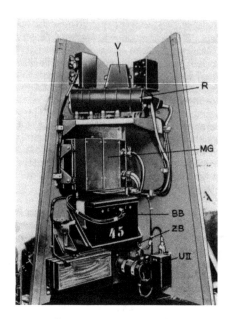

Fig. 3.7. The installed Mischgerät ([TRENKLE 1982, p. 134])

Figure 3.7 is a photograph of the control system's compartment in an A4 rocket. The Mischgerät itself is denoted by *MG*, *BB*, and *ZB* are the main battery and a supplemental battery, while *U II* is the second motor-generator unit powering the gyros, which are installed in one of the other quadrants of this compartment on top of the rocket.

The Mischgerät itself is shown in figure 3.8 – its small size and rugged design are as obvious as they are remarkable. It is built on a base plate holding five plug-in modules, two at the bottom and three at the top. The empty space on the bottom right contains the connectors to the various rocket subsystems.

This device was described by [TOMAYKO 2000, p. 15] as the *"first fully electronic active control system"* and it had great influence on further developments as this quotation from [BILSTEIN 2003, p. 243] shows:

> *"Further work by other Peenemünde veterans and an analog guidance computer devised with American researchers at the Redstone Arsenal culminated in the ST-80, the stabilised platform, inertial guidance system installed in the Army's 1954 Redstone missile [...] The ST-80 of the Redstone evolved into Jupiter's ST-90 (1957) [...]"*

3.1.2 Hoelzer's analog computer

In parallel with his work on the Mischgerät and initially unnoticed by his supervisor, HELMUT HOELZER started development of a fully electronic general purpose

Fig. 3.8. The Mischgerät (photo: ADRI DE KEIJZER, reprinted with permission)

analog computer in 1941. This machine was also based on his idea of using RC combinations to perform integration and differentiation. He was confronted again with many objections and the prejudice by other people. HERMANN STEUDING once commented about HOELZER's ideas:[25]

> "Young man, when I compute something, the results will be correct and I do not need a machine to verify it. By the way, machines cannot do this."

He was even forced by his supervisor to stop fiddling around with this electronic contraption and concentrate on his real work, as he remembers vividly:[26]

> "My boss came to the laboratory where he saw the electronic contraption. He said: 'HOELZER, stop playing with this toy and do your work! I gave the only answer possible under these circumstances: 'Yes, sir!' Next morning, nothing was left in the lab – the computer had been moved to a small, windowless room. [...] Nevertheless, it seemed

---

25 See [TOMAYKO 1985, p. 234].
26 See [HOELZER 1992, pp. 13 f.].

important to me to develop a system that could aid the development of rocket control systems significantly. [...] When it finally worked, everything was forgiven."[27]

Further development on the analog computer continued in secret. After its completion it not only proved to be a valuable tool, but moreover it played a central role in the A4 development process. Without this analog computer, many problems, especially in the area of guidance and control of a ballistic missile like the A4, could not have been solved, at least not in time. In a short span of time many departments in Peenemünde used this machine actively to solve a variety of problems – even the word *analog* used to denote the setup of an analogon became part of engineer's terminology.[28] The influence of this computer cannot be overestimated as [NEUFELD 2007, p. 133] makes clear. He describes this computer as "*a fundamental innovation that really made a mass-produced guidance system possible*".

Since this analog computer was intended to be a research instrument it had to be more precise than the simple RC combinations used in the Mischgerät for integration and differentiation. HOELZER therefore used a controlled current source to charge the capacitor in the case of integration while a controlled voltage source was necessary for the differentiator. Both sources were implemented using AC coupled amplifiers like those being used in the Mischgerät.[29]

Figure 3.9 shows the basic schematic of the differentiator developed by HOELZER for his analog computer.[30] The input values for this computing element are AC voltages that are coupled inductively into the computing element and are amplified by a triode based input stage. A synchronous demodulator in the anode circuit of this tube rectifies the amplified input AC signal. The resulting DC signal is then used to charge the capacitor C while the charging current representing the derivative to be determined, causes a small voltage across resistor r. This voltage

---

27 "*Mein Chef kam ins Labor, sah [den] elektronische[n] Drahtverhau und sagte nur: „HOELZER, hören Sie doch endlich auf mit dieser elektrischen Spielerei und kümmern Sie sich um Ihre Aufgabe. Ab morgen ist das alles weg, verstanden?" Ich sagte das einzige, was man in solcher Situation sagen konnte: „Jawohl, Herr Doktor." Am nächsten Morgen war alles weg und zwar war es jetzt in einem kleinen Raum ohne Fenster [...] Aber mir schien es auch wichtig, ein Gerät zu schaffen, welches, wie man damals dachte, in der Hauptsache für die Entwicklung von Raketensteuerungen von nicht zu überbietender Wichtigkeit war. [...] Als alles funktionierte, wurde mir dann vergeben.*"
28 See [PETZOLD 1992, p. 57].
29 In 1949 EDWIN A. GOLDBERG applied for a patent that described a practical scheme of eliminating the drift effects in DC amplifiers. It was this invention that made the cumbersome AC coupled amplifier approach with the ring modulators, etc., superfluous and the subsequent rapid development of analog computers possible (see section 4.1.2).
30 The symbols in the schematic differ from those used today and are those used in HOELZER's dissertation.

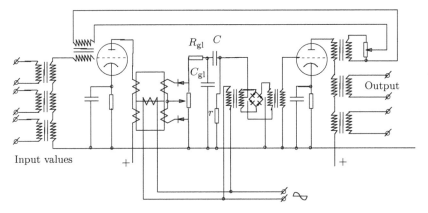

Fig. 3.9. Schematic of HOELZER's differentiator circuit (cf. [HOELZER 1946, fig. I, 13])

is then converted back into an AC voltage by a ring modulator, the output of which driving the grid of the output amplifier stage that in turn feeds several output transformers connected in series.

Apart from this demodulation and modulation scheme there are three noteworthy aspects in this circuit:

1. One of the outputs is used to generate a feedback signal, the amplitude of which can be set with the potentiometer shown in the upper right corner of figure 3.9.
2. The modulator and demodulator are both driven by a common carrier signal.
3. Finally, the differentiator employs a simple low pass filter consisting of R_{gl} and C_{gl}[31] in order to suppress excessive noise.[32]

The integrator circuit differs from the differentiator circuit only in a few respects: The low pass filter consisting of R_{gl} and C_{gl} can be omitted since the operation of integration itself acts as a low pass filter. In addition to that, the output stage is fed directly with the modulated voltage across the capacitor plates, so the resistor r is also no longer required. Figure 3.10 shows the basic schematic of the integrator used in HOELZER's analog computer.

---

[31] The abbreviation *gl* denotes *glätten*, German for smoothing.
[32] Differentiators are nowadays normally avoided in electronic analog computers since they naturally tend to increase the noise of a signal excessively. HOELZER alleviated this problem by the inclusion of this low pass filter.

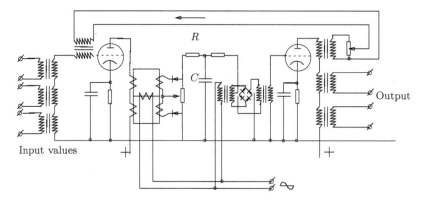

Fig. 3.10. Schematic of HOELZER's integrator circuit (cf. [HOELZER 1946, fig. I, 11])

The analog computer also contained additional computing elements such as multipliers, dividers, and square root function generators.[33]

Especially interesting is a servo circuit suggested by HOELZER to implement division and other operations, since this device is a direct forerunner of the later servo function generators and multipliers described in sections 4.5.1 and 4.6.1. Figure 3.11 shows the basic structure of this computing device. It essentially forms a bridge circuit with a servo motor driving two potentiometers R_1 and R_{out} as its central element. R_1, the fixed resistor R_2, and the two input transformers form the actual bridge circuit. Every imbalance of this bridge will yield an error voltage that is amplified by the Amplifier A, which in turn drives the servo motor canceling out the imbalance by readjusting R_1 according to the current input.

Since R_1 and R_{out} are driven in tandem this self-adjusting bridge yields

$$y = \frac{e_1(t)}{e_2(t)}$$

at the output of the voltage divider R_{out}, which needs an auxiliary supply voltage e_{aux}.

However, this divider circuit was not implemented during war time since the necessary servo motor could not be obtained. Instead HOELZER implemented a divider based on the solution of the differential equation

$$\frac{1}{A}\dot{y} + ax = b$$

---

33 Explicit summers were not necessary since the modulated signals used to transmit values from one unit to another could simply be fed into the multiple transformer coupled inputs of the computing elements. These series connected transformers (see figs. 3.9 and 3.10) then implicitly performed the summing operation.

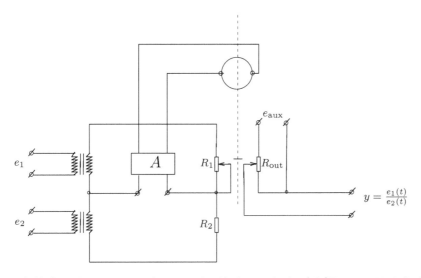

Fig. 3.11. Simplified schematic of HOELZER's self-adjusting bridge (cf. [HOELZER 1946, fig. II, 11])

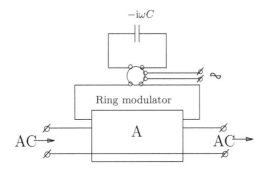

Fig. 3.12. An improved integrator circuit (cf. [HOELZER 1946, fig. II, 11])

where A denotes the gain of the amplifier used in the setup. A should be as high as possible to minimise errors. The multiplication ax was implemented by two ring modulators connected in series.

Figure 3.12 shows an improved integrator circuit that also was proposed by HOELZER. The novel idea here is to simplify the integrator by using only a single AC coupled amplifier instead of two such stages. Since the integration capacitor needs a DC current to be charged, HOELZER's idea was to place the capacitor at one end of a ring modulator while the other side of the ring modulator was used to close a feedback loop over the amplifier. This idea anticipated the basic integrator design that would dominate analog electronic analog computers for the following decades.

Unfortunately, this circuit was also not realized during World War II since integrators based on the original design (figure 3.10) had already been built and worked reliably and satisfactorily.

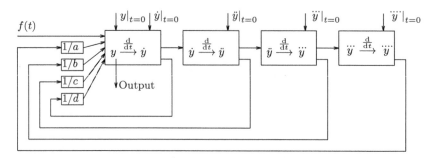

Fig. 3.13. Solution of a differential equation of fourth degree $\ddddot{y} + a\dddot{y} + b\ddot{y} + c\dot{y} + dy = f(t)$ with a feedback circuit employing four differentiators (cf. [HOELZER 1946, fig. II, 2])

HOELZER gives a good example how this first analog computer was programmed in his 1946 dissertation[34] where he outlines three different approaches for solving the following differential equation of fourth degree:

$$\ddddot{y} + a\dddot{y} + b\ddot{y} + c\dot{y} + dy = f(t) \qquad (3.1)$$

Figure 3.13 shows a straightforward approach using four differentiating elements in series to derive $\dot{y}$, $\ddot{y}$, $\dddot{y}$ and $\ddddot{y}$ from y. The required y is generated by adding these derivatives (multiplied by proper coefficients) and an input function $f(t)$.

As straightforward as this solution is, it has a severe drawback: The differentiator chain will increase noise in the variables used, despite the low pass filters employed in each differentiator. Being aware of this fact HOELZER proposed a second approach using four integrators, which is shown in figure 3.14. This setup is the same as any later analog computer programmer would have proposed it. In fact, this computer setup results from the application of the KELVIN feedback technique,[35] which HOELZER probably wasn't aware of.

Unfortunately, this solution was not without problems, either. It solved the inherent noise problem of the purely differentiator based approach but introduced yet another problem due to the non-ideal characteristics of the computing elements, namely the capacitors: Every integrator introduced some amount of drift due to leakage in its integration capacitor. Thus, this circuit was also suboptimal in solving equation (3.1).

To alleviate these problems HOELZER suggested a third setup shown in figure 3.15. This approach is a combination of the integrator or differentiator based

---

[34] See [HOELZER 1946].
[35] See section 7.2.

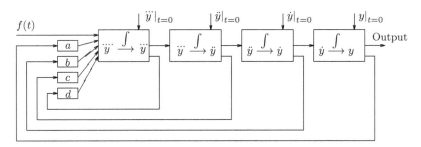

Fig. 3.14. Solution of a differential equation of fourth degree $\dddot{y} + a\dddot{y} + b\ddot{y} + c\dot{y} + dy = f(t)$ with a feedback circuit employing four integrators (cf. [HOELZER 1946, fig. II, 1])

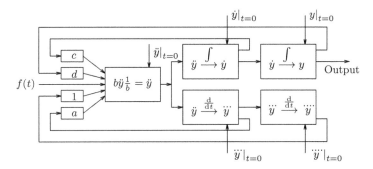

Fig. 3.15. Solution of a differential equation of fourth degree $\dddot{y} + a\dddot{y} + b\ddot{y} + c\dot{y} + dy = f(t)$ with a feedback circuit employing two differentiators and two integrators (cf. [HOELZER 1946, fig. II, 3])

setups and uses two differentiators as well as two integrators thus minimizing the negative effects of each type of computing element.[36]

Figure 3.16 shows the analog rocket motion simulation computer as it was built in the first half of the 1940s. The three top frames contain various computing elements such as integrators, differentiators, a model transmitter, and meters for the various angles, flight time, etc. Below is an electromechanical function generator based on multiple cams mounted on a central shaft driven by a motor which made variable coefficient simulations possible. The bottom compartment contains the power supplies and a motor-generator unit producing the carrier frequency signal for the modulators and demodulators.[37] However, this machine was

---

[36] Later analog computers avoided differentiators wherever possible, but this required some substantial technological advances, most notably very high gain drift stabilised DC amplifiers and high precision computing capacitors.

[37] See [HOSENTHIEN et al. 1962] for a detailed description of this particular system as well as its successors.

Fig. 3.16. HOELZER's analog computer for rocket motion simulation as it was found after World War II (source: NASA, Marshall Space Flight Center)

not a true general purpose analog computer as its computing elements could not be connected freely. Instead it featured a fixed internal structure optimised for its task of simulation rocket motion.

Two of these analog computers were finally built,[38] one of which was taken to the United States where it was used until about 1955.[39] It was used for various early rocket development projects such as the *Hermes* rocket. Based on this machine an improved version was developed under VON BRAUN in 1950 that was

---

[38] In 1993 HELMUT HOELZER started building a replica of this first analog computer, which was completed in 1995. This machine is now part of the collection of the *Deutsches Technikmuseum Berlin*.

[39] This is the machine shown in figure 3.16.

used for about ten years and proved to be a valuable tool in the development of the *Redstone* and *Jupiter* rockets as well as for the design of the first satellite of the United States, *Explorer I*.[40]

3.2 George A. Philbrick's Polyphemus

In 1938 GEORGE A. PHILBRICK,[41] who worked for Foxboro,[42] wrote a proposal describing a novel *simulator*[43] for process control systems. The main goals of the development were stated as follows:[44]

> "We attempt to describe a method for the rapid and easy solution of problems which arise in connection with the technical study of process control. Also included is an electrically operated unit capable of disclosing the behaviour of controlled systems as influenced by their various physical characteristics."

PHILBRICK started with simple analogies like a capacitor modeling a tank, a resistor representing a valve restricting the flow of a medium, etc. Figure 3.17 shows an example used by PHILBRICK in 1938. Depicted are three tanks T_1, T_2, and T_3 connected by pipes P_1 and P_2, which contain valves V_1 and V_2. These series connected tanks can be filled via P_{in} and drained by P_{out}. The latter pipe also contains a valve V_{out}.

At first, he proposed to model this three tank system by means of passive electronic components, with resistors representing the valves and capacitors representing the tanks as shown in figure 3.18.

Of course more complex simulations would require some kind of amplifier so PHILBRICK was faced with pretty much the same problem as HOELZER in Germany: No such devices existed in the late 1930s and state of the art DC amplifiers showed excessive drift, making precise simulations impossible. Consequently, PHILBRICK decided to use AC amplifiers in some of his simulators as did HOELZER[45] but omitted the modulator and synchronous demodulator circuits since his simulator was intended to be used in a mode of operation now known as *repetitive opera-*

---

40 See [TOMAYKO 2000, p. 236].
41 01/05/1913–12/01/1974
42 Founded as *Industrial Instrument Company* in 1908, Foxboro (now owned by Schneider Electric) still is one of the leading brands of industrial control and measurement systems.
43 The term simulator is defined by [HOLST 1982, p. 144] as follows: "*A simulator is a fixed (to a large degree) structure embodying one unique model. [...] A simulator's purpose is specific: to provide the accurate realization of its model for various parameters, stimuli, and operator interactions of interest to its users.*"
44 Cf. [HOLST 1982, p. 143].
45 Cf. [HOLST 1982, p. 149].

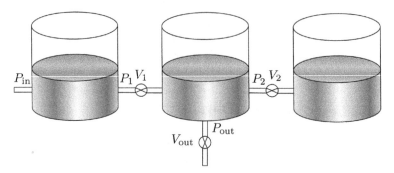

Fig. 3.17. Three interconnected tanks (cf. [HOLST 1982, p. 146])

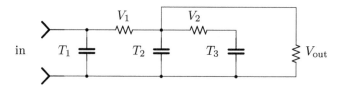

Fig. 3.18. Simulation circuit for the three interconnected tanks of figure 3.17 (cf. [HOLST 1982, p. 146])

tion. In this mode a simulation is run automatically over and over again in quick succession which makes it possible to display the solution of the problem as a flicker-free graph on an oscilloscope. Changing the settings of the potentiometers representing coefficients of the system being simulated causes this graph to change immediately, making this an incredibly powerful tool for investigating the behaviour of dynamic systems.

Based on these ideas PHILBRICK set out to develop a process simulator that was eventually called *Polyphemus* due to its appearance, with a single oscilloscope mounted in the top section of a 19-inch rack holding the simulator's components. With some imagination this setup looked a bit like Polyphemus, the Cyclops of the Odyssey.

Although the basic structure of this simulator was fixed, it could be adapted to different simulation tasks, within limits, by means of a removable cardboard front plate. This plate contained a schematic diagram showing the structure of the process to be simulated and annotations for the various control elements of the simulator like potentiometers, etc. Figure 3.19 shows Polyphemus with a faceplate representing a system consisting of two liquid baths with stirrers and steam and cold water inlets.[46] Other face plates depicting pneumatic controllers, etc., were also available.

---

46 Cf. [HOLST 1982, p. 152].

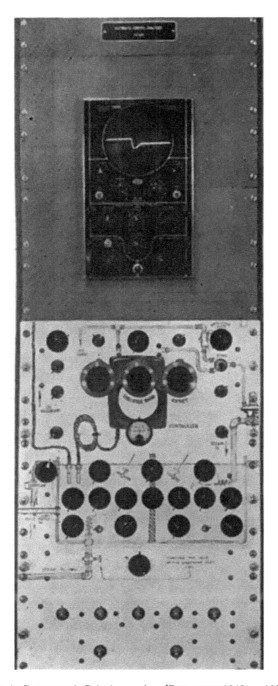

Fig. 3.19. GEORGE A. PHILBRICK's Polyphemus (see [PHILBRICK 1948, p. 108])

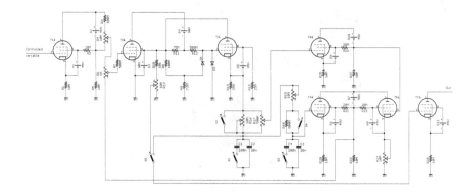

Fig. 3.20. Seven-stage amplifier from GEORGE A. PHILBRICK's laboratory notebook, March 11, 1940 (see [HOLST 1982, p. 154])

Although Polyphemus was intended as a tool for basic research it turned out to be much more versatile and useful. Eventually, it was mainly used as a training simulator and as an aid for sales people that could easily be used to demonstrate the behaviour of a proposed process controller to prospective customers.[47] In retrospective Polyphemus has been described as follows:[48]

> "In those days, an electronic analog machine was a pioneering venture. The Kelvin-Bush differential analysers were mechanical, expensive, and very large. Appropriate electronic techniques, if they existed, were not available. Nevertheless, it was evident that only this medium offered the required flexibility and speed, not to mention economy."

This world's first all electronic process simulator is now part of the collection of the *Smithsonian Institution* – its successor, featuring two oscilloscopes and thus no longer looking like a Cyclops, was in active use at Foxboro well into the 1980s.[49]

To get an impression of the complexity of the circuits developed by GEORGE A. PHILBRICK for his numerous process simulators, figure 3.20 shows the schematic diagram of a seven stage circuit that models a process controller – particularly noteworthy are the two diodes for implementing non-linear responses as well as the DC coupled subcircuits.

In 1946 GEORGE A. PHILBRICK founded a company named *GAP/R*, short for *George A. Philbrick Researches*, that developed and sold products ranging from operational amplifiers[50] to fully fledged analog computers and simulators. These

---

47 See [HOLST 1982, pp. 153 f.].
48 See [GAP/R Evolution].
49 Cf. [HOLST 1982, pp. 155 f.].
50 See section 4.1.

owed much to his early experiences with Polyphemus and its successors. One of the most famous products of his company was the first commercially available operational amplifier, the *K2-W*,[51] which shaped a whole industry. In 1966 GAP/R merged with *Teledyne* to form *Teledyne-Philbrick*.

3.3 Electronic fire control systems

The decline of mechanical fire control systems like those described in section 2.7 had already started before the advent of World War II.[52] This was not due to insufficient accuracy or precision or even slow response times – in fact, the military feared that in the case of a war the sheer complexity of these machines would prevent their production in sufficient quantities. The main obstacles were that the necessary parts and resources would be short in supply in such a situation and the necessary highly skilled craftsmen would not be available due to other war effort demands. To make things even worse the rapid development of radar systems also rendered the classic fire control systems obsolete since they could not be directly coupled with radar stations.

In 1940 DAVID B. PARKINSON, who worked for the *Bell Telephone Laboratories (BTL)*, and his coworkers were developing an *automatic level recorder*, a kind of a strip-chart recorder employing a logarithmic scale. To drive the recorder's pen a servo mechanism based on a feedback scheme was necessary and this inspired PARKINSON to think about using this technique for the automatic control of azimuth and elevation of a gun.[53] His supervisor, CLARENCE A. LOVELL, was enthusiastic about this idea and together they started a thorough investigation of the necessary technology and mathematics.[54]

It was decided to use DC voltages to represent the various variables in a fire control calculation. Coordinates of the target and the gun were represented in a three-dimensional Cartesian coordinate system requiring computer circuits for implementing trigonometric functions, summing, differentiation, etc. Target speed components would be generated by electronic differentiation of the position signals. PARKINSON summarized the main features of the proposed fire control computer as follows:[55]

---

[51] See section 4.1.
[52] See [CLYMER 1993, p. 30].
[53] In [222, p. 135] PARKINSON describes a lucid dream that led him to this development. See also [ZORPETTE 1989].
[54] It is remarkable that *"[n]either* PARKINSON *nor* LOVELL *had any experience in fire control; they did not even know of the existence of mechanical gun directors"* as [222, p. 137] states.
[55] See [MINDELL 1995, p. 73].

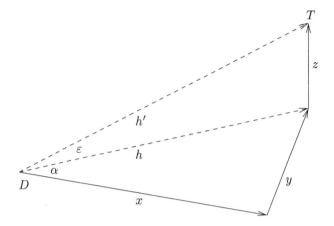

Fig. 3.21. Basic fire control task

> "It required (1) a means of solving equations electrically (potentiometers), (2) a means of deriving rate for prediction (an electrical differentiator), and (3) a means of moving the guns in response to firing solutions."

The patent application[56] states this explicitly:

> "An important form of the invention is a device which, when supplied with electrical voltages proportional to two sides of a triangle, will set itself to indicate an angle of the triangle and to produce a voltage proportional to the other side of the triangle.
> Another form of the invention is a device which, when supplied with voltages proportional to the rectangular coordinates of a point, will set itself to indicate the polar coordinates of the point."

The basic problem of aiming a gun is shown in figure 3.21. The instantaneous Cartesian coordinates of a target T, provided by the tracking radar, with respect to the position D of the gun are given as x, y, and z. Based on these, the gun's azimuth and elevation angles α and ε have to be computed, taking into account the future predicted position of the target and the ballistic characteristics of the shell.

Figure 3.22 shows the structure of a trigonometric function generator developed by LOVELL, PARKINSON, and WEBER, which was used in their electronic fire control computer. The motor on the far left is running with constant speed and drives two bevel gears 3 and 4, which rotate in clockwise and counterclockwise direction respectively. Two electromagnetic clutches, 5 and 6, can be activated by a control signal thus controlling, which gear will be coupled to the vertical shaft, which in turn drives the two potentiometers shown at the top left of the figure.

---

56 See [LOVELL et al. 1946].

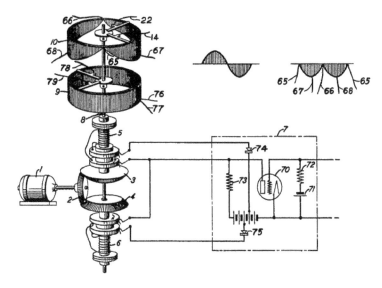

Fig. 3.22. Sine/cosine generator used in the T-10 fire control computer ([LOVELL et al. 1946, fig. 1/2])

The control signals for these two clutches were generated by the comparator circuit shown in the box denoted by 7. This circuit can generate three different control signal combinations: Upper clutch, lower clutch or no clutch at all will be engaged. These states are controlled by the input signal connected to the appropriately biased grid of the triode labelled 70.

It should be noted that the lower of the two potentiometers shown is a linear potentiometer yielding a voltage varying linearly with the angular position of its wiper when connected as a voltage divider, while the upper potentiometer is of a special type, wound to deliver a voltage corresponding to the desired trigonometric function of the wiper angle.

The combination of this circuit with an amplifier finally yields a servo system that can be used to automatically compute $\sin(\alpha)$, etc., given an angle α. This forced LOVELL, PARKINSON, and WEBER to develop an amplifier circuit as shown in figure 3.23, which is a precursor of later *operational amplifiers*.[57]

This amplifier acts as an inverting summer[58] with two inputs connected to the resistors 95 and 96 and a feedback resistor 94. Using this amplifier and the device shown in figure 3.22, a servo system can now be implemented by connecting the output of this summer to the input of the comparator controlling the two clutches. The inputs of the summer are connected to the output of the linear potentiometer

---

[57] See section 4.1.
[58] See section 4.2.

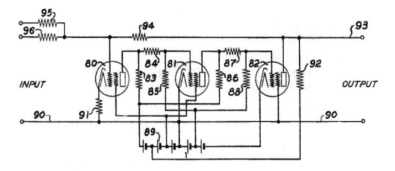

Fig. 3.23. Operational amplifier used in the T-10 fire control computer ([LOVELL et al. 1946, fig. 5])

and a circuit yielding α. Whenever α deviates from the angular position of the potentiometers' wiper the summer will generate a positive or negative output signal thus forcing the comparator to engage either the upper or the lower clutch to minimise the error between the wipers' position and the value α.

LOVELL, PARKINSON, and WEBER realized that the computing elements they proposed were not limited to fire control but could be used to solve a variety of problems as the following quote from [MINDELL 1995, p. 74] shows:

> "A digression from the principal subject is made to comment that the use of servo mechanisms to solve simultaneous systems of equations is feasible and, in a large number of cases, practicable. This fact may lead to the application of this type of mechanism to the solution of many types of problems dissociated from the one in question."

The actual circuit developed for the implementation of a coordinate transformation relies on the observation that

$$x = h \cos(\alpha) \quad (3.2)$$
$$y = h \sin(\alpha) \quad (3.3)$$

given the triangle x, y, h in figure 3.21. (3.2) and (3.3) can be expanded to

$$x \sin(\alpha) = h \cos(\alpha) \sin(\alpha)$$
$$y \cos(\alpha) = h \sin(\alpha) \cos(\alpha)$$

yielding $x \sin(\alpha) - y \cos(\alpha) = 0$, which is used to control the basic servo circuit of the coordinate resolver shown in figure 3.24.

This circuit has two inputs x and y, which are connected to two specially wound potentiometers 12 and 18 with four connections each. The two *polarity reversing repeater* devices are in fact summers with only one input each and gain 1.[59]

---

[59] These could also be called *inverters*.

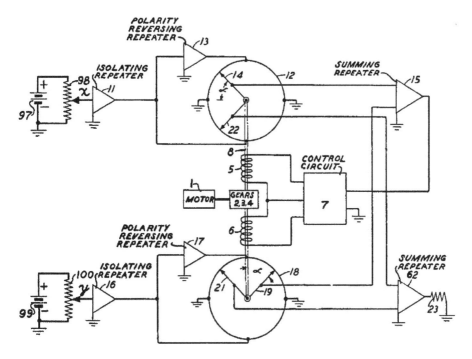

Fig. 3.24. Coordinate resolver used in the T-10 fire control computer ([LOVELL et al. 1946, fig. 4])

Both potentiometers are connected to $-x$, ground (representing the value 0), x, 0 and $-y$, 0, y, 0 respectively. Each potentiometer has two wipers mounted at an angle of $\pi/2$ driven by a common shaft that can be connected to one of the two bevel gears driven by a motor by means of the two clutches denoted by 5 and 6.

The potentiometers are wound in such a way that their resistance represents $\sin(\varphi)$ and $\cos(\varphi)$ with φ denoting the wiper angle (due to the two wipers mounted at a relative angle of π/s with respect to each other, both functions are generated simultaneously using only one such potentiometer). Thus, the wipers 14 and 22 yield $x\sin(\alpha)$ and $x\cos(\alpha)$ while the lower potentiometer generates $-y\cos(\alpha)$ and $y\sin(\alpha)$ at its wipers 19 and 21 respectively. The summing *repeater* 15 now computes the value

$$c = x\sin(\alpha) - y\cos(\alpha), \tag{3.4}$$

which is used to control the comparator 7 driving the two clutches, thus closing the servo loop. As long as c equals zero, both clutches are disengaged. As soon as c deviates from zero, one of the clutches will be activated, thus connecting the central shaft driving the two potentiometers to the motor, which will in turn minimise the error term c until it reaches zero again. Thus, the angular position of the wipers is α, which can be used to control the gun's azimuth.

The second summing repeater, 62, computes $h = x\cos(\alpha) + y\sin(\alpha)$, which is the horizontal range to the target. Using this value and z as inputs for a second, identical circuit, eventually yields h' and the elevation angle ε represented by its potentiometer shaft position. It was even suggested to make the motors driving these two sub-circuits strong enough so that they could actually drive the gun directly to point it to the target T.

The resulting fire control computer *T-10* consisted of four such servo circuits as well as 30 summing amplifiers and five power supplies. Since the voltages representing the Cartesian coordinates of the target with respect to the gun's location often changed quite slowly, the amplifiers had to be DC coupled amplifiers – thus introducing unavoidable drift effects into the calculations. Assuming that amplifier 15 of figure 3.24 exhibits a drift error voltage e this would cause the servo circuit to try making

$$c = x\sin(\alpha) - y\cos(\alpha) + e$$

zero instead of the term (3.4). This, in turn, results in non-negligible errors regarding α, ε, h and h'. Accordingly, the T-10 computer featured additional inputs to its summing amplifiers, which were connected to potentiometers. These had to be adjusted manually to minimise the drift voltage e of each amplifier. In addition to that problem, the T-10 relied on the computation of derivatives thus introducing additional noise into the control signals used to drive the servo circuits. This required the implementation of complex filter networks, further complicating its circuitry.

Despite of these shortcomings, the T-10, which was eventually produced under the designation *M-9* and coupled directly to a *SCR-584* radar system, proved to be greatly superior to the earlier mechanical fire control computers. Due to its easier manufacturing, it was decided on February 12, 1942 to stop all developments of mechanical fire control systems and to concentrate on all electronic devices in the future.[60] The hopes were high regarding the M-9:[61]

> "*The M-9 Director, electrically operated, is, we feel in Ordnance, one of the greatest advances in the art of fire control made during this war, and we anticipate from the M-9 Director very great things as the war goes on.*"

These hopes were not in vain as it turned out – the M-9 was deployed in large numbers[62] to Great Britain to shoot down *V1* German flying bombs. Initially 10 to 34 percent of the incoming flying bombs could be downed automatically and sometimes even higher rates were achieved:[63]

---

[60] See [HIGGINS et al. 1982, p. 225].
[61] See [222, pp. X f.].
[62] Eventually more than 3000 M-9 gun directors were built and deployed.
[63] Cf. [222, p. 148].

"*In a single week in August, the Germans launched 91 V1's from the Antwerp area, and heavy guns controlled by M-9's destroyed 89 of them.*"

Using the M-9, coupled with the use of proximity fuses, brought the number of shells required to hit an enemy aircraft down from many thousand to a mere 100. Hitting an incoming V1 flying bomb required about 200 shells.[64]

Later developments such as the experimental fire control computer *T-15* took many more parameters and effects into account, such as drift of the shell during flight, windage correction, muzzle velocity, air density, etc.[65] The T-15 performed all computations in polar coordinates thus eliminating the coordinate system transformations otherwise necessary.[66]

The technologies developed for these fire control computers formed the basis for following developments like a trainer for aerial gunners[67] that already used continuously operating servo systems in contrast to the controlled clutches of the T-10. Even more complex was the *Flight Training Apparatus* developed in 1948 by DEHMEL,[68] which implemented the basic operation of integration by means of a servo system controlling a potentiometer, the angular wiper position of which corresponds to the integral of the function applied to the servo circuit, etc.

3.4 MIT

In the fall of 1945 A. B. MACNEE set out to develop an all electronic analog computer at the MIT because[69]

"*[...] it was felt that there was considerable need for a differential analyser of somewhat different characteristics from any then in existence or under development. There appeared to be the need for a machine having the following characteristics: (a) moderate accuracy, of perhaps 1 to 10 per cent, (b) much lower cost than existing differential analysers, (c) the ability to handle every type of ordinary differential equation, (d) high speed of operation, (e) above all, extreme flexibility in order to permit the rapid investigation of wide ranges of equation parameters and initial conditions.*"

---

[64] See [ZORPETTE 1989].
[65] [BOGHOSIAN et al. 1950] gives an impression of the complexity of these fire control computers.
[66] Another system that performed all calculations in Cartesian coordinates is described in [BEDFORD et al. 1952].
[67] See [DEHMEL 1949].
[68] See [DEHMEL 1954].
[69] See [MACNEE 1948, p. 1].

These goals summarize the main advantages of an all electronic approach to building analog computers. The high-speed operation[70] mentioned is a very important feature since it allows the behaviour of systems modelled on the analog computer to be explored in a highly interactive way and often even faster than real-time. The operator can change the parameters of a simulation and immediately see the effects these changes have on the system.

This particular system developed by MACNEE consisted of integrators, summers, a high-speed function generator,[71] and a high-speed multiplier.[72] One of the novel ideas employed in this computer was that it ran in a repetitive mode of operation with a repetition rate of 60 Hz between individual computer runs, thus allowing a flicker-free picture to be displayed on an oscilloscope. Apart from the fact that this greatly facilitated interactive operation where an operator can change various parameters of a simulation, it had the additional benefit that AC coupled amplifiers could be used in the computer since even constant values would continuously reset 60 times a second due to the repetitive operation. Thus, it was possible to completely avoid using DC coupled amplifiers with their inevitable drift.

As an example of the capabilities of this early machine, figure 3.25 shows a particular solution for the MATHIEU *equation*[73]

$$\ddot{I} + \omega_0^2 \left(1 + \varepsilon \cos(\omega_m t)\right) = 0 \tag{3.5}$$

obtained by this electronic analog computer. The three curves shown on the oscilloscope's display are I, $-\dot{I}$ and $\ddot{I}$.

No other technology in the 1940s would have allowed the solution of such an equation in only 1/60 of a second. Machines like this MIT analog computer made it crystal clear that the days of mechanical and electromechanical analog computers were gone and the future belonged to high-speed electronic devices. These machines were destined to become crucial elements for the technological progress of the 1950s to the late 1970s.

3.5 The Caltech Computer

Compared with the machines described so far, the *Caltech* (short for *California Institute of Technology*) analog computer was a behemoth – it weight 15 tons

---

[70] Even not too low compared to our modern day stored-program digital computers in many cases.
[71] This function generator was based on optical feedback as described in section 4.5.3.
[72] See section 4.6.2.
[73] See section 8.9.

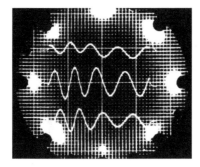

Fig. 3.25. A particular solution of the MATHIEU equation (3.5) obtained with MACNEE's all electronic analog computer, 1948

and occupied a considerable amount of floor space in its final incarnation. This machine was the first development task of the Caltech *Analysis Laboratory*, which was established in 1946. The computer was initially developed by GILBERT D. MCCANN[74] and HARRY E. CRINER and was put into active service as early as 1947 in a partially completed state. Even in this early stage of development the system was successfully applied to solve problems for the *Jet Propulsion Laboratory* (*JPL*), the military, and various aerospace companies.

The Caltech Computer differed substantially in its structure from the machines described previously as it used mainly passive electronic components, such as resistors, inductors, capacitors, and transformers instead of more abstract computing elements based on operational amplifiers as proposed and used by HOELZER, PHILBRICK, and MACNEE. The system was aptly named a *Direct Analogy Electrical Analog Computer* and resembled more of an intricate electric Tinkertoy kit than an abstract mathematical machine.[75] Figure 3.26 shows about half of the computing and control elements of the completed system. A noteworthy device is the forcing function generator visible in the lower left of the picture. It consists of a motor driven axle on which machined disks are mounted, which move the sliders of potentiometers thus generating arbitrary functions, which could be fed into an analog computer setup to explore the behaviour of forced systems.

Figure 3.27 shows a typical setup of the Caltech analog computer. The rhombus shape represents a function generator,[76] the rectangular box below is a multi-

---

[74] MCCANN shifted his research from computers to nervous systems, visual perception, and other biological topics.
[75] See [Caltech 1949]. A wealth of information on the application of such passive analog computers can be found in [PASCHKIS et al. 1968] and [VOLYNSKII et al. 1965].
[76] There were two types of function generators employed in this analog computer: a diode function generator as described in section 4.5.5 and photoformers (see section 4.5.3) denoted by a rhombus sign.

Fig. 3.26. The Caltech analog computer

plier (with implicit sign inversion as denoted by the $-M$), while the small box on the bottom is a summer. The circular component on the left denotes the forcing function generator, which produces a time-dependent forcing function as an input to the computer setup. This setup solves the differential equation

$$F(x) = k_1 \frac{d^2 y}{dx^2} + f(y)\frac{dy}{dx} + k_2 y \qquad (3.6)$$

by implementing its electric equivalent

$$E_0(t) = L\ddot{q} + \frac{\text{AMR}}{\text{C}} f_1(q)\dot{q} + \frac{q}{\text{G}}. \qquad (3.7)$$

It was also possible to solve partial differential equations on the Caltech analog computer by discretising space (or time) along one or more axes resulting in a finite difference mesh-like network of (mostly) passive components. Due to the limited number of such passive computing components these discretisations were pretty coarse but nevertheless useful for a variety of problems. Figure 3.28 shows a typical setup of this machine for a partial differential equation. The system was successfully applied to 4th order partial differential equations as well as to systems of simultaneous partial differential equations.

It is interesting to note that the high degree of interactivity typical of analog computers was put into use and mentioned as early as in 1949. [McCann 1949, p. 507] notes:

"Since circuit element settings and inserted mesh currents can be changed rapidly, iterative methods of solution can be set up in a very short time"

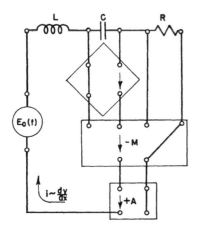 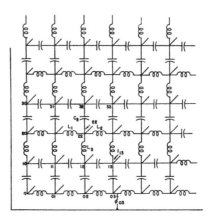

Fig. 3.27. Setup of equation (3.6) transformed into (3.7) on the Caltech analog computer (see [MCCANN 1949, p. 505])

Fig. 3.28. Setup of the Caltech analog computer for solving the finite difference equivalent of a partial differential equation (see [MCCANN 1949, p. 507])

The wealth of applications of this remarkable analog computer is described in [MCCANN 1949] and [MCCANN et al. 1949]. Among these were servomechanisms and automatic control systems, missile, and autopilot related problems, the analysis of mechanical vibrations, the solution of linear and non-linear ordinary differential equations, partial differential equations describing the behaviour of elastic beams and plates, and many more.

RICHARD H. MACNEAL especially remembers the work on beams as follows:[77]

> "[...] MACCANN [...] along with HARRY E. CRINER, had worked out how to model beams with electric analog computers, using transformers as a primary agent. For vibrations, you would use inductors for stiffness and capacitors for mass. [...] In those days, aircraft were analyzed as beams. There was one beam for the fuselage, one beam for the left wing [...] One of the most serious problems in aircraft structural analysis is the flutter problem, which is the instability issue. We got to the point where we could do that fairly well."

Following this brief description of a few of the earliest electronic analog computers the following chapter will now focus on the basic computing elements employed in analog electronic analog computers past, present, and future alike.

---

[77] See [MACNEAL 2002, pp. 5 ff.]

4 Basic computing elements

The following chapters give an overview of the most important computing elements found in a typical electronic analog computer. Some functions, such as multiplication, can be implemented in very different ways. Although often only one implementation variant prevails, other approaches will be also presented to show the wide range of possible implementations.

An analog computer typically relies on voltages[1] for representing and transmitting values between the various computing elements, which constitute a computer setup for a particular problem. These voltages can only take on values within a certain interval. For most vacuum tube based analog computers this interval ranges from -100 V to $+100$ V while most (but not all) transistorised analog computers and later machines used values in the range of $-10\ldots10$ V.[2] To simplify things it is common to think of variables occurring in an analog computer setup to be in the interval $[-1, 1]$ regardless of the actual voltages used in a particular machine. The bounds of this interval are called *machine units* and correspond to the actual minimum and maximum voltages for any given machine. Thus, a variable represented by $+50$ V on a ±100 V machine is said to have the value 0.5. The same value would describe a variable represented by 5 V on a ±10 V transistorised system. All calculations are thus with respect to these machine units of ±1 and never with respect to the actual range of voltages used by a particular analog computer.

All computing elements described in the following sections will work on normalized values in the range of ±1 machine unit.

4.1 Operational amplifiers

The central element of every electronic analog computer is the *operational amplifier*, which forms the heart of most computing circuits such as summers, integrators, etc. This central role of the operational amplifier led to the following remark by [TRUITT et al. 1960, p. 2-58]: "*The Operational Amplifier is the King of Analog Computing Components*". It was clear from the beginning of the development of

---

[1] Most classic analog computers use voltages except machines made by Applied Dynamics, which used currents. Modern implementations of analog computers on integrated circuits will mainly use currents to represent values.

[2] These two voltage ranges are not the only ones to be found in actual machines – some tube based circuits work with values in the range of ±50 V and some transistorised analog computers even allow values between $+100$ and -100 V while others work with voltages ranging from -5 V to $+5$ V.

electronic analog computers that high gain amplifiers capable of working on DC voltages with negligible error voltages caused by drift effects and the like were required for usable general purpose machines. HOELZER, PHILBRICK, MACNEE and all of the other pioneers were plagued by the deficiencies of their respective amplifiers. HOELZER overcame these problems by resorting to AC coupled amplifiers operating on modulated AC signals, which required modulator and demodulator circuits, while MACNEE solved this problem by building a machine that computed at such high speed that he could use drift free AC amplifiers in his machine.

It was not until 1949 that a practical drift correction technique was developed[3] by EDWIN A. GOLDBERG. This made the implementation of high precision operational amplifiers with extremely high gain and very low drift possible, which led to a revolution in the development of analog electronic analog computers.[4]

The term *operational amplifier* was coined by JOHN RAGAZZINI[5] in 1947 and defined as follows:[6]

> "As an amplifier so connected can perform the mathematical operations of arithmetic and calculus on the voltages applied to its input, it is hereafter termed an 'Operational Amplifier'."

Figure 4.1 shows the symbol used today to denote an operational amplifier in a schematic. Typically such an amplifier has two inputs, one inverting and one non-inverting, and one output that represents the amplified difference between the values applied to the inputs. In most analog computers, the non-inverting input is not available for use in a computer setup and is either connected to ground – representing the value zero – or used for drift compensation. Thus, differing from the standard symbol used in today's electronics – an *open* operational amplifier, i. e., an amplifier that has no elements connecting its output to its input, is normally denoted by the symbol shown in figure 4.2 in an analog computer setup diagram. This computing element always inverts the sign of its input signal.

In the early 20th century amplifiers were normally operated in a rather straightforward mode of operation since they did not use any kind of feedback to stabilise their behaviour. Thus, these amplifiers were quite unstable and difficult to use since even the slightest changes regarding their components resulted in massively different behaviour of the overall amplifier. Then, in 1927, HAROLD

---

[3] See section 4.1.2.
[4] A comprehensive account of the history of operational amplifiers can be found in [JUNG 2006], which is very worthwhile to read.
[5] 1912–11/22/1988
[6] See [RAGAZZINI et al. 1947]. It should be noted that the developments on which this paper is based were mostly done by LOEBE JULIE, (12/10/1920–06/07/2015) who was working under the auspices of GEORGE A. PHILBRICK at Bell Labs (cf. [JUNG 2006, p. 779]).

Fig. 4.1. Symbol of an operational amplifier

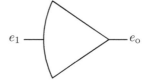

Fig. 4.2. Symbol of an *open* or *uncommitted* amplifier

STEPHEN BLACK[7] invented the *negative-feedback amplifier* that should become one of the most influential basic circuits in electronics.[8] He was elected as a member of the "Electrical Engineering Hall of Fame" for his seminal work on amplifiers (see [BRITAIN 2011]). The idea of negative-feedback was so novel and unusual that it took nine years from Black's patent application to its issuance (see [JUNG 2006, p. 767]).[9]

At the time, BLACK was working for the Bell telephone laboratories and his negative-feedback scheme that is taken as granted today allowed the Bell telephone system to successfully implement carrier telephony, which was until then hampered by the poor stability and high distortion of existing amplifiers without negative feedback; these were prone to errors caused by varying supply voltages, aging of circuit elements (especially the tubes), etc.

Figure 4.3 shows the basic concept of BLACK's invention: An amplifier consisting of an odd number of inverting stages (resulting in an inverting amplifier) is fed with two signals that are applied to its inverting input: The desired signal E itself, which is connected via a series resistor C and a feedback signal originating from the output of the amplifier. This signal is connected to the input of the amplifier be means of a feedback resistor f. Figure 4.4 shows this principle using today's standard symbols.

The idea of feeding back some amount of the output signal of an inverting amplifier to its input allows to trade gain for stability as the following consideration shows: With A denoting the gain of the amplifier and e_{SJ} representing the voltage at the summing junction at its inverting input, as in figure 4.4, its output voltage is

$$e_\text{o} = -Ae_{\text{SJ}}$$

---

[7] 04/14/1898–12/11/1983
[8] A detailed account of BLACK's work is given in [KLINE 1993] and [MINDELL 2000].
[9] It should be noted that BLACK was not the only developer working on negative-feedback schemes – others included PAUL VOIGT, 12/09/1901–02/09/1981, (in memoriam [KLIPSCH 1981]), ALAN BLUMLEIN, 06/29/1903–06/07/1942, and BERNARDUS DOMINICUS HUBERTUS TELLEGEN, 06/24/1900–08/30/1990 (cf. [JUNG 2006, p. 767]).

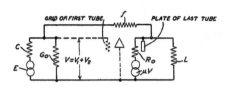

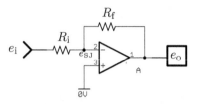

Fig. 4.3. The basic principle of negative-feedback according to [BLACK 1937]

Fig. 4.4. Operational amplifier with negative feedback

yielding

$$e_{\text{SJ}} = -\frac{e_o}{A}. \tag{4.1}$$

The currents at the summing junction are

$$i_i = \frac{e_i}{R_i} \quad \text{(input current through } R_i\text{)},$$

$$i_f = \frac{e_o}{R_f} \quad \text{(feedback current through } R_f\text{), and}$$

$$i_- \approx 10^{-9} \text{ A} \quad \text{(current flowing in the inverting input)}.$$

Applying KIRCHHOFF's law yields

$$i_- = i_i + i_f = \frac{e_i i - e_{\text{SJ}}}{R_i} + i\frac{e_o - e_{\text{SJ}}}{R_f}. \tag{4.2}$$

Since i_- is negligibly small, a few nA[10] at most, equation (4.2) can be rearranged

$$\frac{e_i - e_{\text{SJ}}}{R_i} = -\frac{e_o - e_{\text{SJ}}}{R_f}.$$

Replacing e_{SJ} according to (4.1) yields

$$\frac{e_i + \dfrac{e_o}{A}}{R_i} = -\frac{e_o + \dfrac{e_o}{A}}{R_f}.$$

Rearranging this equation yields

$$e_o \left(\frac{1}{AR_i} + \frac{1}{R_f} + \frac{1}{AR_f} \right) = -\frac{e_i}{R_i}$$

and thus

$$e_o = \frac{-\dfrac{e_i}{R_i}}{\dfrac{R_f + AR_i + R_i}{AR_i R_f}} = \frac{-\dfrac{e_i}{R_i} AR_i R_f}{R_f + AR_i + R_i}.$$

---

[10] 1 nanoampére = 10^{-9} A.

Dividing by R_i and A and rearranging finally yields

$$e_o = \frac{-\dfrac{R_f}{R_i}e_i}{1 + \dfrac{1}{A}\left(\dfrac{R_f}{R_i}+1\right)}. \tag{4.3}$$

Since the gain A is usually very large – some amplifiers used in analog computers reached values up to $A = 10^8$ and even $A = 10^9$ – and since the ratio R_f/R_i is normally quite small, rarely exceeding 10, equation (4.3) can be further simplified to

$$e_o = \frac{R_f}{R_i}e_i, \tag{4.4}$$

so the behaviour of a high gain amplifier with negligible input current at its inverting input is completely determined by the circuit elements in the feedback loop and at its input.

4.1.1 Early operational amplifiers

This section focuses on some early and influential operational amplifiers designs. Historic references that might be of interest in this context are [GRAY 1948] and [CHANCE et al. 1947].

An early dual channel operational amplifier is depicted in figures 4.5 and 4.6. This amplifier was used in the ground based analog guidance computer of a *Nike* surface to air missile system.[11] The first version of this missile system was deployed from 1954 on well into the early 1960s. This amplifier still shows its roots, which can be traced back to the M-9 fire control computer. In fact, the amplifiers used for Nike were identical to those used in the *M-33* fire control computer, a successor of the M-9.[12]

The schematic diagram for a single channel of this operational amplifier is shown in figure 4.7. It features a now classic differential input stage consisting of both halves of a dual triode. The interstage-coupling is done via the common cathode resistor R1. The non-inverting input is explicitly used as an offset input for (automatic) drift compensation in this application. The following stages are no longer differential and consist of a pentode section and a simple triode output stage for a total of three amplification stages. Thus, the overall amplifier is an inverting amplifier.

Since the output stage is not of the push-pull type, an external pull-up resistor is necessary. This resistor would be connected between the output triode's anode

---

[11] See section 13.15.8.
[12] The schematic of the M-9 operational amplifier can be found in [JUNG 2006, p. 781].

Fig. 4.5. Top view of a typical dual operational amplifier used in the Nike computer

Fig. 4.6. Bottom view of the same amplifier

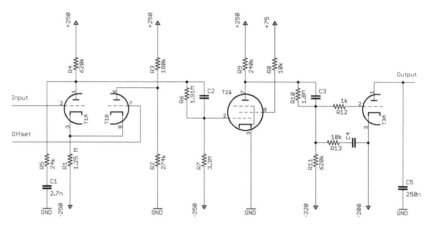

Fig. 4.7. Operational amplifier used in the Nike computer

and the $+320$ V supply. All in all this amplifier needed five different supply voltages of -320 V, ± 250 V, -200 V and $+75$ V and has an output voltage range of ± 100 V, although the full range can only be maintained up to about 100 Hz,[13] which is not that impressive but is more than enough to steer a surface to air missile.

A far simpler yet commercially very successful operational amplifier is shown on the left hand side of figure 4.8: The famous *K2-W* amplifier developed by GAP/R, the company GEORGE A. PHILBRICK founded after his early process simulator work at Foxboro.[14] The K2-W amplifier only requires two supply voltages of ± 300 V and has an open-loop gain of about 15,000 with an output voltage

---

[13] See [PEASE 2003].
[14] As [PEASE 2003] notes, this is not the first "modern" operational amplifier – there were earlier developments, but none could match the K2-W in terms of practicability and commercial success.

Fig. 4.8. K2-W and K2-X operational amplifiers (GAP/R)

range of ±50 V at a maximum output load of 1 mA. Rise time is about 2 μs, the input impedance is roughly 100 MΩ resulting in an input current of only about 10 nA.[15]

Figure 4.9 shows the schematic of this pluggable operational amplifier. It is based on two 12AX7 dual triodes and had its market debut in 1953. The first 12AX7 acts as differential input stage with a common cathode resistor.[16] This input stage is followed by a single triode inverter and a cathode follower driving the output, which is tied by three paralleled pull-down resistors to the negative supply voltage. An interesting detail is the Neon light bulb in series with a 680 kΩ resistor. This circuit acts as a level shifter.[17]

---

15 Not bad for an operational amplifier – even from today's perspective.
16 This configuration is called a *long-tailed pair*
17 Nowadays, a ZENER *diode* would be used instead of a Neon tube. In early specimen of the K2-W the level shifter consisted of a *thyrite* device – today this would be called a *varistor*. These are made from silicon carbide and were not only used as level shifters in the early days of electronic analog computers but also as function generators in applications where precision was not of prime importance. Today such devices are commonly used as surge arrestors.

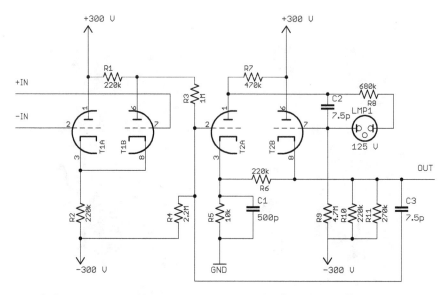

Fig. 4.9. Schematic of the K2-W operational amplifier (cf. [GAP/R K2W])

Figure 4.8 shows on the right an improved version of the K2-W, the *K2-X* operational amplifier. This Device has a significantly faster output signal rise time than the K2-W and features an output voltage swing of ± 100 V at 3 mA.[18]

4.1.2 Drift stabilisation

All of these early DC coupled operational amplifiers were plagued by drift, i.e., small error signals that propagate through the various amplifier stages and show up with a reasonable amplitude at the output, causing significant errors in computations. Recalling figure 4.4 and taking into account that the input current even of early vacuum tube based operational amplifiers is negligibly small, it is clear that the voltage e_{SJ} at the summing junction would be zero in the case of a theoretically drift-free amplifier since the input current i_i flowing through R_i and the feedback current i_f through R_f cancel each other.

Any drift caused by intrinsic effects of the amplifier result in an output signal different from 0 V even if $e_i = 0$. Thus, all drift effects in an amplifier can be modelled by a drift voltage e_d acting on the summing junction of the amplifier via a large input resistor. If one could inject a commensurate voltage with reverse sign at the non-inverting input of the amplifier, it would cancel out the drift error.

---

[18] The schematic of the K2-X operational amplifier can be found in [RUSSELL 1962, p. 6-20].

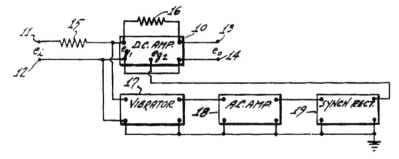

Fig. 4.10. Drift stabilisation scheme according to Edwin A. Goldberg and Jules Lehmann (cf. [Goldberg et al. 1954])

This is the basic idea Edwin A. Goldberg describes in his 1949 patent.[19] If one could measure the error voltage due to drift, e_d at the summing junction and amplify it without introducing further errors with a drift-free amplifier, the resulting output signal could be used to cancel out the drift of the main amplifier. Figure 4.10 shows the setup proposed by Goldberg.

The DC amplifier shown in the upper half of the picture is the DC coupled main amplifier whose drift is to be eliminated. To achieve this, the error voltage at the input of this amplifier, the summing junction, is converted to an AC voltage by means of the electromechanical chopper relay[20] in figure 4.10.

The output signal of this chopper[21] is of rectangular shape swinging between 0 V and the error voltage e_d picked up from the summing junction of the main amplifier. This AC voltage can now be amplified with next to no additional drift error by a separate AC coupled amplifier. The output signal of this amplifier is then rectified (typically by means of a synchronous demodulator driven by the same voltage feeding the chopper) yielding a DC signal that can then be used to compensate for the drift of the main DC amplifier.[22]

---

[19] See [Goldberg et al. 1954].
[20] These relays are also called *vibrators*, especially when used in power supplies for battery powered mobile vacuum tube devices such as car radios, etc.
[21] These choppers were typically driven by a 60 Hz or a 400 Hz signal resulting in a very characteristic audible hum emitted by such amplifiers.
[22] The output voltage of the drift compensation stage is often also used to detect amplifier overload. If an amplifier is overloaded it cannot maintain the necessary feedback signal to keep the summing junction at ground potential, which is reflected by an excessively high output voltage from the AC amplifier, which can be easily detected. Such overloads may result from improper setup of the analog computer typically due to scaling errors regarding the underlying equations. Since an overloaded amplifier will yield an erroneous result, overload conditions are normally indicated visually or acoustically. Ideally a computation run is halted in case of an overload so that its cause can be determined by checking which amplifier went into overload.

This ingenious scheme was quickly put into use in nearly all operational amplifiers used for analog computers. Apart from the automatic drift compensation this *chopper stabilisation* had a second advantage: For low frequency input signals the gains of the main DC amplifier and the stabilising AC amplifiers essentially add up yielding a very high gain overall amplifier. As equation (4.3) shows, a high open loop gain[23] is essential for keeping the errors of a circuit with a negative feedback path low. The amplifier developed by GOLDBERG had an impressive overall DC gain of 150,000,000 corresponding to roughly 163 dB.[24] Later developments achieved even higher gains for low frequency input signals.

Electromechanical choppers were used well into the 1970s to stabilise precision operational amplifiers because they exhibit an extremely low resistance when the contacts are closed and nearly infinite resistance in the open state. This came at a price: Most of the commercially available devices had a short life-span. In addition to this, the contact pairs often acted as parasitic thermocouples thus introducing another error in the drift-compensation circuit. The second problem was finally overcome by using arcane metal compositions for the contact pairs, such as Pt/Rh.

Figure 4.11 shows a typical 400 Hz chopper made by the German manufacturer *KACO* for use in Telefunken analog computers. The 400 Hz excitation voltage is supplied through a top connector (not visible here) to minimise injection of noise into the amplifier circuit. The top half of the chopper contains the coil, which excites the middle contact to oscillate between the two contacts on the left and right at the bottom of the relay. The chopper system itself is suspended from a mounting frame by a rubber band, thus minimising vibration of the overall enclosure.

Some analog computers used a single AC coupled amplifier for drift stabilisation of a large number of operational amplifiers by employing a round robin mode of operation. This reduced cost substantially as the complexity of a single operational amplifier was basically cut in half but required a motor driven rotary switch to connect the single stabilisation amplifier to each operational amplifier in sequence. The downside of this approach is that it sacrifices the extremely high gain of a typical chopper-stabilised amplifier. In addition to this, the recovery time from an overload condition is substantially increased.

An example for this technique can be found in the analog computer used in the ground based guidance system for Nike missiles. Here a rotating switch connected one AC coupled amplifier with its associated input chopper and output rectifier sequentially to a number of amplifiers. This required an additional capacitor at each of the operational amplifier inputs used for drift compensation to store the

---

[23] Open loop gain is the gain exhibited by an amplifier without any external negative feedback path in place.
[24] See [JUNG 2006, p. 780].

Fig. 4.11. Typical electromechanical chopper

signal generated by the AC amplifier for duration of one revolution of the rotary switch.

Figure 4.12 shows a typical vacuum tube operational amplifier with chopper stabilisation. The chopper is the gray rectangle visible in the upper right part of the picture. This particular amplifier is used in a Solartron *Minispace* analog computer.

The schematic of the first fully transistorised chopper-stabilised operational amplifier developed at Telefunken in 1959 is shown in figure 4.13. Its main developers were HANS OTTO GOLDMANN and GÜNTER MEYER-BRÖTZ.

It is worthwhile to take a closer look at remarkable circuit. The main amplifier shown in the upper half of the schematic has two differential amplifier stages in series. The first stage, T1 and T2, is the input stage with the base of T1 – this is the inverting input of the operational amplifier – connected to the summing junction by C1. The two antiparallel diodes D1 and D2 limit the voltage at this point. It should be noted that this main amplifier is AC coupled to the summing junction, so a pure DC signal cannot pass. The base of T2 – the non-inverting input of the amplifier – is fed by the AC coupled amplifier shown in the lower half of the schematic.

T1 and T2 drive a second differential amplifier stage consisting of T3 and T4, which then feeds T5 and finally the differential output stage with the power transistors T6 and T7 (mounted on a heat sink).

The amplifier depicted in this schematic is wired as an inverter, i.e., a summer with a single input. $R_f = 200k$ together with the high-pass T-filter consisting of

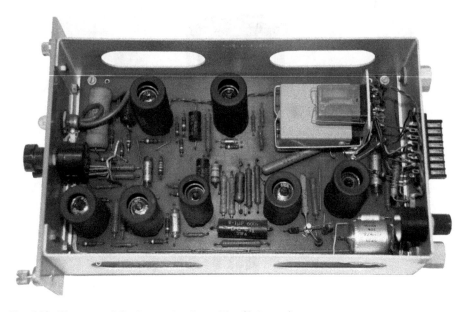

Fig. 4.12. Chopper-stabilised operational amplifier (Solartron)

two 600 pF capacitors and a 100k resistor connected to ground is the feedback path while the 200k resistor connected to the summing junction is an input with weight 1.

The summing junction is also connected to the AC coupled amplifier shown in the lower half of figure 4.13 by means of the low-pass T-filter consisting of R1, R2, and C1 and the Airpax chopper. The chopper relay is driven by a 400 Hz square wave signal and repetitively shorts the output of the low-pass to ground thus creating an AC signal, which is filtered by another high-pass T-filter (C2, C3, R3). This signal is then amplified by the AC coupled amplifier built around the transistors T1, T2, T3, and T4.

The output AC signal at R17 is the rectified by the synchronous rectifier consisting of D3, R23, D4, and R24. This subcircuit is also driven by a 400 Hz square wave signal, which is in a fixed phase relationship to the chopper excitation voltage. A static drift correction voltage can be set by means of the potentiometer labelled "0 Punkt" (zero setting), which is added to the output signal of the AC coupled amplifier. The resulting signal is then fed to the non-inverting input of the main amplifier.

A signal entering this combination of two amplifiers thus takes two paths: Its DC and very low frequency parts are mainly amplified by the AC coupled amplifier and then enter the main amplifier through its non-inverting input while higher-frequency components are amplified by the DC coupled main amplifier only. Thus, for low frequency and DC signals the overall open loop gain is the product of the gains of both amplifiers.

The output of the AC coupled amplifier is also fed into the circuit consisting of T5, D5, T6 with their associated passive components. This controls the relay coil shown on the far right. In case of an overload, i.e., the summing junction can no longer be held at 0 V by the DC/AC amplifier combination, the relay will be activated and will signal the overload condition to the control circuits of the analog computer.

The production version of this amplifier (which differs only slightly from this schematic) is shown in figure 4.14. It was used for more than 10 years in the various fully transistorised Telefunken analog computers. The circuit card on the left side contains the DC coupled main amplifier with its two large germanium output transistors mounted on a cooling plate visible on the top. The card on the right is the drift stabilising AC coupled amplifier fed with the chopped error signal obtained from the summing junction of the main amplifier and provides the drift correction signal. Not shown is the chopper relay. The overall combined amplifier has a DC gain of up to 10^9 for DC to low frequency signals.[25]

Figure 4.15 shows the successor of this amplifier using an all electronic chopper thus greatly maximising MTBF[26] and simplifying its overall construction. This amplifier was used in the Telefunken *RA 770* precision analog computer and has a DC gain of $3 \cdot 10^8$ for low frequency input signals. The electronic chopper and synchronous rectifier is under the shielding shown in the right upper part of the picture. The output stage transistors with their heat sinks are visible in the lower right part of the picture.[27]

4.2 Summers

One of the most basic computing elements of an analog computer is the *summer* or *summing amplifier*, which is basically a direct consequence of equations (4.3) and (4.4). Figure 4.16 shows the simplified schematic of one of the earliest summing amplifier developed by KARL DALE SWARTZEL[28] in 1941 at *Bell Laboratories* for application in a fire control computer.[29]

---

[25] [MEYER-BRÖTZ et al. 1966] contains a detailed description of a follow-up development of this amplifier based on Silicon transistors.
[26] Mean Time between Failure
[27] The single integrated circuit, a LM709 operational amplifier forms the AC coupled amplifier of this circuit.
[28] 06/19/1907–04/23/1998
[29] Cf. [JUNG 2006, pp. 777 f.].

86 — 4 Basic computing elements

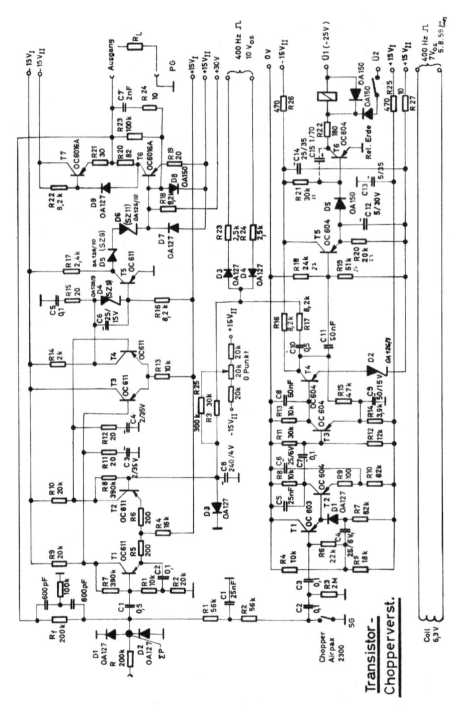

Fig. 4.13. Schematic of the first chopper-stabilised transistorised operational amplifier developed at Telefunken in 1959

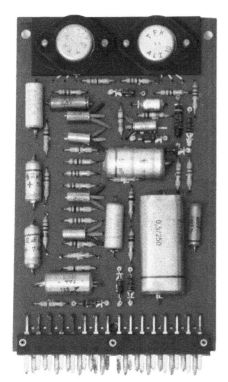
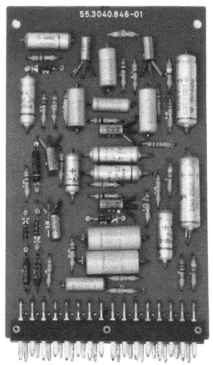

Fig. 4.14. Early fully transistorised chopper-stabilised amplifier

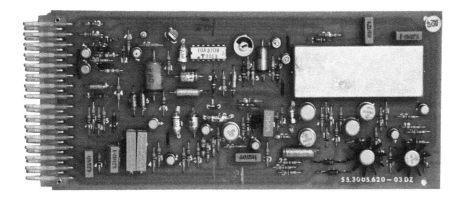

Fig. 4.15. Late chopper-stabilised amplifier with all electronic chopper

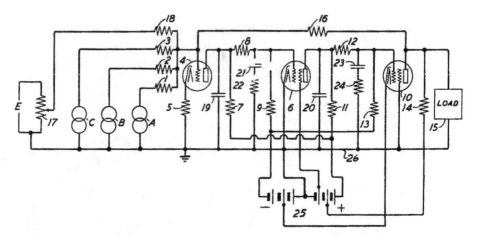

Fig. 4.16. Schematic of the "Summing Amplifier" developed by KARL D. SWARTZEL (cf. [SWARTZEL 1946])

This particular amplifier has three inverting stages employing one triode in the input stage and two tetrodes for the remaining two stages, so the overall amplifier is of the inverting type making a negative feedback scheme as developed by BLACK possible. The summing junction at the grid of the triode input stage (4) is connected to four input resistors 1, 2, 3 and 18 as well as to the feedback resistor 16. Of these four inputs only three are used for signals to be processed while the fourth input is reserved for a manual drift compensation circuit consisting of a voltage source E and a potentiometer 17. This amplifier requires supply voltages of ± 350 V, -135 V, $+250$ V and $+75$ V and has an overall gain of about 60,000. The various RC networks between the three amplifier stages stabilise the circuit.

Figure 4.17 shows an equivalent setup based on an operational amplifier. In this case three inputs e_1, e_2 and e_3 and a connection to the summing junction[30] as well as an output connection are available. The simplified symbol for such a summer is shown in figure 4.18.

Given sufficient open loop gain A of the operational amplifier used, the operation performed by this circuit is basically only determined by the passive elements at its input and in the feedback loop.

---

[30] Many analog computers allow direct connections to the summing junction of the operational amplifiers used in summers and integrators. Typical applications include (biased) diodes for amplitude limiting and little capacitors between the output of an amplifier and its summing junction to suppress parasitic oscillations, which can occur in setups like square rooting, etc.

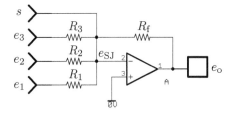

Fig. 4.17. Schematic of a three-input summer

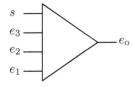

Fig. 4.18. Symbol of a summer

Extending equation (4.4) by additional inputs yields

$$\sum_{i=1}^{n} \frac{e_i}{R_i} = -\frac{e_o}{R_f} \qquad (4.5)$$

for n inputs with input resistors $R_1, \ldots, R_n$ and a feedback resistor R_f. Introducing weighting coefficients $a_i = R_f/R_i$ for all inputs, the output voltage of this circuit is

$$e_o = -\sum_{i=1}^{n} a_i e_i.$$

Thus, a typical summing amplifier yields the negative of the sum of its weighted input voltages.[31] Typical values for a_i are 1 and 10 although in some cases even weights 4 or 5 can be found.[32] If other weights are required in a computation, inputs can be paralleled or external resistors or potentiometers can be connected to the summing junction input.

Figure 4.19 shows a typical early summing amplifier – a model K3-A *adding component* made by GAP/R. This device computes the sum of four input voltages e_1, e_2, e_3 and e_4 plus an adjustable constant value e_0 that can be set manually using the potentiometer in the top half of the device. It delivers two output signals with opposite signs satisfying $\pm ke_0 + e_1 + e_2 + e_3 + e_4$ with $-1 \leq k \leq 1$ and $e \in \{0, 5, 50\}$ V.

4.3 Integrators

Replacing the feedback resistor R_f in a summer circuit with a feedback capacitor C yields a circuit capable of integration as shown in figure 4.20. The feedback

---

[31] This is a source of confusion for most beginners in analog computer programming.
[32] If an input's weight differs from 1 it is written next to the input in the triangle of the simplified summer symbol shown in figure 4.18. If no weight is noted, the corresponding input is weighted with 1.

Fig. 4.19. Model K3-A adding component (GAP/R), front and interior view

current i_f flowing through the capacitor C is determined by $i_\mathrm{f} = C\dot{e}_\mathrm{o}$. Accordingly, equation (4.5) can be changed to

$$\sum_{i=1}^{n} \frac{e_i}{R_i} = -C\dot{e}_\mathrm{o}. \tag{4.6}$$

Redefining the weighting coefficients as $a_i = 1/R_i C$ and solving (4.6) for e_o yields

$$e_\mathrm{o} = \int_0^t \sum_{i=1}^{n} a_i e_i \, \mathrm{d}t + e(0)$$

where $e(0)$ denotes the *initial condition* of the integrator.

Figure 4.21 shows the symbol used to denote an integrator in an analog computer setup diagram. Since such an integrator is obviously working with time being the free variable it needs some means to reset, or even better preset, the integrating capacitor C to an initial value. In addition to this, additional circuitry is necessary to start and stop the operation of the integrator. Figure 4.22 shows the simplified schematic of a more realistic integrator as it is used in a typical analog computer.

The central control elements are the two switches labelled IC, short for *initial condition* and OP, short for *operate/halt*. The integrator shown in this figure is currently in its IC mode: The inputs e_1, e_2 and e_3 are grounded through the operate/halt switch while the initial condition switch effectively turns the circuit

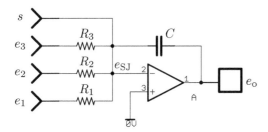

Fig. 4.20. Basic structure of an integrator

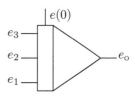

Fig. 4.21. Symbol of an integrator with three inputs e_1, e_2, e_3, and initial condition input $e(0)$

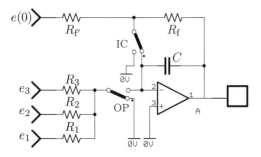

Fig. 4.22. More detailed schematic of an integrator showing the control switches h and c

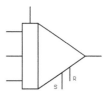

Fig. 4.23. Integrator symbol showing control inputs S and R for IC and OP switches

into a simple summer with only one input $e(0)$. The integration capacitor C is switched in parallel with R_f. Since the summing junction at the inverting input of the operational amplifier is always maintained at 0 V, the capacitor will be charged to the voltage at the output of the amplifier. This, in turn, is determined by the input signal at $e(0)$ and the ratio $R_f/R_{f'}$, which is typically equal to 1. The capacitor is thus pre-charged to the initial value $-e(0)$.[33]

To switch the integrator into the OP mode both switches change their state. This grounds the connection between $R_{f'}$ and R_f effectively disabling the input

---

[33] Some analog computers from the 1950s employed a simpler circuit allowing the integration capacitor to be discharged only prior to switching the integrator into OP mode. Some circuits used Neon tubes to short circuit the integration capacitors. Little Neon tubes were enclosed in a coil that could be energized by a RF signal, which in turn caused the tube to conduct. Since this does not provide any means to preset an integrator to $e(0)$ this actual initial condition value had to be introduced later in the computer setup by a summer or the like. A schematic of a Philbrick K2-J integrator with a Neon reset tube triggered by an RF generator can be found in [RUSSELL 1962, p. 6-24].

$e(0)$ and connects the input resistors R_1, R_2 and R_3 to the summing junction of the amplifier. The circuit now performs an integration with respect to time as its free variable.

If the switch OP is reset the integrator enters *halt mode*. In this mode the last value of the integration will be stored.[34] This mode is normally used to pause a computation to read out values of interest from the various elements of the current setup.

In most cases the control switches IC and OP of all integrators in an analog computer setup are controlled by a common set of signals. In some cases it is necessary to control individual integrators or groups of integrators. This makes it possible to solve equations with one group of integrators based on initial conditions generated by a second group. Another typical application for such a more sophisticated control setup is the use of integrators as analog memory cells.

If an integrator is to be operated under individual control the symbol shown in figure 4.23 is used, which shows the actual inputs controlling the switches IC and OP of the integrator.

A realistic integrator circuit is still more complex than the schematic shown in figure 4.22: High-speed integrators use electronic switches in place of the relay contacts shown here. In addition to this, most integrators feature more than one integration capacitor C thus allowing to select different ratios a_i for their operation.[35] Some machines offer as many as four or five different values for C (usually powers of ten). Typically, these are selected by shorting certain jacks of an integrator or from a central control panel.

Figure 4.24 shows a dual integrator made by *Electronic Associates Inc.*, *EAI* for short. The chassis contains the passive elements for two integrators – the associated operational amplifiers are contained in a second drawer and are not shown here. Each integrator channel contains four capacitors implementing four different time-scale factors.

4.4 Coefficient potentiometers

The fixed input weights a_i of summers and integrators are not sufficient for representing general coefficients in a computer setup. This done using *coefficient potentiometers*. In the simplest case these are just potentiometers connected as

---

[34] Due to small leakage currents in the capacitor and even the tiniest drift effects the output of the integrator will slowly change during halt mode. Accordingly, this mode should not be active for extended periods of time.

[35] These are called *time-scale factors*, since they allow faster or slower operation of integrators with respect to integration time.

Fig. 4.24. EAI dual integrator used in an EAI 580 analog computer

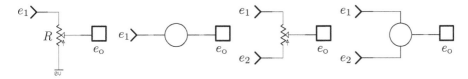

Fig. 4.25. Coefficient potentiometers – basic implementation and representation

voltage dividers as shown in figure 4.25. In an analog computer program such a voltage divider is typically represented by a circle with one input and one output connection.[36] In some cases it is necessary to have a potentiometer with two inputs, which is shown in the right half of the figure.

With α denoting the position of the potentiometer's wiper, the overall resistance R is partitioned into an upper and a lower resistor R_u and R_l connected in series:[37]

$$R_\mathrm{u} = (1-\alpha)R$$
$$R_\mathrm{l} = \alpha R$$

Without load applied to the wiper of the potentiometer the operation of such a coefficient potentiometer can be described by

$$\frac{e_\mathrm{o}}{e_1} = \frac{R_\mathrm{l}}{R}.$$

---

36 Input and output of a coefficient potentiometer are not explicitly denoted in this representation form although they cannot be used interchangeably. In fact, connecting the output of a coefficient potentiometer to the output of a computing element can and most often will damage the potentiometer in a classic analog computer, so care should be taken.

37 A linear potentiometer is assumed here.

In a traditional analog computer this is a hypothetical case since the following computing elements normally act as a non-negligible load.[38] A computing element connected to the wiper of a potentiometer acts as a resistive load R_{load} connected in parallel to R_1. Thus,

$$R_1 \parallel R_{\text{load}} = \frac{R_1 R_{\text{load}}}{R_1 + R_{\text{load}}}.$$

The total resistance of the coefficient potentiometer with a load connected is then

$$R_{\text{total}} = R_u + R_1 \parallel R_{\text{load}} = R_u + \frac{R_1 R_{\text{load}}}{R_1 + R_{\text{load}}}.$$

The loaded potentiometer can thus be described by

$$\frac{e_o}{e_1} = \frac{R_1 \parallel R_{\text{load}}}{R_{\text{total}}} = \frac{\frac{R_1 R_{\text{load}}}{R_1 + R_{\text{load}}}}{R_u + \frac{R_1 R_{\text{load}}}{R_1 + R_{\text{load}}}} = \frac{R_1 R_{\text{load}}}{R_u(R_1 + R_{\text{load}}) + R_1 R_{\text{load}}}$$

$$= \frac{R_1}{\frac{R_u R_1}{R_{\text{load}}} + R_u + R_1} = \frac{\alpha R}{(1-\alpha)R \frac{\alpha R}{R_{\text{load}}} + R}$$

$$= \frac{\alpha}{(1-\alpha)\alpha \frac{R}{R_{\text{load}}} + 1}. \tag{4.7}$$

Figure 4.26 shows the effect of potentiometer loading with respect to the value set by α. It is obvious that setting a potentiometer manually by just looking at the value displayed on its dial will not yield the desired value since the load exerted by the computing elements connected to the potentiometer has to be taken into account.[39]

To overcome this problem most analog computers have a special mode of operation, called *potentiometer set (Pot Set)*. In this mode, which can be selected from the central control panel of the computer, all integrators and summers have their input resistor networks grounded, so no actual computation takes place while every computing element exhibits its actual load to its signal sources – including the various coefficient potentiometers. To set a potentiometer to a precise value in this mode, small and simple systems offer a push button next to each potentiometer that will connect its input to +1 machine unit and connect a precision

---

[38] Later and more modern implementations used buffer amplifiers for the coefficient potentiometers as shown below.

[39] In some cases load-correction graphs were used to give the required offset of α for some typical loads.

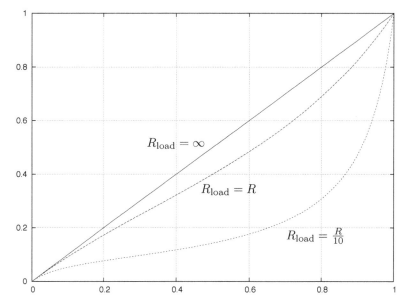

Fig. 4.26. Coefficient potentiometer with $R_{\text{load}} = \infty$, $R_{\text{load}} = R$ and $R_{\text{load}} = R/10$

voltmeter – either a compensation voltmeter[40] or a digital voltmeter – to its wiper when depressed. Turning the potentiometer's, dial the desired coefficient value can now be set very precisely by means of the voltmeter.

When cheap integrated circuit operational amplifiers became available some analog computers employed simple buffer stages to decouple the coefficient potentiometers from the load of the following computing element, thus vastly simplifying the setup procedure.[41] Figure 4.27 shows two typical buffering schemes: The circuit on the left uses an operational amplifier as a simple non-inverting output buffer for the potentiometer R.

The circuit shown on the right in figure 4.27 is more versatile and is used for example in the *EAI 2000* hybrid computer. In contrast to a simple voltage divider with buffer amplifier this circuit allows coefficient values to be set in the range $-1\ldots 1$ as the following consideration shows: If α is set to 0, the potentiometer's wiper is at its grounded end, thus the non-inverting input of the operational am-

---

40 A compensation voltmeter basically consists of a reference voltage source, i. e., a precision potentiometer connected as a voltage divider between ground and +1 machine unit, and a null voltmeter connected between the output of this voltage divider and an input connection. Adjusting one of the inputs so that the null voltmeter is no longer deflected guarantees that there is no load acting on either the input source or the voltage divider generating the comparison voltage, thus values can be set very precisely with little effort.

41 In addition, an output buffer ensures that a potentiometer cannot be damaged by feeding a voltage into its output.

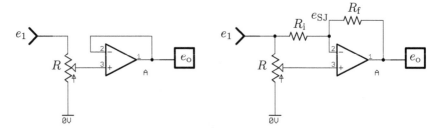

Fig. 4.27. Buffered coefficient potentiometers

plifier is grounded and the amplifier acts as a simple inverting amplifier with gain R_f/R_i. Typically $R_f = R_i$ is chosen so that the overall gain is 1. Setting the potentiometer to $\alpha = 1$ causes the wiper to be at the upper end of the potentiometer thus connecting the non-inverting input of the amplifier with the input e_1. Since the amplifier will maintain the voltage e_{SJ} at its inverting summing junction at 0 V, it follows that $e_o = e_1$. So this circuit can be described by[42]

$$\frac{e_o}{e_1} = 2\alpha - 1.$$

Large installations often used servo driven potentiometers that could be set from a central control panel or automatically from an attached paper tape reader or even an attached stored-program computer, thus allowing rapid setup of coefficients. Figure 4.28 shows a typical servo potentiometer from an *EAI 580* analog computer. The 10-turn wire wound potentiometer is visible on the left with the motor driving a reduction gear visible in the upper right half. The small circuit board below the motor holds the relay used to select the potentiometer for readout. Later hybrid computers[43] often replaced classic potentiometers by digitally controlled attenuators.[44]

Using only summers, integrators and coefficient potentiometers, an analog computer would be restricted to the solution of simple linear differential equations of the form

$$\sum_{i=0}^{n} a_i y^{(i)} = 0.$$

To overcome this restriction additional computing elements are required and these are described in the following sections.

---

42 Cf. [BADER 1985, pp. 76 f.].
43 See section 9.
44 These are basically multiplying *digital to analog converters (DAC)*.

Fig. 4.28. EAI servo potentiometer

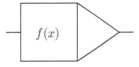

Fig. 4.29. Symbol of a function generator

4.5 Function generators

Another basic task is the generation of a general function $f(x)$. Ideally such a function could be derived as solution of differential equation but in many cases functions are required, which are defined by values obtained from experiments or other sources. In these cases special computing elements called *function generators* are required. Generating arbitrary functions with purely analog electronic means is not a trivial task and often requires an large amount of circuitry.[45] Even more difficult, at least in the common case, is the generation of functions like $f(x,\dots)$ of more than one variable by analog electronic means. The symbol for representing a function generator with a single input is shown in figure 4.29.

A plethora of different approaches to generate arbitrary functions have been developed over time. The following sections describe some of the more widely used or technologically more interesting implementations.

4.5.1 Servo function generators

The earliest function generators used by LOVELL et al. in their fire control computer were based on a servo circuit controlling a motor driving at least two potentiometers mounted on a common shaft. One of these potentiometers is used to generate a feedback value for the servo circuit while the other(s) generate the function(s) desired depending on the angular position of their respective wipers. Figure 3.22 (section 3.3) shows such a setup used to generate trigonometric functions.

---

[45] This was one of the problems, which led to the development of *hybrid computers* (see section 9). Using a stored-program digital computer coupled to an analog computer, the digital system can be used to generate functions by suitable algorithms or table lookups.

Other implementations of servo based function generators use cams driven by the servo controlled motor with a feeler pin riding on the cam's surface driving an output potentiometer. Another interesting implementation of a cam potentiometer is described in [LOVEMAN 1962].

These techniques have the disadvantage that specially wound potentiometers are required and these are delicate instruments and difficult to manufacture. The same is true for machined precision cams. Consequently early *tapped potentiometers* were employed for function generation. These are based on standard 10-turn wire wound potentiometers that have been modified to feature multiple taps connected equidistantly to the resistive element. Using these taps, the potentiometer can be used as a linear interpolator between the voltages applied to its various taps. The taps' positions correspond to the breakpoints while the voltage applied to the tap represents the desired function value at this particular breakpoint. Using a servo system driving the potentiometer's shaft according to an input value x such a setup yields a function $f(x)$, which is the result of linear interpolation between the values set at the taps.[46]

4.5.2 Curve followers

Another servo-based function generator is the *curve follower*, which is basically a servo circuit fed by a pickup connected to a xy-plotter as shown in figure 4.30.

The basic idea is as simple as it is elegant: The function to be generated is painted or plotted on a sheet of paper with conductive ink. This graph is then placed onto the table of a modified xy-plotter that has a pickup coil mounted in place of the pen normally used. An external sine generator is then connected to the conductive trace of the function graph on the paper. A simple servo circuit now feeds the y-input of the plotter in such a way that the pickup coil will always be directly over the conducting function graph, while the x-input of the plotter is fed with the argument of the function to be generated. The y value used to drive the y-input of the curve follower now represents the desired value $f(x)$.

Other curve followers used an optical pickup to follow the function graph instead of the conductive ink and inductive pickup setup. The basic principle of operation remains the same.

---

[46] See [LOVEMAN 1962, pp. 3-36 ff.] for more information about tapped potentiometers.

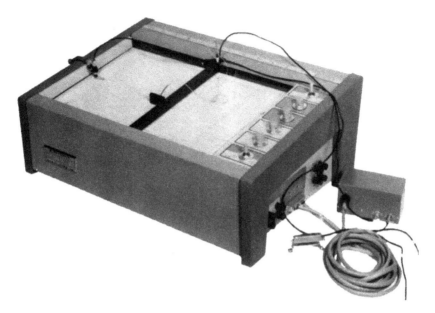

Fig. 4.30. EAI curve follower

4.5.3 Photoformers

As elegant as these electromechanical function generation techniques are, their main disadvantage is that they cannot cope with rapidly varying function arguments x due to the relatively large moving masses. Fortunately the basic curve follower idea can be easily implemented using an all electronic approach like the one shown in figure 4.31.

The function to be generated is represented by an opaque mask placed in front of an oscilloscope tube. The input to the x-deflection plates of the display tube is fed by the function argument. Using a photomultiplier and a simple servo circuit a signal for driving the y-deflection plates is now generated so that the spot of light on the face of the display will always be placed at the boundary of the opaque mask; this system effectively generates $y = f(x)$.[47]

---

[47] Cf. [MACNEE 1948, Sec. IIIE], [HAUG 1960, p. A3] and [KORN 1962, pp. 3-68 ff.] for more detailed information. Even a WILLIAMS-tube like construction was once employed. This proved useful in repetitive operation of high-speed analog computers when the result of one run had to be used as the input for the next run, see [HAUG 1960, p. A3].

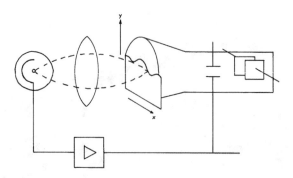

Fig. 4.31. Optical function generator (cf. [HAUG 1960, p. A3])

4.5.4 Varistor function generators

A different approach to function generation is the application of *varistors*.[48] A varistor is a current sensitive non-linear resistor implementing a basic function of the form[49]

$$e_o = ae_1^n$$

with n being determined during production. Typical values are $n = 2$ or $n = 3$. The precision of these varistors is quite good. [KORN 1962, p. 3-75] quotes an accuracy of 0.4 V for the *Quadratron*,[50] a varistor made by *Douglas Aircraft Company*.

Of course having square and cube functions does not give the flexibility of a photoformer but these varistor circuits were extremely cheap. This made it possible to use these devices to generate functions by truncated power series.[51]

4.5.5 Diode function generators

The availability of reliable and cheap germanium, and later silicon diodes gave rise to another technique of implementing function generators, the *diode function generator* or *polygon function generator*. The basic structure of such a circuit is shown in figure 4.32.

---

48 See [KORN 1962, pp. 3-73 ff.].
49 Cf. [KORN 1962, p. 3-75].
50 See [GAP/R 1959].
51 Cf. [KORN et al. 1964, pp. 6-14 f.] for more details.

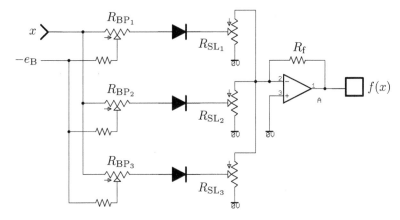

Fig. 4.32. Principle of operation of a diode function generator

An ideal diode acts like a switch conducting current only in one direction but not in the reverse. So a simple ideal diode might be used to implement a function such as

$$f(x) = \begin{cases} ax & \text{if } x > 0 \\ 0 & \text{otherwise} \end{cases}$$

where the slope a can be set by a series resistor. If the voltage at which the diode starts conducting could be shifted arbitrarily a function could be generated by a piecewise linear approximation using several of these circuits.

The circuit shown in figure 4.32 is based on this idea: Each of the three diodes starts conducting at a voltage defined by the setting of its associated bias potentiometer R_{BP_i}, which effectively sets a breakpoint position. The voltages at the cathode of each of these diodes are then summed by an operational amplifier with variable input weights defined by the setting of R_{SL_i} and the feedback resistor R_f. R_{SL_i} defined the *slope* of the particular partial function.

Of course, a practical diode function generator features many more than just three of these biased diode circuits.[52] A typical implementation of a function generator made by Telefunken is shown in figure 4.33. This particular example generates $f(x) = x^2$ for values $x \geq 0$ and was used in the various transistorised Telefunken analog computers.

A more versatile variable diode function generator, a *VDFG*, made by EAI is shown in figure 4.34. It features ten diode segments with coarse and fine control for the respective breakpoints and slopes. The possibility of defining the breakpoints had the advantage that the breakpoints could be set to match the function

---

[52] More information can be found in [KORN 1962, pp. 3-66 ff.] and [HAUG 1960, p. A4] for example.

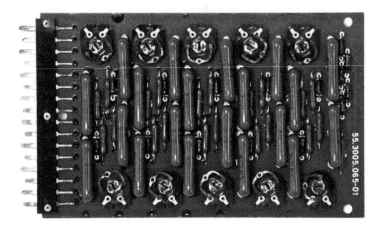

Fig. 4.33. $f(x) = x^2$ for $x \geq 0$ (Telefunken)

being generated as well as possible, thus minimising errors. On the other hand, it greatly complicated the overall manual setup procedure. Consequently, other manufacturers implemented diode function generators with 21 equally spaced fixed breakpoints with only the slopes of the individual partial functions to be set manually.

The last and maybe most remarkable successor of these diode function generators is the AD639 integrated circuit, which implements the basic trigonometric functions and their inverses with a high degree of precision.[53]

4.5.6 Inverse functions

Many functions can be readily implemented by generating the inverse of another readily available function. Using a function generator in the negative feedback path of an operational amplifier circuit yields its inverse function.[54] Figure 4.35 shows the basic setup required to generate an inverse function $f^{-1}(x)$ using a function generator $f(y)$ and an *open amplifier*.[55]

---

[53] See [GILBERT 1982] and [Analog Devices].
[54] At least as long as there is no change of sign and the feedback function is reasonably smooth – in most cases it is advisable to add a small capacitor, about several ten pF, between the output of the amplifier and its summing junction to stabilise the circuit.
[55] An operational amplifier with no (implicit) feedback path is called open amplifier.

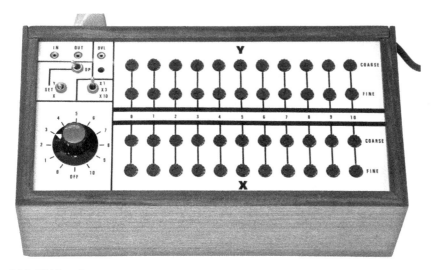

Fig. 4.34. EAI function generator

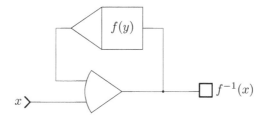

Fig. 4.35. Generating an inverse function using an open amplifier

4.5.7 Functions of two variables

Generating functions of more than one variable, the case with two variables being the most common, is difficult using only analog electronics.[56] Some approaches to solve this problem involve the use of three-dimensional cams[57] with appropriate servo motor circuits. Another technique is based vaguely on the curve follower idea: An xy-plotter with its pen replaced by a pickup stylus is controlled by its x- and y-inputs moving a pickup over a specially prepared conductive paper or plastic sheet. On this sheet equipotential contour lines are drawn with conductive ink. These lines are then each connected to adjustable voltage sources. The conductive sheet

---

[56] Some basic information about this topic can be found in [KORN 1962, pp. 3-78 ff.] and [HAUG 1960, pp. A9 f.].
[57] Cf. figure 2.7.

Fig. 4.36. Contour map function generator

then performs a two-dimensional linear interpolation between these equipotential lines.

A variation of that scheme is the electrolytic tank. In the simplest case it is just a rectangular bath filled with a conductive fluid. A pickup electrode is then moved through this trough by an xy-plotter. Using wires embedded in the bath, equipotential lines can be implemented as above. Even more complicated functions can be represented using a non-flat bottom plate. In some cases even a three-dimensional movement of the pickup electrode has been employed yielding functions of the form $f(x, y, z)$.

Similar to the mechanical barrel cam[58] is the *contour map* function generator shown in figure 4.36.[59] The function is represented by a three-dimensional "landscape", which is probed by a tip being moved in x- and y-direction as controlled by the two function arguments. The actual height of the tip touching the contour then corresponds to the desired function value $f(x, y)$.

Yet another possibility is to use a couple of function generators of one variable and feed a tapped potentiometer function generator with their respective outputs, thus linearly interpolating between the outputs of the individual function generators. In cases where the response times achievable with these servo-based techniques is not sufficient, the use of an oscilloscope tube, a translucent mask, and a photomultiplier has been proposed. The main problem here is the manufacturing

---

[58] See section 2.5.1.
[59] See [GALLAGHER et al. 1957, p. 87].

of a precise translucent mask representing the function $f(x, y)$ by means of the extinction of light.

4.6 Multiplication

The seemingly trivial operation of multiplying two time varying variables is quite complicated to implement. Accordingly, a plethora of different approaches were suggested and used in the past to build usable multipliers for analog computers.[60] Since multiplication can be seen as a function $f(x, y) = xy$ of two arguments, many of the ideas described in the preceding section can be used to implement this operation.

4.6.1 Servo multipliers

An obvious way to implement a multiplier is based on a servo function generator driving a number of linearly wound potentiometers mounted on a common motor driven shaft. The input variable to control the servo circuit is the multiplier x while the ends of the free potentiometers, i.e., those that are not needed for the servo-feedback, are connected to the multiplicands $+y_i$ and $-y_i$.[61] If x and y can take on values in the interval $[-1, 1]$, the multiplier is said to be capable of *four-quadrant operation*.

Figure 4.37 shows a five channel servo multiplier made by Solartron: The left half of the device houses the power supply transformers.[62] The power output transistors of the servo circuit driving the motor are mounted on the two large black heat sinks visible in the upper middle of the picture.

At the heart of the servo multiplier are the six precision potentiometers mounted on a common shaft, visible on the right of the picture. One of these potentiometers is used to generate the feedback signal for the servo circuit while the remaining five potentiometers are available for multiplication. Thus, with one input x, this servo multiplier can generate five products $xy_1, xy_2, \ldots, xy_5$.

Apart from their mechanical complexity servo multipliers have the inherent drawback that the rate of change of the x is limited due to the inertias involved. If x changes too rapidly, the servo circuit will be unable to follow introducing errors

---

[60] A comprehensive classification of analog multipliers can be found in [CELINSKI et al. 1964].
[61] Typically, these potentiometers feature a grounded middle tap, thus enforcing that $0 \cdot y$ is as close to the value zero as possible regardless of any unavoidable little non-linearities in the potentiometer itself.
[62] Servo multipliers typically feature dedicated power supplies for the actual servo circuit to minimise noise coupling into the precision supplies of the analog computer.

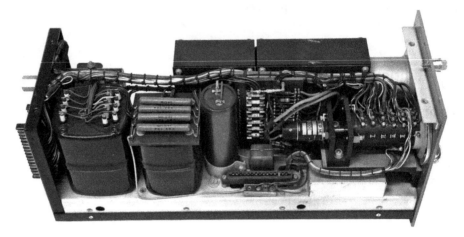

Fig. 4.37. Solartron servo multiplier

in the products. Nevertheless, in many applications this restriction turns out to be not a severe one – the advantage of generating a number of products based on a common multiplicator often outweighs it. Additionally, the rate of change of the y_i is not limited.

4.6.2 Crossed-fields electron-beam multiplier

One of the first all electronic multiplication schemes is the *crossed-fields electron-beam multiplier* described by [MACNEE 1948, Sec. IIIA]. The central element of this type of multiplier is an oscilloscope tube with its x- and y-deflection plates and an additional deflection coil on the tube's neck.[63]

The x-deflection plate pair is controlled by the x-input voltage causing the electron beam to move horizontally with a velocity $|v_x| \sim x$. The y input signal drives a power amplifier, which in turn feeds the deflection coil wound around the neck of the tube. This coil generates a magnetic field $B \sim y$ thus deflecting the electron beam in the vertical direction by an amount xy.

The screen of the display tube is divided into an upper and lower half by an opaque mask. Using a photodetector[64] a servo circuit like the one outlined in section 4.5.3 can now be setup. This circuit drives the y-deflection plates of the oscilloscope tube in such a way that the spot of light will always stay on the edge

---

[63] Cf. [HAUG 1960, pp. A6 f.].
[64] Alternatively, a shield dividing the tube screen into two equal halves with two photodetectors has also been used successfully.

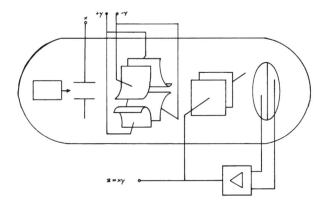

Fig. 4.38. Structure of a hyperbolic field tube (cf. [HAUG 1960, p. A7])

of the opaque mask. Thus, the output voltage of the servo circuit is in proportion to xy.

4.6.3 Hyperbolic field multiplier

An even more arcane multiplication scheme based on a very special vacuum tube is the *hyperbolic field multiplier*, which was invented at the university Darmstadt by FRIEDRICH-WILHELM GUNDLACH[65] in the early 1950s as a high speed multiplier.[66]

Figure 4.38 shows the structure of this hyperbolic field tube. It has three sets of deflection plates – from left to right these are a vertical deflection plate pair, four plates generating a hyperbolic electrostatic field, and another pair of deflection plates to deflect the electron beam horizontally. On the far right are two target plates connected to a difference amplifier.

The vertical deflection plates are driven by the multiplicator input x while the multiplicand y is fed in a crosswise fashion to the hyperbolic plates. The field created by these is proportional to $y\eta\xi$ with η and ξ being the coordinates of a coordinate system with its origin in the middle of these four deflection plates.[67]

An electron beam, which has not been deflected by the vertical deflection plates, will cross the hyperbolic field undisturbed since $\eta = \xi = 0$. If the beam

---

[65] 02/02/1912–01/27/1994
[66] See [GUNDLACH 1955], [SCHMIDT 1956] and [KLEIN 1965].
[67] See [HAUG 1960, p. A7].

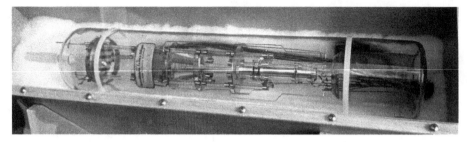

Fig. 4.39. Hyperbola tube

has been deflected due to $x \neq 0$ experiences an additional deflection due to this hyperbolic field resulting in a additional horizontal deflection proportional to xy.[68]

The electron beam now hits the target plates causing a tiny voltage, which is amplified by the difference amplifier connected to these plates. The output voltage of this amplifier drives the horizontal deflection plates so that the electron beam will always be exactly in the middle of the two target plates, so this voltage will by xy.

This type of multiplier is capable of operation at high frequencies with respect to both input voltages x and y due to its purely electrostatic deflection system. [HAUG 1960, p. A7] mentions a dynamic multiplication error of only about 1% at signal frequencies of 80 kHz – a remarkable value for its time. The complete schematic of a multiplier based on this special tube can be found in [DHEN 1960, p. 31].[69] Figure 4.39 shows the only known surviving hyperbolic field tube.[70]

4.6.4 Other multiplication tubes

The 1950s saw several other vacuum tube based approaches to multiplication apart from the hyperbolic field tube. One of these devices was developed by ALEXANDER SOMERVILL. Figure 4.40 shows the principle of operation. An electron beam with square cross section is deflected by two sets of deflection plates in x and y direction before hitting a target consisting of four square collector plates. The current

---

[68] This tube is not too dissimilar to the crossed-fields electron-beam multiplier described before but without the external deflection coil, which has been replaced by the hyperbolic deflection plates.

[69] This multiplier was part of a large analog computer, *ELARD*, developed and built at the Technical University of Darmstadt by WALTER DHEN, see [DHEN 1959].

[70] This specimen is part of the *Informatiksammlung* of the *Darmstadt University of Applied Sciences*.

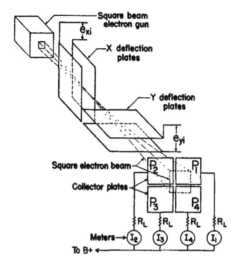

Fig. 4.40. Electron beam multiplication tube (see [SOMERVILLE, p. 145])

measured by each of the meters $I_{1...4}$ depends on the position of the deflected electron beam.

If the beam is not deflected, i.e., $x = y = 0$, it hits the four plates $P_{1...4}$ in a symmetrical way and the currents, taking the different signs of the plates into account, cancel out. If, for example, $x = y = 1$, current will only flow through P_1 corresponding to the product $xy = 1$, etc. This basic idea can be further improved by employing feedback circuitry to linearize the multiplier.[71]

A similar device developed by E. J. ANGELO in 1952 had a bandwidth of 70 kHz and a precision of about 2% of full-scale output. This device used an electron beam with a circular cross section and four square target plates.

4.6.5 Time division multipliers

Another approach to multiplication is the *time division multiplier*,[72] which is a particular implementation of a larger family of *modulation multipliers*.

As described before, a servo multiplier is essentially just a voltage-controlled voltage divider using potentiometers to yield the desired product values. Voltage division can also be achieved by modulating a square wave signal. The amplitude

---

[71] Unfortunately, SOMERVILLE was unable to perform dynamic tests due to a tube failure (see [ANGELO 1952, p. 3]).

[72] See [KORN et al. 1964, pp. 7-10 ff.]. Other terms for this device are *mark-space multiplier* or *pulsed-attenuator multiplier*.

of this signal is controlled by the y input, while x controls the duty cycle.[73] $x = \pm 1$ corresponds to a signal being constantly at the value $\pm y$ while $x = 0$ would result in a duty cycle of 1 : 1. Generating the average of this square wave signal by means of an RC low-pass filter yields an output voltage of xy.

As in the case of the servo multiplier, this scheme can be readily extended to the computation of multiple products with a common multiplicator x. Accordingly, most time division multipliers contain one mark-space modulator circuit driving several amplitude modulators, which are controlled by individual inputs y_i.[74]

[KORN et al. 1954] describes the simple relay-based multiplier shown in figure 4.41. The variable x is fed with opposite signs to the circuit and one SPDT relay contact switches between $+x$ and $-x$. This signal is fed to a ripple filter consisting of two T-filters, yielding the desired output value $x_o = \frac{xy}{2z}$.

At the heart of the circuit is an integrator consisting of a DC coupled amplifier with two inputs and a capacitor in its feedback path. The output of this integrator drives the relay. One of the integrator inputs is connected to the term y while the other input E_z is connected to $+z$ and $-z$ by means of a second SPDT relay contact. z is typically the machine unit of the computer. If $y = 0$ the relay will toggle between $+z$ and $-z$ with a 50% duty cycle resulting in $x_o = 0$. When $y \neq 0$ it biases the integratorm which will in turn result in a different duty cycle of the relay operation so that the integrator input value E_z will cancel out y yielding the desired product value at x_o.

A setup like this has some drawbacks such as limited bandwidth, mainly determined by the maximum frequency at which the relay can operate reliably, and limited life time due to wear of the relay contacts.[75] A variation of this idea is the *triangle-integration* multiplier as described in [HARTMANN et al. 1961].

4.6.6 Logarithmic multipliers

An obvious implementation of a multiplier is based on adding logarithms to achieve a multiplication – the very same principle that slide rules depend on. Figure 4.42 shows the basic setup of a *logarithmic multiplier*, which uses three function generators, two yielding the logarithm of an argument and one yielding the inverse logarithm.[76] An interesting linear-to-logarithmic circuit based on standard vac-

---

[73] This is the ratio of the mark- and space-times of the signal.
[74] An early and very detailed circuit description of such a time division multiplier can be found in [LILAMAND 1956]. A modern implementation might use a switched capacitor building block as described in [Linear Technology, p. 9].
[75] Using modern integrated circuit analog switches very low cost multipliers can be implemented using this basic approach.
[76] Cf. [MORRILL 1962, pp. 3-41 f.].

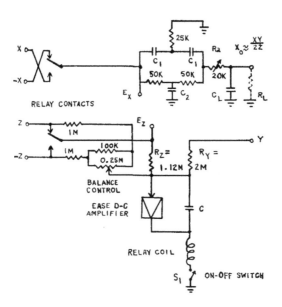

Fig. 4.41. Relay based time-division multiplier (see [KORN et al. 1954, p. 980])

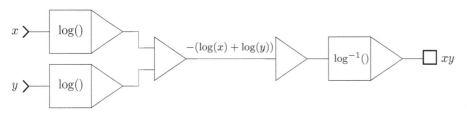

Fig. 4.42. Logarithmic multiplier

uum tubes, such as the dual triode 12AX7, can be found in [HOWARD et al. 1953] while [SAVANT et al. 1954] describes a practical multiplier based on this idea.

The main problem of this type of multiplier is the fact that the two variables x and y are restricted to positive values unless special precautions are taken, such as generating absolute values and restoring the sign of the multiplication result.

4.6.7 Quarter square multipliers

By far the most widely used classic multiplier circuit is the *quarter square multiplier*. It is based on
$$xy = \frac{(x+y)^2 - (x-y)^2}{4},$$

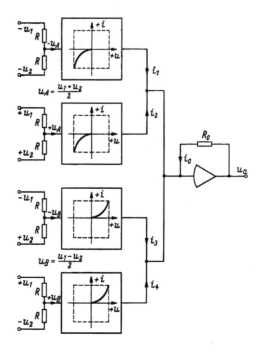

Fig. 4.43. Principle of operation of a quarter square multiplier (cf. [GILOI et al. 1963, p. 92])

which only requires the computation of two square functions,[77] a difference and a division by 4, which can be easily done with a voltage divider or a suitable feedback resistor of the output amplifier.[78] Figure 4.43 shows the basic circuitry of such a multiplier.

A particular requirement of this type of multiplier is that both variables, the multiplicator as well as the multiplicand, have to be supplied with both signs, which typically is not a problem in most analog computer setups due to the sign changing behaviour of summers and integrators. The main advantages of this multiplier are its high bandwidth as well as its relatively low price. The only

---

[77] The function generator card shown in figure 4.33 is actually part of a quarter square multiplier.

[78] An interesting variation of this multiplication technique has been used by Hitachi in the 505E analog computer. Its multipliers are based on

$$xy = \frac{1}{2}\left(\frac{(x+y-1)^2}{2} - \frac{(x-y-1)^2}{2}\right) + y,$$

see [Hitachi 505E, p. 16].

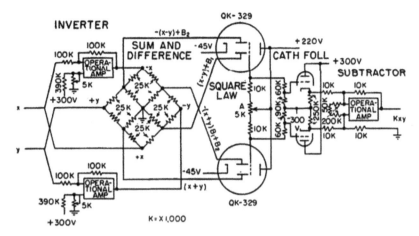

Fig. 4.44. Quarter square multiplier based on two QK-329 beam-deflection tubes (see [MILLER et al. 1955, p. 163])

drawback is that there is no way to generate multiple products of a common multiplicator as in the case of servo and time division multipliers.

In the late 1940s Raytheon started development of special non-linear beam-deflection tubes for analog multiplication. The first of these tubes, the *QK-256* used an electron beam with rectangular cross section that could be deflected before passing through a specially-shaped mask. Depending on the amount of deflection, the beam passing through this mask changed its cross section in a non-linear way. Based on this tube, the *QK-329* was developed, which allowed the construction of very high bandwidth – up to 80 Mhz – quarter square multipliers.[79]

Figure 4.44 shows a typical quarter square multiplier based on two QK-329 tubes. The bandwidth of this setup is primarily limited by the operational amplifiers used in the input and output stages and not by the beam-deflection tube implementing the square function.

Quadratrons, as described in section 4.5.4, were also used for low bandwidth (up to about 400 Hz[80]) and low precision quarter square multipliers.[81]

---

[79] See [DASSOW 2015], [MILLER et al. 1954], and [MILLER et al. 1955].
[80] [CELINSKI et al. 1964, p. 152] reports up to 5 kHz bandwidth for these multipliers.
[81] [GUL'KO 1961] describes an interesting multiplier circuit using such varistors with increased passband.

4.6.8 Other multiplication schemes

Over the years many other multiplications schemes have been developed and implemented. There are *dynamometer multipliers* balancing the forces of multiple electromagnets against each other[82], *strain-gauge multipliers* balancing a strain-gauge with voice coils,[83] *heat-transfer multipliers*,[84] HALL *effect multipliers*,[85] and many more.[86]

An interesting hybrid multiplier technique for analog computers using binary counters and very early digital-analog-converters is described in [GOLDBERG 1951].

In the late 1960s analog transconductance multipliers were developed.[87] These are based on the effect that the gain of a transistor depends on its collector current, so that base and collector current can be multiplied by a differential transistor pair followed by a difference amplifier.

Modern analog multipliers are typically based on GILBERT *cells* developed by BARRIE GILBERT[88] in the late 1960s.[89]

4.7 Division and square root

Division and square root are normally implemented in a straightforward fashion using a multiplication circuit and an open amplifier to generate the inverse function of multiplication. Figure 4.45 shows how a division operation can be implemented using this technique by employing a multiplier[90] in the feedback loop of an open amplifier. Connecting the y input of the multiplier also to the output of the amplifier sets up a square root function.

It should be noted that circuits using open amplifiers with active elements in the feedback path may tend to unstable operation i.e., oscillation. To stabilize such circuits it is common to connect a small capacitor in the range of several 10 pF between the summing junction and the output of the amplifier.

---

[82] See [MORRILL 1962, p. 3-40].
[83] Cf. [MORRILL 1962, p. 3-41].
[84] See [SAVET 1962].
[85] Cf. [HAUG 1960, p. A7].
[86] [EDWARDS 1954] contains a comprehensive overview of various historic multiplication schemes used in analog computers.
[87] See [STATA 1968].
[88] 06/05/1937–01/30/2020
[89] See [Analog Devices 2008]. [MORTON 1966] describes a multiplier capable of operation up to 10 MHz, which is quite similar to a GILBERT cell mutliplier.
[90] The multiplier is also often denoted with a "Π" instead of the "×" symbol.

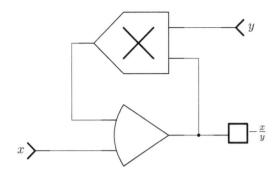

Fig. 4.45. Division as inverse function of multiplication

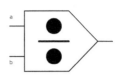

Fig. 4.46. Symbol for a divider

4.8 Comparators

Some problems require a means of implementing step functions or to switch a variable from one value to another, etc. This is typically done using a *comparator* in conjunction with analog switches in an analog computer. Basically, a comparator is just an operational amplifier with biased diodes in its feedback loop thus limiting its output typically to two values such as $+1$ and -1.

Figure 4.47 shows the general principle of the operation of a comparator.[91] At its heart is an amplifier featuring typically two inputs with resistors R_1 and R_2.[92] The feedback loop of this amplifier consists of two diodes biased by two voltage sources e_{b_1} and e_{b_2} depicted by batteries in this schematic.[93]

In the simplest case the output of the amplifier controls a relay. If the sum of the two input signals e_1 and e_2 is greater than zero the output will be driven to its lower limit, while the upper limit will be reached for a sum less than zero. Thus, a comparator can be used to compare the voltages of two (or more) signals.

Small early tabletop computers often did not provide dedicated comparators – the operational amplifiers were too precious to be committed to a specific function like this. Instead their patch panels often featured a couple of free diodes, which can be used to setup a comparator based on an open amplifier. Comparators directly driving a relay are often called *relay comparators*.

Larger and more complex analog computer typically contain a number of comparators, which do not control a relay directly. Instead the comparators yield a

---

[91] Cf. [GILOI et al. 1963, p. 80]. The summing junction is denoted by S.
[92] Usually it is $R_1 = R_2$.
[93] Of course these bias sources are not actual batteries in a real implementation – they can be readily implemented by either using coefficient potentiometers introducing a bias voltage or the two diodes can be altogether replaced by ZENER diodes.

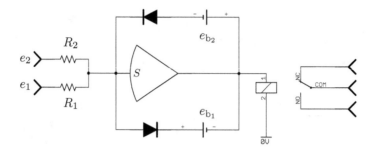

Fig. 4.47. Principle of operation of a classic comparator

Fig. 4.48. Symbol of a relay comparator   **Fig. 4.49.** Symbol of a comparator

digital output signal, which can be used to control a relay driver or to signal certain events to a connected digital computer, etc. Such comparators are typically called *electronic comparators*. Figures 4.48 and 4.49 show the symbols for a relay comparator and an electronic comparator yielding a logic output signal respectively.

4.9 Limiters

Limiters are used to limit the value of a variable. This is often necessary in simulations of mechanical systems where moving parts can hit stops, etc. The basic circuit of a limiter is similar to that of the amplifier part in a relay comparator. By using appropriately biased diodes forming a feedback path of the operational amplifier, its output voltage can be limited by an upper and lower bound. Figure 4.50 shows the structure of a basic limiter.[94]

---

[94] Sometimes this simple circuit is insufficient since the slope of the limited portion of the signal is not zero. See [GILOI et al. 1963, pp. 207 ff.] and [KORN 1962, pp. 3-62 ff.] for implementation variants yielding more precise results.

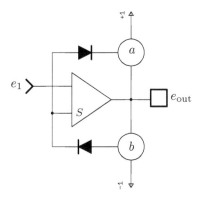

Fig. 4.50. Simple limiter circuit

4.10 Resolvers

Resolvers are among the most complex individual computing elements in an analog computer. Typically only large systems contain resolvers as stand-alone units. The purpose of a resolver is the conversion of polar to rectangular coordinates and vice versa. In addition to this, a resolver can also yield the sine and cosine values of a given input angle. Resolvers also allow the rotation of coordinate systems, which is an especially useful operation in aerospace related problems.

One of the earliest resolvers described in literature is the device developed by LOVELL et al.[95] This device is a typical *servo resolver* – this type of resolver was used widely in aerospace applications where coordinate transformations are abundant and the rates of change of the variables to be transformed are normally sufficiently slow to be suitable for an electromechanical approach. Later analog computers also featured all electronic resolvers allowing high-speed operation.

Resolvers are quite complicated and there are too many implementation variants to be described in detail here.[96]

4.11 Time delay

Many problems require some form of time delay to model transport, diffusion effects, etc. An ideal time delay unit would act as shown in figure 4.51: An input

---

[95] See chapter 3.3.
[96] [CARLSON et al. 1967, pp. 58 ff.], [GILOI et al. 1963, pp. 95 ff.], [LOVEMAN 1962, pp. 3-2 ff.], [MORRILL 1962, pp. 3-56 ff.], [KLEY et al. 1966], and [VOGEL 1977] contain details and implementation examples.

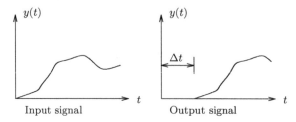

Fig. 4.51. Behaviour of an ideal time delay

signal is delayed by a variable amount Δt. Given an input signal $y(t)$ the output $e_o(t)$ of the time delay unit is then $e_o(t) = y(t - \Delta t)$.[97]

There have been many different approaches to achieve a near ideal time delay for use in an analog computers. The most basic idea from today's point of view is to digitize the input signal, store it in a digital memory system and read it out with the desired delay time Δt. The digital values read from memory are then fed to a *digital-analog converter*, a *DAC*. Although this yields very good results and is basically not affected by the form of the input signal,[98] this approach became feasible only in the late 1970s with the advent of cheap *random access memory* integrated circuits (*RAM*). Figure 4.52 shows a typical digital time delay unit made by EAI in the 1970s.

Earlier implementations were based on other technologies such as tape recording systems[99] or capacitor storage systems. The latter are based on a wheel or drum housing a number of capacitors oriented radially and sharing a common connection at the axis with the other terminal of the individual capacitors connected to contacts on the surface of the drum or wheel. The segmented circumference of the rotating assembly (essentially a commutator) is electrically connected to two brushes: One for storing a value into a capacitor, the other for reading out values, which are then fed to an amplifier. The time delay Δt is determined by the angular positions of the two brushes as well as the angular velocity of the capacitor wheel.

The main disadvantage of such systems is the very limited number of storage elements, effectively limiting the number of points where the function to be delayed is sampled.[100] Yet another approach is based on the curve follower.[101] The function to be delayed is plotted by a strip-chart recorder. The paper containing the plot is

---

[97] During start-up time, $t < \Delta t$, the output is usually zero.
[98] Apart from questions regarding the sampling period and the resolution of the converters.
[99] Cf. [KENNEDY 1962, pp. 6-10 ff.].
[100] [KENNEDY 1962, pp. 6-8 ff.].
[101] Cf. section 4.5.2.

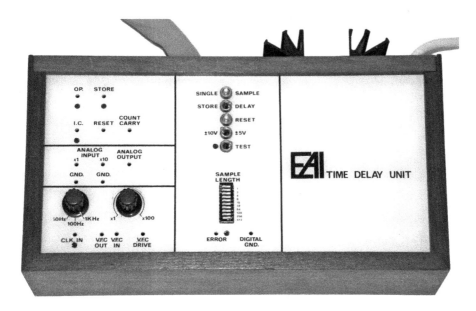

Fig. 4.52. EAI time delay unit

then continuously fed into a one-dimensional curve follower, which in turn delivers the delayed function.[102]

In cases where the substantial cost of a dedicated time delay system could not be justified, delay functions can be set up using PADÉ *approximations* (named after HENRI PADÉ[103]) or a STUBBS-SINGLE *approximation*. In LAPLACE-transform notation delays are characterized by the following general transfer function:

$$\frac{Y(s)}{X(s)} = \mathrm{e}^{-Ts}$$

Figure 4.53 shows the implementation of a 4$^\text{th}$-order STUBBS-SINGLE approximation, which is based on the transfer function

$$S_4(s) = \frac{1 - \dfrac{1}{2}Ts + \dfrac{15}{134}T^2s^2 - \dfrac{13.55}{1072}T^3s^3 + \dfrac{1}{1072}T^4s^4}{1 + \dfrac{1}{2}Ts + \dfrac{15}{134}T^2s^2 + \dfrac{13.55}{1072}T^3s^3 + \dfrac{1}{1072}T^4s^4}.$$

These delay functions tend to become quite complicated with increasing expectations regarding the fidelity of the delay. Their setup is also quite time consuming due to the many coefficient potentiometers required. More details

---

102 See [CARLSON et al. 1967, p. 225].
103 12/17/1863–06/09/1953

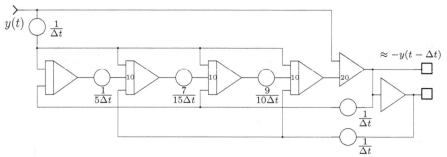

Fig. 4.53. 4th-order STUBBS-SINGLE approximation for time delay

and examples regarding PADÉ and STUBBS-SINGLE approximations can be found in [KENNEDY 1962, pp. 6-5 ff.], [CARLSON et al. 1967, pp. 225 ff.] and [STUBBS et al. 1954]. A comprehensive description of electronic time delays can be found in [ULMANN 2020/1, pp. 95–108].

4.12 Random noise generators

Many real-world problems to be analysed by analog computers involve random processes. Accordingly, sources of random noise are sometimes required in an analog computer.[104] Typical applications include the analysis of correlation functions and delay errors in signal networks, mechanical system analysis, the measurement of spectral densities, the study of optimisation processes by sequential random perturbations and many more.

Typical random sources employed in analog computers were based on physical systems exhibiting a random behaviour such as the decay of radioactive isotopes, resistor noise, or recombination noise on PN-junctions in bipolar semiconductors. Often complex demodulation schemes and special function generators are employed to yield the desired probability distribution of the random signal.[105] It is also possible and often advisable to use digital feedback shift registers and digital-analog-converters to produce random voltage signals of certain properties.

---

104 See [RIDEOUT 1962, pp. 12-5 f.], [KORN et al. 1964], [GILOI et al. 1963, pp. 357 ff.], [LANING et al. 1956], and [KORN 1966] for application examples and programming techniques.
105 See [KORN 1962, pp. 81 ff.].

Fig. 4.54. *Brush* six channel recorder

4.13 Output devices

Since an analog electronic analog computer typically represents values by voltages or currents, typical classic output devices are strip-chart recorders as shown in figure 4.54, xy-plotters (see figure 4.30), and oscilloscopes.

In cases where a graphical output of some solution is not required *digital voltmeters* (*DVM*s) are used to display values variables in a simulation run with high precision. Large analog computers often feature a built-in digital voltmeter with an automatic selector.[106] To get a snapshot of the current state of a computer run, all integrators are typically placed into halt mode first. Then the output value of every computing element of interest is read out by the digital voltmeter and displayed or printed in decimal form. Additionally, these values can be transmitted to an attached stored-program digital computer or can be stored on some digital medium.

Figure 4.55 shows the display of a typical precision digital voltmeter – in this case the precision DVM of a Telefunken RA 770 analog computer. The left half of the display[107] shows the address of the computing element selected for readout.

---

[106] This selector was often based on stepping relays in early analog computers, which were then replaced by readout relays located at every computing element. These readout relays could then connect the output of an individual computing element to a central readout bus, which is connected to the input of the central digital voltmeter.

[107] This particular display is of the projective type, i. e., every place of the display contains up to ten incandescend bulbs. Between each bulb and the screen of that particular digit a stencil with some symbol or digit is placed. Powering one of these tiny light bulbs will then project the desired shape.

Fig. 4.55. Readout of integrator I000 on a Telefunken RA 770 analog computer

Fig. 4.56. EAI 6200 digital voltmeter

In this case it is I000 denoting integrator number 000. The output value of this element is currently +0.9998. It should be noted that most of the digital voltmeters used in analog computers display their values scaled with respect to the machine unit of the system, so the above value corresponds to +9.998 V since the RA 770 is a ±10 V system.

A typical stand-alone digital voltmeter is shown in figure 4.56. This 3.5 digit instrument was built by Electronic Associates Inc. and was typically used in conjunction with small tabletop analog computers such as the TR-10 or TR-20. Its construction is quite remarkable as it consists of a frequency counter module and a DVM module. The instrument can be used as a frequency counter or as a digital voltmeter. In the latter mode the frequency counter becomes part of the DVM circuitry.

The following chapter will now focus on the anatomy of a typical classic medium sized analog electronic analog computer giving an impression of the arrangement of computing elements described in the preceding sections.

5 The anatomy of a classic analog computer

The system chosen for this section on the anatomy of classic analog computers is the medium sized *EAI-580*, built by Electronic Associates Inc. in the late 1960s and shown in figure 5.1. Although labelled as an "ANALOG/HYBRID COMPUTING SYSTEM" by EAI's marketing department, it is just a basic analog computer with provision to interface it to a digital computer if required in which case suitable external analog to digital and digital to analog converters (*ADC*s and *DAC*s) must be added.[1]

5.1 Analog patch panel

The most prominent feature of any medium or large classic analog computer is its central *(analog) patch panel*. It contains hundreds, or even thousands, of jacks, which connect to all inputs and outputs of the various computing elements of the system. A program is set up by means of *patch cables*, which are plugged into these jacks, to configure the computer to solve the problem. Patch cables for medium precision computers are typically simple unshielded cables with banana plugs on their ends. *Precision analog computers* often use shielded patch cables with special plugs and intricate shielded patch panels to achieve up to four decimal places of precision.

Since the process of patching an analog computer is quite time consuming, the patch panels of all but the smallest analog computers are exchangeable. Otherwise the times to switch from one program to another would be unreasonably high.

Figure 5.2 shows the analog patch panel of an EAI 580 analog computer. This panel is of the modular type, using insets, which represent the different computing elements. It contains four rows of computing elements each consisting of 15 inserts. Each of these has 2 columns and 12 rows of connectors. Such a computer typically contains the following types of computing elements:

- Dual operational amplifiers, usually paired with dual integrator circuits,[2]
- quadruple amplifiers,
- x^2-function generators, which are used to setup quarter square multipliers,

---

[1] See section 9.

[2] In contrast to other systems, the EAI 580 requires the programmer to setup integrators by explicitly interconnecting operational amplifiers with dedicated integrator circuits. This approach has the distinct advantage of giving maximum freedom to the programmer. On the other hand, this approach breaks with the idea of a mathematical machine as some knowledge of the inner workings of the system is required in order to program it.

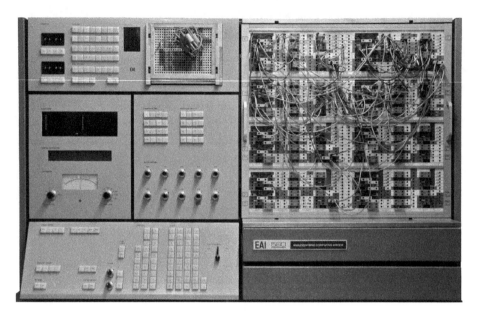

Fig. 5.1. EAI 580 analog computer

- coefficient potentiometers,
- comparators with electronic analog switches,
- function generators, and
- *trunk lines*,[3] etc.

5.2 Function generators

Located below the analog patch panel are two drawers, which each contain four diode function generators. Each of these function generators has ten diode segments with adjustable breakpoint and slope.[4] Figure 5.3 shows the extended lower drawer.

As noted before, there are two basic variants of diode function generators: With and without adjustable breakpoints. The former implementation has the advantage that the breakpoints for a function approximation can be clustered in regions where the function to be implemented is "interesting" (i.e., changes

---

3 A trunk line is a direct connection between a patch panel connector and a connector on the back of the machine where input/output-devices may be connected in a convenient way without further cluttering the analog patch panel.

4 Cf. figure 4.32.

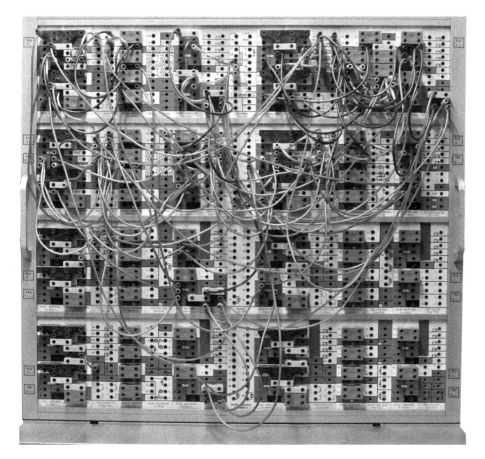

Fig. 5.2. EAI 580 analog patch panel

rapidly in a certain region). This approach has the disadvantage of a tortuous setup procedure, as the breakpoints and slopes must be calculated and set manually. The second implementation variant, which is typically used in Telefunken analog computers, speeds up the setup considerably by omitting the variable breakpoints but this can sacrifice precision of the function approximation. It is mainly a matter of personal preference, which implementation is preferred.

5.3 Digital patch panel and controls

Located in the top half on the left of the EAI 580 is a small digital subsystem consisting of a tiny patch panel, two thumb wheel switches to preset digital counters/dividers and a number of switches, which can be used to set or reset individual bits of these counters. These switches also display the current state of each associ-

Fig. 5.3. EAI 580 function generators

ated bit. The remaining switches control individual flip-flops and a timer. Figure 5.4 shows this subsystem.

The various digital elements (simple logic gates, inverters, counters, and dividers) can be interconnected by means of patch cables plugged into the small removable digital patch panel. Although many applications of an analog computer do not require any digital elements at all, there are more complex tasks such as parameter optimisation, etc., which benefit greatly from the flexibility offered by a digital control system. A typical example is the simulation of an automatic transmission for a car: The behaviour of the gears, the torque converter, etc., can be simulated readily by the analog part of the computer. The control of these transmission parts, the gear changing decisions under varying loads, etc., is then implemented using the digital part of the computer.

The inputs to these digital elements are typically clock signals generated by a central timing unit, outputs of comparators used in the analog setup, manual switches, etc. Normally, the digital outputs are used to control electronic switches

Fig. 5.4. EAI 580 digital controls and patch panel

on the analog computer[5] or to control the operation of individual integrators or groups of integrators. Some tasks such as parameter optimisation often require at least two groups of integrators, which operate in an alternating fashion: One integrator group is halted and delivers initial conditions for a second group, which will perform some computation based on these values. This second group is then halted and can be used to update the first integrator group, etc. Alternatively, the first integrator group can compute a new set of initial conditions while the second group is still running.[6]

5.4 Readout

Located below the digital control unit is the overload display and readout system on the left and manual coefficient potentiometers and switches associated with comparators and function relays on the right. Figure 5.5 shows these subsystems of the EAI 580 analog computer in detail.

The overload display on the upper left has one incandescent light bulb per amplifier in the system to display overload conditions. There are two basic types of overload that may occur in an operational amplifier, in both of which it will be unable to keep its output at the level required to satisfy the condition that the inverting summing junction of the amplifier must be maintained at 0 V. Taking a summer with two inputs e_1, e_2 and associated weights $a_1 = a_2 = 1$ as an example,

---

[5] Using such switches differing gear ratios can be selected in the automatic transmission example.
[6] Some machines like the Telefunken RA 770 feature an intricate control and timing system that allows operating modes like this to be implemented without having to resort to individually patched digital control elements.

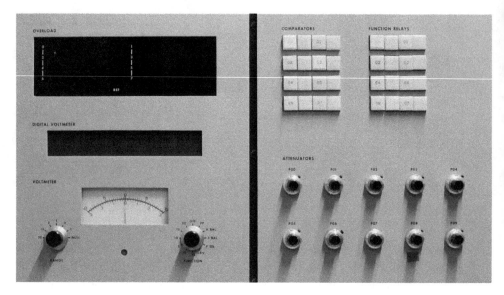

Fig. 5.5. EAI 580 readout, overload display, manual potentiometers and manual switches

this error condition could result from the attempt to add two input voltages of $+1$ each. The output voltage of an amplifier cannot raise above or go below its positive and negative supply voltages. In this example the output, which should be -2, would be limited by the amplifier's output range to a lower value such as -1.4. Consequently, the current flowing through the feedback resistor can no longer cancel the input currents at the summing junction.

A similar problem would result from too high a load current at the output of an amplifier, in which case it could not maintain the proper output voltage. Connecting the output of an amplifier to ground or to one of the reference voltages would result in such an overload condition, as would be feeding too many computing elements from a single operational amplifier's output.

When chopper-stabilised amplifiers are used, the detection of such overload conditions is easy: In each case the inverting summing junction can no longer be maintained at 0 V. This will in turn yield an excessive error correction signal at the output of the stabilising amplifier, which can then be used to light the corresponding overload indicator for this amplifier.

Since an overload condition normally renders a computation useless, it will in most cases also place the analog computer into halt mode. The programmer can then check the program to determine and remedy the reason for the overload.

Below the overload indicator panel is the display of a precision digital voltmeter with four digits of resolution. This is used to read out program variables.[7] Digital voltmeters like this can often be connected to a printer and a selector unit, which allows automatic readout of a number of computing elements.

On the bottom left is an analog voltmeter, which can also be used to determine the value of program variables as well as for a quick checkout of the computer. By means of the rotary switch on its right the various power supply voltages can be checked easily. In addition to this, the voltmeter is used for manual balancing of the operational amplifiers.[8]

The right half of this panel contains a couple of switches connected to the comparators and function relays of the analog computer. Below these are ten manual ten-turn coefficient potentiometers, called *attenuators* on this machine. The machine has many more coefficient potentiometers, which are set by servo-controlled motors, as it proved impractical to make users manually set up to several hundred coefficient potentiometers.

5.5 Control

The unit on the lower left of the computer is the central control panel, which is used to control the overall operation of the computer and for setting the servo coefficient potentiometers. Figure 5.6 shows this control panel in detail.

Located on the far left are, from bottom to top, the power switches, four control switches to select the mode of operation of the analog computing elements and similar switches for the digital part. The available analog operating modes are (from left to right[9]):

PP: Short for *Program Panel*, this mode transfers the overall control of all integrators to the digital unit on the top left.
IC: Place all integrators into initial condition mode.
HD: All integrators are placed into halt mode.
OP: Switch all integrators into operate mode.

The logic mode switches are used to

---

[7] Normally the computer is placed into halt mode for readout.
[8] In some cases the drift of an amplifier is so excessive that the chopper stabilisation stage cannot compensate it completely. Consequently, each amplifier has a potentiometer for manual coarse zeroing.
[9] The key labels have faded away due to the age and usage of the particular machine depicted.

Fig. 5.6. EAI 580 control panel

- transfer the control of the digital elements to the digital patch panel (PP),
- to clear (C),
- stop (S),
- and run (R) the digital circuit setup on the digital patch panel.

The switches labelled 10^6, 10^5, 10^1 and STP select the basic digital clock rate, one pulse per microsecond being the default cycle time, or place the digital subsystem into single step mode. The RMT switch places the digital unit into remote mode, which allows several EAI 580 systems to be coupled together or to couple an EAI 580 analog computer with a stored-program computer, basically forming a hybrid computer.[10]

The two switches labelled DIS and ENG are used to disengage and engage the large removable analog patch panel. Due to its many connections, it is locked into place by a motor driven mechanism, which is controlled by these two switches.

Above these two switches is a row of three buttons labelled SP, ST and RMT. These place the machine into the potentiometer-set mode (SP), a check mode called *static test*[11] or into remote mode, which transfers the control of the analog portion to an external system such as a memory programmed computer.

Next to these are two switches, which are used to select the time base of the system. By default all integrators are set to a time-scale of 1 second, so that an input voltage corresponding to one machine unit will yield -1 at the output of the integrator after one second. Pressing the button labelled 2MS switches all integrators to a second set of integration capacitors with 1/500 of the capacity of

---

10 See section 9.
11 This mode is quite similar to the initial condition mode, but instead of initial conditions static check voltages can be applied to integrators allowing a check of a program without actually running it.

the normal feedback capacitors, thus speeding up the computation by a factor of 500.

The two ten-turn potentiometers control timers for *repetitive operation*. In this mode a computation will be performed over and over again, cycling the integrators through their initial condition and operate modes automatically. The keys labelled .1, 1 and 10 set a global multiplcative factor for the timers controlling the repetitive operation of the system.

The next three columns of switches are labelled ADDRESS and allows the selection of any computing element of the analog section of the EAI 580 for readout or setup in the case of servo potentiometers. Each element has an element type and a two digit decimal address. The output value of an addressed element is displayed on the digital voltmeter, together with a one letter function code denoting the type of the element.[12]

The next block of switches is labelled RDAC and forms a simple, yet precise resistor-based analog-digital converter. Its main purpose is to set servo controlled coefficient potentiometers: First, using the ADDRESS switches, a potentiometer to be set is selected. In the next step the desired value is keyed into the RDAC keys and the SET key is pressed. This causes the servo motor of the selected potentiometer to position its wiper to the desired value. In some cases it is desirable to control a servo potentiometer manually, which can be done with the big lever on the far right. Turning it to the left will cause the servo motor of the selected potentiometer to decrease the coefficient's value; turning it to the right increases its value.

5.6 Performing a computation

A computation on the EAI 580 analog computer is performed as follows:

1. The computer is switched on by pressing the ON button.
2. The programs prepared on the analog and digital panels have to be inserted onto the machine. While the digital patch panel is engaged and disengaged manually by means of a lever, the analog patch panel is unlocked or locked by pressing the DIS or ENG switch on the control panel.
3. The various servo driven coefficient potentiometers and the manually controlled attenuators must be set.
4. If required for this computation, the diode function generators must be set up manually.

---

[12] "A" denotes an amplifier, "P" a potentiometer, etc.

5. If required, initial conditions for the digital computing elements have to be set by pushing the switches associated with the counters, flip-flops, etc.
6. The time-scale is set if repetitive operation is desired.
7. The analog and digital subsystems of the analog computer can then be put into the required mode of operation. In the simplest case this will just involve setting the analog part of the machine into IC mode, followed by OP.

6 Some typical analog computers

So many companies made so many different models of analog computers that it is impossible to describe them in detail here.[1] The following sections describe some typical analog computers.

6.1 Telefunken RA 1

In the early 1950s Dr. ERNST KETTEL, an engineer working for Telefunken, a world-renowned manufacturer of radio transmitters, receivers, RADAR sets, etc., started thinking about building an analog computer to do research in his main field of interest – communications equipment. Eventually, this project transformed Telefunken into a leading analog computer manufacturer in Germany and Europe. It seems reasonable to assume that Dr. KETTEL met HELMUT HOELZER during his work in Peenemünde, where he developed radio control and telemetry systems for the rocket development projects. He had probably seen HOELZER's general purpose analog computer in operation there, which would explain his general interest in this topic.

In 1953 actual work on this machine started with the help of Dr. ADOLF KLEY. The machine became known as the *RA 1*[2] and is shown in figure 6.1.

[FEILMEIER 1974, p. 18] describes the RA 1 as a typical system of the *heroic age* of analog computing, which spanned from about 1945 until 1955. The machine is of a completely modular design. Three racks hold a variety of computing and support elements. The right hand rack contains the power supplies delivering the highly precise machine voltages of ± 100 V, the main supply voltages of ± 200 V, a bias voltage of -450 V, etc.

The rack on the left holds the following subsystems from top to bottom:

– A vacuum diode function generator with 21 fixed breakpoints and 21 precision potentiometers and switches to set the slope of the polygon,

---

[1] In fact, many influential systems, such as the EAI 580, the EAI 680, the large systems EAI 7800 and EAI 8800, as well as many vendors like Beckman, COMCOR, Dornier, Simulators Inc., Goodyear, Solartron, etc., are not described. This should not imply that their machines are inferior to those described here.

[2] *RA 1* is the abbreviation for *Rechner, Analog* number one. "Rechner" is the German term for computer. Telefunken followed a simple scheme for naming its analog computers, which were all denoted by *RA* followed by a number. Later tabletop machines were labelled with an additional *T* to distinguish these from the larger systems.

Fig. 6.1. The Telefunken RA 1 as it is currently preserved in a private collection

- four quarter square multipliers[3],
- eight summers,
- an oscilloscope of rather interesting design with two display tubes,
- a collection of limiters and the like,
- some *free diodes*,[4],
- relays to set up comparator circuits,

---

3 Of these eight units, seven use dual diode vacuum tubes EAA 91 (equivalent to the 6AL5) for the square functions while one module is of a clearly experimental nature: This left most multiplier uses unlabelled germanium diodes instead of vacuum tubes.

4 Many analog computers contain *free* or *uncommitted* diodes, which are very useful to limit the output of computing elements, which can be used to implement functions such as absolute value, etc.

- five coefficient potentiometers,
- a selectable integrator/summer/amplifier,
- and a summer.

The rack in the middle contains (top to bottom)

- a second diode function generator,
- four quarter square multipliers,
- eight switchable integrators/summers/amplifiers,
- 16 coefficient potentiometers,
- another eight switchable integrators/summers/amplifiers,
- eight coefficient potentiometers and
- a drawer containing the central timing and control module as well as six switchable integrators/summers/amplifiers.

It is quite obvious that the RA 1 was built as a piece of laboratory equipment and not as the prototype of a product to be sold to customers. All computing elements are interconnected with shielded patch cables but there is no central patch panel, there is no means to simplify setting up coefficient potentiometers, there is no built-in voltmeter, etc.

It must have come quite as a surprise for Telefunken that customers from industry who saw this machine during visits to the Telefunken laboratory in Ulm expressed their interest in buying such a machine. Forced by this unexpected customer demand, Telefunken agreed to build a batch of ten systems, which were slightly improved over the prototype RA 1.[5] A very large configuration of this production model, named *RA 463/2*[6] is shown in figure 6.2. Figure 6.3 shows a typical medium-scale installation.[7]

A typical RA 463/2 system, consisting of only three racks with no operator console table, was priced at 158,000 DM in 1958 (see [Telefunken 1958]) corresponding to about 410,000 US$ today. Since the first batch of these machines was sold even before the computers were even built, additional systems were manufactured – reluctantly at first – and thus Telefunken entered the analog computer business.

---

[5] A contemporary description of this machine with potential application can be found in [HERSCHEL 1957].
[6] See [Telefunken 1958].
[7] Sitting on the left is Prof. Dr. OTTO FÖLLINGER with Dr. WERNER AMMON on the right.

Fig. 6.2. A large Telefunken RA 463/2 analog computer installation (see [AMMON, p. 1])

6.2 GAP/R analog computers

The analog computer pioneer GEORGE A. PHILBRICK and founder of *GAP/R* (*George A. Philbrick Researches*) not only brought operational amplifiers into widespread commercial use in 1952 but also built and sold highly modular electronic analog computers.[8]

These computers are built from "black boxes", which implemented various computing functions such as the K3-A adding component shown in figure 4.19 in section 4.2. These modules typically employed a number of operational amplifiers such as the K2-W[9] as well as the necessary passive elements to implement a certain function. These components were grouped into linear devices such as the K3-A adding component, the K3-C coefficient component, integrators, differentia-

---

[8] A short history of these developments in the years from 1938 to 1957 can be found in [COULTER].
[9] See section 4.1.1.

Fig. 6.3. Another Telefunken RA 463/2 analog computer

Fig. 6.4. Typical GAP/R analog computer setup from 1957

tors, etc. Non-linear components included limiters, backlash, multipliers, square functions, square root, etc.

These systems were primarily intended for high-speed, repetitive operation.[10] A remarkable feature was the *Electronic Graph Paper*, which basically displayed a grid on an oscilloscope, thus giving *"instantaneous and automatic calibration of both voltage and time on the oscilloscope screen itself"*.[11]

Figure 6.4 shows a typical GAP/R analog computer installation from 1957. The large oscilloscope display in the middle is part of the electronic graph paper capability. The overall construction is highly modular and flexible.

6.3 EAI 231R

The Electronic Associates *EAI 231R* analog computer is one of the machines that became archetypal for analog computers during the 1960s. It was introduced in

---

10 See [GAP/R EAC, p. 3].
11 See [GAP/R EAC, p. 3].

Fig. 6.5. EAI 231R analog computer

1959 and already had a long ancestry, which made this system far superior to the Telefunken RA 1 and its production model RA 463/2 with respect to its overall layout, the available computing elements, and their precision. The 231R and its predecessors were part of the *PACE* (short for *Precision Analog Computing Equipment*) series including the earlier 16-24A, 16-24D, 16-31R and 16-131R models.

A typical EAI 231R analog computer is shown in figure 6.5. This system became one of the most widely-used analog computers and systems of this type were still in use in the late 1970s. The system is dominated by the large patch panel on the right side containing 3450 plug sockets. These sockets as well as the patch cables are completely shielded, to minimise cross talk and noise injection. Below the patch panel is the operator control panel containing 20 manually operated coefficient potentiometers on the right and the timing control, analog voltmeter,

Fig. 6.6. Removing the patch field from an EAI 231R computer (see [EAI PACE 231R, p. 6])

and computing element selection circuitry on the left. The two large bays on the top of the system can hold either additional 80 manual coefficient potentiometers or servo potentiometers.[12] Located above the operator panel are a random noise generator (recognizable by its two circular analog instruments), a digital voltmeter, two servo multipliers and a printer connected to the DVM.[13]

The bays below the operator console contain 24 modules, each with four operational amplifiers and two power supplies. All input and feedback resistors and capacitors are contained in a temperature controlled oven to minimise drift effects due to temperature variations. Figure 6.6 show the removable patch field. The system depicted has manual coefficient potentiometers installed in the top enclosures and an extension rack mounted to its right.

A large EAI 231R installation is shown in figure 6.7. It consists of two fully equipped EAI 231R analog computers, each with one extension rack to its left and one to its right. A large flatbed plotter can be seen in the left foreground. Visible in the middle of the picture is a smaller EAI *Variplotter* and on the right are two multi-channel recorders and an *ADIOS* (short for *Automatic Digital In-*

---

12 The system shown here is equipped with servo potentiometers. These could be set automatically from an attached setup system or a digital computer.
13 Short for *Digital Voltmeter*.

Fig. 6.7. Installation of two EAI 231R with various plotters and an ADIOS console

put Output System).[14] console, which can be used to automatically set up servo potentiometers under paper tape control.

The 231R was such a successful system that EAI developed an enhanced version, the *EAI 231RV*, that was put on the market in 1964. In contrast to the 231R, it contains a sophisticated digital control system that allows control of individual integrators and easy implementation of complex control sequences. In addition, the integrators featured additional time-scale factors. These systems found widespread use in nearly all branches of science and engineering but most prominently in the field of aerospace engineering.

Figure 6.8 shows a *HYDAC 2000* system, consisting of an EAI 231RV analog computer with several expansion racks and an eight channel recorder on the right. On the left a DOS 350 digital console can be seen. This consists of a plethora of simple digital circuit elements such as individual gates, flip-flops, monoflops, etc., that can be interconnected by means of a central digital patch panel similar to that used in the analog computer. Using these elements, quite complex control circuits can be implemented.

Such a HYDAC system, costing over one million Dollars,[15] was acquired by General Dynamics in 1964 for the development of the F-111 jet fighter.[16]

6.4 Early transistorised systems

Telefunken was the first company that successfully developed a fully transistorised analog computer. After the rather unexpected commercial success of the vacuum tube based RA 463/2, the company's management decided that the era of vacuum tubes was over – a pretty bold decision in the mid 1950s when transistors were still exotic devices with widely varying performance characteristics. A project

---

14 Some background information about ADIOS can be found in [VAN WAUVE 1962].
15 Adjusted for inflation this corresponds to roughly 9.2 million USD in 2022.
16 Cf. [N. N. 1964/2].

Fig. 6.8. EAI HYDAC-2000 system (see [N. N. 1964/1])

to develop a transistorised high-performance operational amplifier with chopper stabilisation was started.[17] This development was undertaken by GÜNTER MEYER-BRÖTZ, who just finished university,[18] and HANS OTTO GOLDMANN.

A small tabletop analog computer was built as proof-of-concept. The resulting system is shown in figure 6.9. It is clearly a direct descendant of the RA 1 and RA 463/2 – there is no central patch panel, all computing elements are housed in separate modules and the machine features a built-in oscilloscope.

It was decided to develop a line of transistorised analog computers based on this prototype. The first such machine was the *RAT 700*.[19] The operational amplifier used in this first fully transistorised analog computer has been shown in chapter 4, figure 4.14. The RAT 700 was first sold in 1959 and became the progenitor of a long and commercially successful series of analog tabletop computers built by Telefunken. Figure 6.10 shows one of the early machines sold in 1959.[20]

---

[17] See section 4.1.2 for a description of this amplifier. The much simpler amplifiers in the RA 1 and its corresponding production model RA 463/2 had no such drift compensation. Accordingly, these machines were capable of repetitive operation only with a maximum integration time of 110 seconds. Longer runs would compromise the results too much due to drift effects.
[18] Prof. Dr. MEYER-BRÖTZ told the author that he was hired without having any real experience in the field of amplifier development because he did not know that it was considered impossible to build a fully transistorised precision operational amplifier. Nearly everybody in the 1950s considered transistors to be well suited for switching operation but not up to the task for precision analog computer applications.
[19] Short for *analog tabletop computer* (German *Rechner Analog Tisch*).
[20] Later models had a different colour scheme and a slightly different enclosure.

6.4 Early transistorised systems — 143

Fig. 6.9. First fully transistorised tabletop analog computer made by Telefunken (see [ERNST 1960, p. 255])

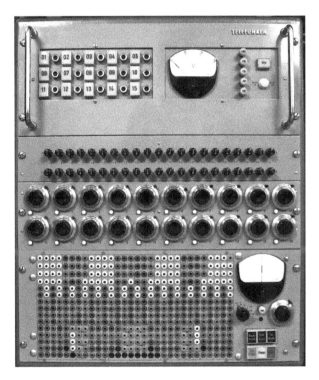

Fig. 6.10. The RAT 700 tabletop analog computer

In contrast to the RA 1 and RA 463/2 this tabletop computer features a central patch panel[21] located in the lowest of four chassis-mounted modules, which contains the input and feedback networks of the summers and integrators and additional computing elements such as free diodes, relays, etc. The operator control panel can be seen on the right hand side of this module. The middle section of the machine holds two half-height drawers containing 20 precision coefficient potentiometers[22] and two diode function generators with fixed breakpoints. The top drawer holds the power supplies (± 15 V, -25 V for relays, and 6 V for lamps, and the highly stabilised machine units of ± 10 V) on the right and 15 chopper-stabilised precision operational amplifiers with their corresponding overload indicators on the left.

A fully expanded RAT 700 contains 19 operational amplifiers,[23] four quarter square multipliers, two diode function generators, 20 coefficient potentiometers, eight free diodes and two relays that can be used to set up comparators in conjunction with a summer.

The RAT 700, which sold remarkably well, was in fact just a by-product of the development of a large precision analog computer that was completed in 1960 and introduced to the public at the Hanover trade show in the same year. This machine, the *RA 800*, shown in its basic configuration in figure 6.11, was living proof that a precision analog computer with a machine unit of ± 10 V could be built with transistors instead of the then ubiquitous vacuum tubes. Around this time, other companies started their own developments of transistorised analog computers to compete with the RA 800.

The left rack contains, from top to bottom, a power supply, the central removable patch panel, 50 precision coefficient potentiometers, two servo resolvers and two drawers each with four time division multipliers. The right rack contains, top to bottom, a power supply, two drawers each containing four diode function generators, ten manual switches, a digital voltmeter, the central control panel with various timers, a compensation voltmeter and computing element readout selec-

---

21 Even a removable patch panel was offered as an option due to customers requesting some way to quickly change the setup from one program to another.

22 To protect these expensive potentiometers against damage caused by patching errors, a small incandescent bulb is placed in series with the wiper. Under normal circumstances only a tiny current flows through the wiper and the small resistance of the bulb in its cold state does not introduce any noticeable error into the computation. In case of a erroneous connection resulting in excessive current flowing through the potentiometer's wiper, the bulb lights up and limits the current to a value small enough so that will not damage the potentiometer.

23 Eight of these operational amplifiers are committed to switchable integrators/summers, seven are part of dedicated summers while the remaining four are needed for the multipliers and function generators.

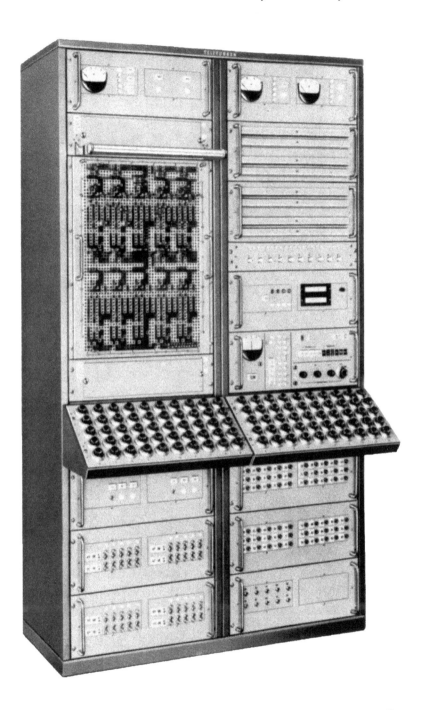

Fig. 6.11. A basic RA 800 precision analog computer (see [Meyer-Brötz 1960, p. 176])

Fig. 6.12. The EAI TR-10 tabletop analog computer

tion logic, 50 additional potentiometers, two drawers each holding 30 operational amplifiers, and ten electronic comparators.

At the same time, EAI was developing a transistorised operational amplifier that would be the heart of its first tabletop analog computer, the *TR-10*, which was introduced to the market in 1960. Figure 6.12 shows the front view of a TR-10. Instead of a central patch panel, it exposes its various components to the programmer.[24] Interestingly, this machine – like several other later systems made by EAI – does not offer integrators and summers as abstract computing components but requires the user to set up these elements explicitly by interconnecting operational amplifiers with suitable passive networks.[25]

---

24 A removable patch panel was later offered as an option.
25 This is in sharp contrast to other systems such as all of the Telefunken made machines, which hide the inner workings of their computing elements and only offer mathematical operations as the basic elements for a computer setup.

From top to bottom, this particular TR-10 contains 20 coefficient potentiometers, twelve computing networks to build summers, integrators, etc., ten dual operational amplifiers that have to be connected to the computing networks, and the control panel at the bottom.

6.5 Later analog computers

In the mid 1960s analog computers were increasingly often coupled with digital computers, forming *hybrid computers*.[26] Accordingly, most manufacturers started development of these systems by augmenting existing analog computers with the necessary hardware interfacing components such as analog-digital and digital-analog converters, etc. A typical example of this strategy is the EAI HYDAC 2000 system shown in figure 6.8. Telefunken used a similar approach by evolving the RA 800 into the *RA 800H*, the *H* denoting its hybrid capability. In fact, this machine was basically still a RA 800 system but had a digital control system similar to the DOS 350 of the HYDAC-2000.

Figure 6.13 shows a RA 800H system. The rack on the left holds the digital extension, the *DEX 802*.[27] This is basically just a collection of simple digital elements like logic gates, counters and timers. In addition to this, there are additional drawers containing analog function generators, an electronic resolver, etc. Thanks to the seemingly simple DEX 802 the RA 800H was a much more versatile system than its RA 800 predecessor. It was possible to solve directly complex tasks such as parameter optimisations on the RA 800H.

The last Telefunken analog computer is the RA 770, unveiled in 1966, which was a significant improvement over the RA 800H. Shown in figure 6.14, this system achieves an overall precision of 10^{-4} thanks to improved operational amplifiers employing silicon transistors and fully electronic choppers.[28] Using integrated circuits in the control system and other improvements made the RA 770 a physically much smaller and yet more capable system, with a wider complement of computing elements, than the RA 800H.

Both the RA 770 and the RA 800H can also be coupled with a stored-program digital computer by means of a *hybrides Koppelwerk*[29] *HKW 860* to form a true hybrid computer. Unfortunately, Telefunken stopped the development of analog

---

[26] See section 9.
[27] Short for "Digital-Experimentierzusatz", which translates roughly to "digital extension for experiments".
[28] These amplifiers, described in [MEYER-BRÖTZ et al. 1966], and shown in figure 4.15, chapter 4, have an open-loop gain of $3 \cdot 10^8$.
[29] German for *hybrid coupler*.

Fig. 6.13. A RA 800H system (see [ADLER 1968, p. 273])

computers in the early 1970s with no successor to the venerable RA 770. Its analog computer branch went out of business in the 1970s while other manufacturers, especially EAI, were still bringing new and enhanced systems to the market.

An example of a late EAI system, a *PACER 500* hybrid computer is shown in figure 6.15.[30] From left to right, the console typewriter for the stored-program digital processor, the digital processor itself, the dominating analog patch field, the operator control panel, including a small digital patch field, and a small cathode ray tube display can be seen. Systems like these were in widespread use in many scientific and engineering branches well into the 1980s.

---

[30] This system was based on the earlier EAI 580 analog computer, now equipped with the necessary interface electronic for attaching a digital computer.

6.5 Later analog computers

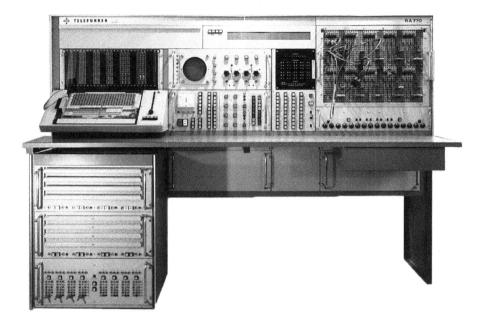

Fig. 6.14. Telefunken RA 770 precision analog computer

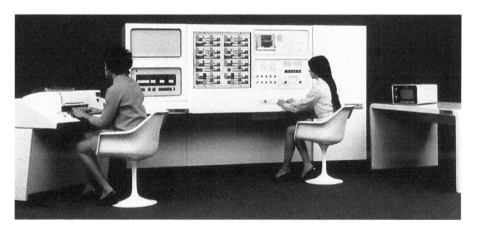

Fig. 6.15. An EAI PACER 500 hybrid computer system

Fig. 6.16. Front view of THE ANALOG THING

6.6 THE ANALOG THING

A noteworthy recent development[31] is *THE ANALOG THING (THAT)*, a small-scale analog computer ideally suited for educational and hobbyist use.[32] Shown in figure 6.16 this little analog computer contains five integrators, each with two time-scale factors, four summers, four sign-inverters, eight coefficient potentiometers, two multipliers, two comparators with electronic switches, several free elements such as diodes, capacitors, and resistor networks. It also supports manual operation as well as repetitive operation so that solutions can be displayed as steady figures on an oscilloscope.

One especially interesting feature of this system is that it is an open hardware project, so its schematics are publicly available and can be used as a starting point for other developments.

---

[31] This computer was introduced in 2021.
[32] See https://the-analog-thing.org.

7 Programming

Programming an analog computer differs substantially from the algorithmic approach used for stored-program digital computers. Programming an analog computer can be considered being easier and maybe even more "natural" than developing and implementing an algorithm on a digital computer. This is especially true for engineers, physicists and mathematicians, since the analog computer is, due to its nature of forming an analog for a problem to be solved, closely related to their realms of expertise and their ways of thinking about complex dynamical systems.

From the perspective of a contemporary programmer, programming an analog computer will look like an arcane task, as it bears no resemblance at all to today's ubiquitous algorithmic approach. Since analog computing is not part of the standard curriculum in a computer science class, programming such a machine is considered a curiosity while the much more abstract algorithmic approach has become natural.

The following sections are an introduction to analog computer programming and are based on practical examples of increasing complexity.[1] Since programming, be it analog or algorithmic, is to a certain degree an art form, becoming an expert analog computer programmer requires a lot of hands-on experience.

One of the biggest hurdles for aspiring analog computer programmers is to overcome the deeply ingrained algorithmic approach to programming, which has been the predominant approach to programming since the late 1970s.

7.1 Basic approach

Figure 7.1 shows the basic approach to programming an analog computer: First of all a thorough analysis of the system to be simulated is required, which typically results in a differential equation, or systems of coupled differential, or partial differential, equations.[2] Based on these equations a *program* for the analog computer is then devised. This program is a description of the connections required between the analog computer's various computing elements to form an electronic analog of the underlying problem. Since the variables within such a program are limited to the interval $[-1, 1]$ a *scaling* or *normalization* step is typically necessary to ensure

---

[1] A much more comprehensive account of analog (and hybrid, for that matter) computer programming can be found in [ULMANN 2020/1].

[2] Of course, this first step is also required for developing an algorithmic solution for a given problem.

Fig. 7.1. Typical workflow in analog simulation (cf. [TRUITT et al. 1960, p. 1-108])

that no variable within the program will exceed the machine units.[3] This can often be quite tricky. Scaling a problem will not only change the values of coefficients but also introduce additional coefficients into the computer setup.

Given sufficient initial conditions for the integrators in the computer setup, the system of which an *analog* has been implemented can now be simulated.

There are two main approaches to derive an analog computer program from a set of equations, either the *feedback technique*[4] developed by Lord KELVIN in

---

[3] The value representation within an analog computer is thus similar to fixed point binary arithmetic or the mantissa part of a floating point number.
[4] Cf. [SCHWARZ 1971, pp. 25/165 ff.].

1875/1876 or the *substitution method*.[5] Both methods start with a set of differential equations of the general form

$$\frac{d^n}{dx^n}y(x) + a_{n-1}\frac{d^{n-1}}{dx^{n-1}}y(x) + a_1\frac{d}{dx}y(x) + a_0y(x) = f(x) \tag{7.1}$$

or shorter

$$y^{(n)}(x) + a_{n-1}y^{(n-1)}(x) + \cdots + a_1y'(x) + a_0y(x) = f(x)$$

with initial conditions

$$y(0), y^{(1)}(0), y^{(2)}(0), \ldots, y^{(n-1)}(0)$$

representing the system to be solved with the analog computer. Derivatives with respect to time t or the scaled *machine time* τ, such as $\frac{dy}{dt}$ or $\frac{dy}{d\tau}$, will be written here as $\dot{y}$.[6]

In an analog computer program differentials are typically avoided in favour of integrations since a (time) derivative of signal amplifies noise (a differentiator basically is a high-pass filter) while an integration filters noise out (low-pass filtering).[7]

7.2 Kelvin's feedback technique

KELVIN's feedback technique, which has also been called the *classical differential analyser technique*,[8] is based on the idea of a series of integrations starting with the highest derivative of a differential equation, successively yielding all lower derivatives necessary to represent the equation in question. As an example the following equation with initial conditions $\ddot{y}(0)$, $\dot{y}(0)$ and $y(0)$ is to be solved with an analog computer:

$$\dddot{y} + a_2\ddot{y} + a_1\dot{y} + a_0y = 0 \tag{7.2}$$

First, the equation is rearranged so that its highest derivative is separated on the left hand side of the equation sign:

$$\dddot{y} = -a_2\ddot{y} - a_1\dot{y} - a_0y. \tag{7.3}$$

---

[5] See [SCHWARZ 1971, pp. 168 ff.].
[6] It should be noted that an electronic analog computer can only integrate with respect to time, while classic mechanical or electromechanical differential analysers as well as their digital counterparts (see chapter 10) do not have this restriction. [JOHNSON et al. 1961, pp. 2–13] describe an electronic generalized integrator.
[7] See section 3.1.2.
[8] See [KORN et al. 1964, p. 1-5].

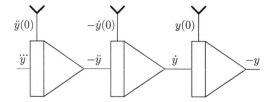

Fig. 7.2. Kelvin's feedback technique, 1st step

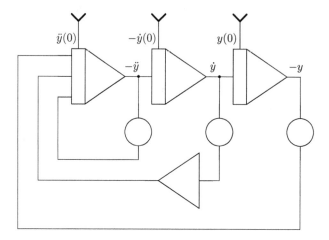

Fig. 7.3. Kelvin's feedback technique, 2nd step

Using three integrators connected in series as shown in figure 7.2 the lower derivatives $-\ddot{y}$, $\dot{y}$ and $-y$ can be computed.[9]

Combining these derivatives, the right hand side of equation (7.3) can be obtained as shown in figure 7.3. The coefficients a_0, a_1 and a_2 are implemented using coefficient potentiometers while an additional summer is necessary to yield $-\dot{y}$. Since every integrator (and summer) performs a sign reversal, the three initial conditions have to be supplied with alternating signs in this setup, too.

The computer setup derived by this simple method is now an analog of the system described by the differential equation (7.2). Depending on its scaling, i.e., the coefficients a_0, a_1 and a_2 and the time- scaling factors of the integrators, the desired solution can be typically obtained in the same period of time the original system would behave. It is also possible to use increase or decrease the time-scale factors yielding the solution faster or slower than real-time. While many problems, especially those with hardware- or humans-in-the-loop, need to run in real-time

---

[9] The alternating signs are caused by the sign-changing nature of practical integrators.

and do not therefore require time scaling, other problems benefit considerably from stretching or compressing time.[10]

Summarising the KELVIN feedback technique, an analog computer setup can be derived from a set of equations by executing the following steps:

1. Rearrange the equations so that the highest derivatives are separated on the left hand side.
2. Generate all lower derivatives that are part of the equations by successive integrations.
3. Using these lower derivatives, the terms on the right hand sides of the equations can be generated.
4. Since these right hand sides are equal to the highest derivatives used in the first step, the feedback loops can now be closed by feeding these values into the inputs of the respective first integrator stages.

7.3 Substitution method

Application of the *substitution method*[11] is best shown by deriving a computer setup for solving a differential equation like[12]

$$\ddot{y} + a_1\dot{y} + a_0 y = f(t). \tag{7.4}$$

To simplify matters, the initial conditions $\dot{y}(0)$ and $y(0)$ are assumed to be zero.

The basic idea is to transform this differential equation into equations of first degree by repeated substitutions. Each substitution decreases the degree of the remaining differential equation by one. The substitution

$$\eta = \dot{y} + a_1 y \tag{7.5}$$

yields

$$\dot{\eta} = \ddot{y} + a_1 \dot{y} \tag{7.6}$$

as its first derivative with respect to time. Substituting (7.6) into (7.4) yields the nonhomogenous differential equation

$$\dot{\eta} + a_0 y = f(t). \tag{7.7}$$

Solving (7.7) for its highest derivative yields

$$\dot{\eta} = f(t) - a_0 y,$$

---

10 Typical examples are the simulation of processes such as neutron dynamics in a nuclear reactor or problems in population dynamics.
11 Also known as *partial feedback method*.
12 Cf. [SCHWARZ 1971, pp. 168 ff.].

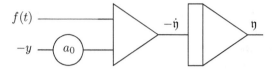

Fig. 7.4. First sub-circuit derived by the method of substitution

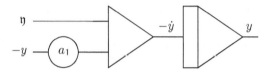

Fig. 7.5. Second sub-circuit derived by the method of substitution

which corresponds to the circuit shown in figure 7.4. Now a second circuit representing equation (7.5) has to be set up. Solving this equation for its highest derivative yields

$$\dot{y} = \eta - a_1 y,$$

which corresponds to the circuit shown in figure 7.5. η is readily available from the circuit shown in figure 7.4.

Combining these two sub-circuits in a straightforward fashion yields the setup shown in figure 7.6. Realizing that every summer and integrator performs a sign inversion, this circuit can be simplified by removing two summers.[13] The resulting simplified setup is shown in figure 7.7.

A major drawback of the substitution method is the complex treatment of non-zero initial conditions. While the initial condition $y(0)$ can be used directly for the last integrator in a program derived by this method, all other initial values have to be transformed according to the previously executed substitutions. So the initial value $\dot{y}(0)$ has to be transformed into an equivalent initial value

$$\dot{\eta}(0) = \dot{y}(0) + a_1 y(0).$$

η can now be used to initialize the corresponding integrator. This initial value transformation makes the application of the substitution method quite cumbersome in real applications. Consequently, the following sections will use the simpler KELVIN method.

---

[13] Since every operation causes some inevitable errors due to noise or drift, analog computers programs should always use the minimum number of computing elements.

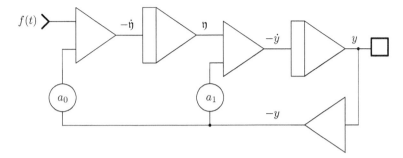

Fig. 7.6. Non-optimised circuit derived by the substitution approach

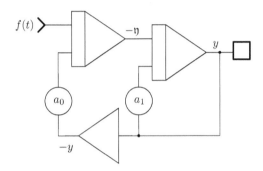

Fig. 7.7. Optimised circuit

In general, both techniques, KELVIN's feedback method as well as the substitution method, can be applied to linear and non-linear and linear differential equations alike.[14]

7.4 Partial differential equations

Many problems are readily described by partial differential equations containing derivatives with respect to more than variable.[15] Since an analog electronic analog

---

[14] In the case of non-linear differential equations the handling of initial conditions can be quite tricky. Sometimes special computer setups are used. Cf. [SOUDACK 1968] and [BROWN 1969] for more information. Some special classes of non-linear differential equations can be tackled with a method different from those described here, see [WHITE 1966].

[15] Most often derivatives with respect to time t and Cartesian coordinates (x, y, z) or spherical coordinates (t, φ, θ) are encountered.

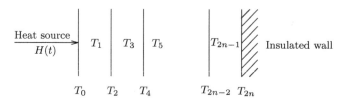

Fig. 7.8. One-dimensional heat-transfer

computer can only use time as the free variable of integration,[16] partial differential equations (*PDE*s) are normally solved by either employing a quotient of differences approach or by separation of variables.[17]

7.4.1 Quotient of differences

The *quotient of differences*-method is analogous to simple numeric algorithmic approaches. Derivatives with respect to another variable than time t are approximated as quotients thus discretising the underlying problem.[18] The derivative

$$\frac{\mathrm{d}f(x)}{\mathrm{d}x} = \lim_{\Delta x \to 0} \frac{f(x + \Delta x) - f(x)}{\Delta x}.$$

is approximated by a quotient of differences:

$$\frac{\Delta f(x)}{\Delta x} \approx \frac{f(x + \Delta x) - f(x)}{\Delta x}$$

As simple as this method is, it usually requires a large number of computing elements, which – especially on classic analog computers – may outweigh the advantage of simplicity. Figure 7.8 shows a simple one-dimensional heat-transfer problem being solved this way.[19]

---

[16] Using a multiplier an integration can be performed with respect to another variable than time according to $\int_{x(0)}^{x(t)} y(x)\,\mathrm{d}x = \int_0^t y(t)\frac{\mathrm{d}x}{\mathrm{d}t}\,\mathrm{d}t$, which is basically a STIELTJES integral.
[17] See [HOWE et al. 1953], [HOWE 1962], or [FORBES 1972].
[18] In some cases a problem variable other than t is identified with machine time τ. In this case all derivatives with respect to t must be discretised accordingly.
[19] [BRYANT et al. 1962, pp. 37–48] and [EAI 7.3.8a 1965] describe a very similar problem.

A heat conducting rod is connected to a time dependent heat source $H(t)$ on its left end while the right end is placed against an insulating wall.[20] This problem can be described by the PDE

$$\frac{\partial^2 T}{\partial x^2} = k\dot{T}.$$

The rod is now divided into slices of equal width as shown in figure 7.8. Each of these slices represented by a difference quotient. The heat source on the left and the insulator on the right side yield the following boundary conditions:

$$\dot{T}_0 = \frac{1}{k\Delta x} H(t) \qquad (7.8)$$

$$\dot{T}_{2n} = \frac{1}{k\Delta x} (T_{2n-1} - T_{2n-2}). \qquad (7.9)$$

The resulting quotient terms with indices $2m$ with $1 \leq m < n$ are

$$\frac{\partial^2 T_{2m}}{\partial x^2} \approx \frac{T_{2m-1} - 2T_{2m} + T_{2m+1}}{\frac{1}{4}(\Delta x)^2} \qquad (7.10)$$

with the interjacent elements being

$$T_{2m+1} = \frac{T_{2m} + T_{2m+2}}{2}.$$

T_{2m+1} represents the mean temperature on the edges of the corresponding slice in figure 7.8. Starting with equation (7.10), a system of differential equations involving only derivatives with respect to time can be derived. Assuming $n = 3$ and taking the boundary conditions (7.8) and (7.9) into account yields the following set of differential equations:

$$\dot{T}_0 = \frac{1}{k\Delta x} H(t)$$

$$T_1 = \frac{T_0 + T_2}{2}$$

$$\dot{T}_2 = \frac{4}{k(\Delta x)^2} (T_1 - 2T_2 + T_3)$$

$$T_3 = \frac{T_2 + T_4}{2}$$

$$\dot{T}_4 = \frac{4}{k(\Delta x)^2} (T_3 - 2T_4 + T_5)$$

$$T_5 = \frac{T_4 + T_6}{2}$$

---

[20] Additional examples can be found in [SEYFERTH 1960, pp. 3-66 ff.], [TRUITT et al. 1960], [AMELING 1963, pp. 284 ff.], [SYDOW 1964, pp. 103 ff.], [MACKAY 1962, pp. 243 ff. and pp. 293 ff.], [Telefunken 1963/1], [GILOI et al. 1963, pp. 255 ff.], and [ULMANN 2020/1, pp. 162 ff.]. A more complex gas flow example is described in [MAHRENHOLTZ 1968, pp. 143 ff.].

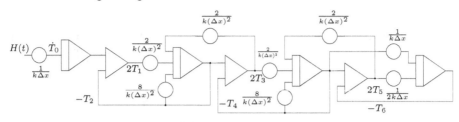

Fig. 7.9. Circuit for one-dimensional heat-transfer

$$\dot{T}_6 = \frac{1}{k\Delta x}(T_5 - T_4)$$

These equations can be readily transformed into the computer setup shown in figure 7.9. The large number of computing elements is obvious.[21] This method has the advantage that the coefficient potentiometers can be set up in groups, thereby simplifying the setup task significantly.

When the derivatives not depending on t are of higher degree, the method of difference quotients becomes rather cumbersome. In these cases, a different approach is often used, based on TAYLOR series approximations for the higher derivatives. Cf. [AMELING 1964, pp. 270 ff.] and [GILOI et al. 1963] for more information regarding this approach.

7.4.2 Separation of variables

Another method for the solution of partial differential equations with an analog computer is based on the separation of variables.[22] While the method described above discretises all variables except t, the separation of variables allows all variables of a PDE to take on continuous values. As a result, computer setups based on the separation of variables often require substantially fewer computing elements. Nevertheless, this technique has a major drawback: It is much more complicated to derive a computer setup by using the separation approach than by the method of difference quotients.

As an example the one-dimensional heat-transfer based on

$$\frac{\partial^2 T}{\partial x^2} = k\dot{T} \qquad (7.11)$$

---

21 It was not uncommon for some classic computer setups, especially in the area of chemical engineering or aerospace technology, to require hundreds of integrators and summers to model heat transfer problems with a reasonable fine discretisation of space. [REIHING 1959] describes an interesting time-sharing approach allowing to map the same set of analog computing elements to small sections of a large grid over and over again thus trading time against discretization granularity.

22 See [STEPANOW 1956, pp. 12 ff.].

will be treated again using this method.[23] The task is the determination of the temperature $T(x,t)$ at a certain position x and a certain time t. The separation of the variables t and x yields

$$T(x,t) = f_1(x)f_2(t). \tag{7.12}$$

From equations (7.11) and (7.12)

$$\frac{\mathrm{d}^2 f_1(x)}{\mathrm{d}x^2} f_2(t) = k\dot{f}_2(t) f_1(x)$$

follows. Separating x and t yields

$$\frac{\frac{\mathrm{d}^2 f_1(x)}{\mathrm{d}x^2}}{f_1(x)} = k\frac{\dot{f}_2(t)}{f_2(t)}. \tag{7.13}$$

Splitting equation (7.13) and setting both sides equal to $-\lambda_n$ yields the eigenvalue problem

$$\frac{\frac{\mathrm{d}^2 f_1(x)}{\mathrm{d}x^2}}{f_1(x)} = -\lambda_n \quad \text{and} \tag{7.14}$$

$$k\frac{\dot{f}_2(t)}{f_2(t)} = -\lambda_n. \tag{7.15}$$

Solving (7.14) and (7.15) for their highest derivatives results in

$$\frac{\mathrm{d}^2 f_1(x)}{\mathrm{d}x^2} = -\lambda_n f_1(x) \quad \text{and} \tag{7.16}$$

$$\dot{f}_2(t) = -\frac{\lambda_n}{k} f_2(t), \tag{7.17}$$

which are readily transformed into the circuits shown in figure 7.10 and 7.11 where the variable x is identified with t.

Initially, for $t = 0$, the rod is at temperature $T(0) \neq 0$. Then its ends located at $x = 0$ and $x = 1$ are forced to a temperature 0 so that $T(0,t) = T(1,t) = 0$. Using the circuit shown in figure 7.11 suitable values λ_n are determined. This is normally done using a manual trial-and-error process by varying the coefficient potentiometer between successive runs of the analog computer with $0 < t \leq 1$.[24] This can be, of course, done automatically by an attached digital computer – one of the advantages of a hybrid computer setup.

---

[23] Similar examples can be found in [SYDOW 1964, pp. 107 f.] and [AMELING 1963, pp. 280 ff.].
[24] Cf. [AMELING 1963, p. 281].

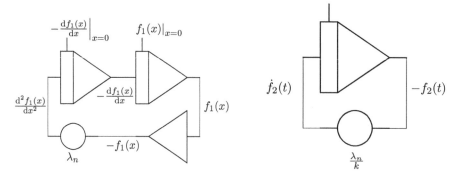

Fig. 7.10. Circuit corresponding to (7.16)   **Fig. 7.11.** Circuit corresponding to (7.17)

With all λ_n determined, a linear combination of the eigenfunctions is sought that approximates $T(x,t)$ to the required degree of precision:[25]

$$T(x,t) = \sum_{i \in \mathcal{I}} \lambda_i T_i \qquad (7.18)$$

Figure 7.12 shows the mechanization of this equation. Using this setup it is either possible to hold t at a fixed value while x is running from 0 to 1 or vice versa. These two variables are controlled with the two switches labelled t_{start} and x_{start}.

7.5 Scaling

Deriving a circuit corresponding to a given set of differential equations is only one step towards the solution of a given problem using an analog electronic analog computer. Since all variables are bounded by ±1 machine unit, in most cases it is necessary to *scale* the variables in a computer setup. An unscaled variable v representing some property of the underlying problem is associated with a *machine variable* $\mathfrak{v}$, which must satisfy

$$-1 \leq \mathfrak{v} \leq 1 \qquad (7.19)$$

at all times.

To achieve this, a set of scaling factors α_v has to be determined so that all $\mathfrak{v} = \alpha_v v$ satisfy (7.19). Determining these α_v can be a time-consuming and difficult process since the overall behaviour of the underlying differential equations has to

---

[25] T_i denotes the ith eigenfunction, while $\mathcal{I}$ denotes a suitable index set.

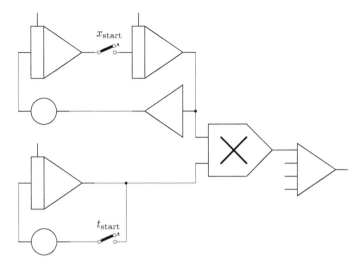

Fig. 7.12. One-dimensional heat-transfer using separation of variables

be considered.[26] When $\max(|v|)$ is known with regard to a unit U then α_v is determined by[27]

$$\alpha_v = \frac{1 \text{ machine unit}}{\max(|v|)} \left[\frac{V}{U}\right].$$

Determining these α_v is further complicated by the technical requirement of using as much of the allowable interval $[-1, 1]$ in order to minimise errors due to inevitable inaccuracies of the actual computing elements.[28] Scaling a problem variable v with a scaling factor α_v yields a machine variable $\mathfrak{v}$, which is measured in Volts (or Amperes in case of a current-coupled computer). Machine variables are often denoted by square brackets in the literature. Considering a problem variable $\ddot{x}$ representing an acceleration, which can assume a maximum value $\max(|\ddot{x}|) = 2 \cdot 9.81 \text{m/s}^2$ and a machine unit of 10 V, results in a practical scaling factor $\alpha_{\ddot{x}} = 1/2$. The resulting machine variable $\alpha_{\ddot{x}}\ddot{x}$ is then denoted by $\left[\frac{\ddot{x}}{2}\right]$.

Time can be scaled, too. *Problem time*, often called *real-time*, is normally denoted by t while machine time is often represented by τ. The time-scaling factor is denoted by λ, yielding $\tau = \lambda t$. Since the integrators are the only computing elements depending on time, time-scaling is either done by changing the time-

---

[26] This is even more complex a problem in the case of non-linear differential equations.
[27] Units are noted in square brackets.
[28] In particular, classic analog electronic multipliers tended to exhibit large errors for small products.

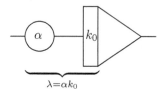

Fig. 7.13. Time-scaling

scale factors of the integrators[29] or by changing the input factors of the integrators. Often both scaling approaches are used together.

Integrators typically feature a number of time-scale factors k_0, which can take values of the form $k_0 = 10^n$ with $0 \leq n \leq N$. N can be as large as 6 in some high-speed analog computers. A time-scale factor $k_0 = 1$ means that the integral over a constant input value -1 will yield $+1$ at the integrator's output after one second. With $k_0 = 10^3$ this output value will be reached in one millisecond.

Any coefficient potentiometers at the inputs of an integrator will influence the overall time-scale factor based on k_0 of the respective integrator. Figure 7.13 shows this effect with a coefficient α.

By changing λ it is possible to run computations slower or faster than real-time. Large analog computers often feature a switch labelled "10×" that exchanges the feedback capacitors of all integrators by components with one-tenth of the normal capacity at the selected value k_0 thus speeding up a computer run by a factor of 10 at the press of single button.[30]

Many methods were devised to facilitate the scaling process, including numerical solutions of a problem by means of a stored-program digital computer to determine the maximum values of the problem variables involved.[31] Other approaches used hybrid computers[32] where the digital part generates scaling values for the analog part.[33]

---

[29] This is done by selecting appropriate capacitors in the feedback path of the integrators.
[30] Some systems even allow this button to be depressed during a computer run without the need of restarting the computation.
[31] A system using digital optimisation techniques to generate scaling factors for analog computers is described in [CELMER et al. 1970]. [SCHWARZE 1972] describes a hybrid computer approach to automatic scaling. A modern implementation of such a tool can be found in https://github.com/bernd-ulmann/DEQscaler.
[32] Cf. chapter 9.
[33] See [HALL et al. 1969] and [GILOI 1975, pp. 129 ff.].

8 Programming examples

Learning how to program an analog computer requires practice but is not too difficult. The following sections contain some programming examples, ranging from simple to difficult, all of which were implemented and run on real analog computers.

8.1 Solving $\ddot{y} + \omega^2 y = 0$

Generating a sine-/cosine signal pair, also called a *quadrature signal pair* on an analog computer is a frequent requirement, since such signals can be used for various tasks such as filter design, research on vibrating mechanical systems, displaying figures on an oscilloscope, etc. On a stored-program digital computer polynomial approximations or even table lookups could be used to generate a function such as $\sin(\omega t)$. On an analog computer a different approach is typically chosen: Solving a differential equation, which yields the desired function as result:

$$\ddot{y} + \omega^2 y = 0 \qquad (8.1)$$

with initial conditions

$$y(0) = a\sin(\varphi) \quad \text{and}$$
$$\dot{y}(0) = a\omega\cos(\varphi).$$

Prior to applying KELVIN'S feedback method, equation (8.1) is solved for its highest derivative yielding

$$\ddot{y} = -\omega^2 y. \qquad (8.2)$$

Since this differential equation is satisfied by any linear combinations of sine and cosine it can be readily used to generate a sine signal, given suitable initial values. In this case, additional scaling of the problem variables is not necessary since $\max(|\sin(\omega t)|) = 1$. Figure 8.1 shows the basic circuit based on equation (8.2). This setup is still missing its initial values and would yield the null-function.[1]

The frequency ω of the resulting sine signal influences the time-scaling process. Figure 8.2 shows the time-scaled circuit. Using two coefficient potentiometers in conjunction with the time-constant k_0 of the associated integrators, $\omega = \alpha k_0$ can be set.[2] Thus, time is identified with τ in this setup. To generate a sine signal the second integrator is fed with an initial value of 0.

---

[1] Over time inevitable errors due to drift and noise would build up forcing the circuit to eventually oscillate even with all initial conditions set to 0.

[2] Some machines, like the Telefunken RA 770, even have mechanically coupled coefficient potentiometers to facilitate setups like this where two potentiometers must be set in lock step.

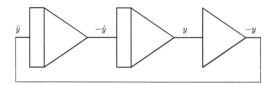

Fig. 8.1. Basic circuit for $y = \sin(\omega t + \varphi)$

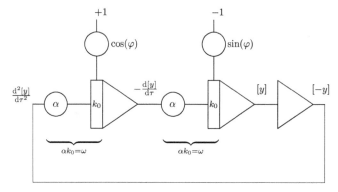

Fig. 8.2. Basic circuit for $y = \sin(\omega t + \varphi)$

Figure 8.3 shows one period of the output signal $[y]$ generated by this circuit. Using the time control system of the analog computer used, the integrators were placed into *operate* mode for the time it takes to generate one full wave. It should be noted that the output amplitude of a setup like this will not stay constant but instead decrease or increase over time. This is mainly due to little differences in the time-scaling factors k_0 of the integrators. If a sine/cosine signal with a stable amplitude is required over an extended period of time, additional measures have to be taken.[3]

8.2 Sweep generator

Since $\omega = \alpha k_0$ controls the output-signal's frequency, the circuit shown in figure 8.2 can be easily extended to a sweep generator. Swept signals are frequently employed to analyse transmission systems, the response of vibrating systems, etc.

---

[3] See section 8.6.

Fig. 8.3. A single sine period generated by the circuit shown in figure 8.2

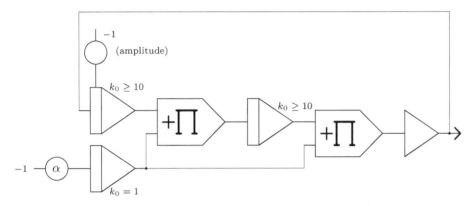

Fig. 8.4. Basic circuit for a sweep generator

Here, the two coefficient potentiometers in front of the two integrators are replaced by multipliers sharing a common factor.[4]

Figure 8.4 shows the basic setup of such a sweep generator.[5] The frequency controlling signal is derived from integrating over a constant value α. To achieve a sufficient slow frequency change two time-scale factors for the integrators involved are used. The integrator with α as its input has a time-scale $k_0 = 1$ while the two

---

[4] In applications like this where two or more multiplicands are to be multiplied with a (slowly changing) multiplier, a servo multiplier or a time division multiplier could be used with advantage. If one of these multipliers is inverting the sign of its output value, the sign inverting summer in figure 8.4 is not required. In a classic analog computer the sign-inversion can be easily achieved with a quarter square multiplier. Since these multipliers require both input signals with both polarities, $\pm x$ and $\pm y$, the sign of the result can be inverted by interchanging the connections to one of these signal pairs.

[5] It should be noted that this circuit can be used to generate the functions $-\sin(\varphi)$ and $\cos(\varphi)$ if the first derivative $\dot{\varphi}$ is available as input. In this case, $\dot{\varphi}$ is fed to the two multipliers directly instead of the output signal of the integrator on the lower left.

Fig. 8.5. Patch panel setup for the sweep generator (THE ANALOG THING)

integrators forming the basic oscillator loop run with a time-scale ten times (or more) as fast ($k_0 = 10$). The patch panel setup for this sweep generator is shown in figure 8.5. Figure 8.6 shows the output of a typical sweep run with linearly increasing frequency.

8.3 Mass-spring-damper system

The next problem is a bit more complex. A mass-spring-damper system consisting of the parts shown in figure 8.7 is to be modelled and simulated with an analog computer. These parts are a mass m exerting a force $m\ddot{y}$, a spring with stiffness s resulting in a force sy and a damper with damping coefficient d exerting a force $d\dot{y}$. Since the forces in a closed physical system add up to 0 the mass-spring-damper system shown in figure 8.8 can be described by the following differential equation:

$$m\ddot{y} + d\dot{y} + sy = 0. \tag{8.3}$$

Rearranging so that the highest derivative is on the left side of the equal sign yields

$$\ddot{y} = \frac{-(d\dot{y} + sy)}{m} \tag{8.4}$$

8.3 Mass-spring-damper system

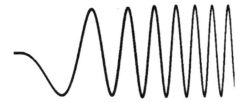

Fig. 8.6. Sine oscillation with variable frequency

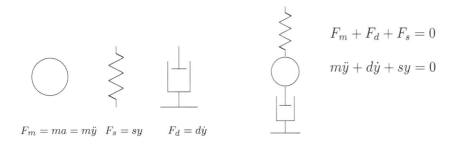

Fig. 8.7. Parts of a mass-spring-damper system

Fig. 8.8. Mass-spring-damper system

as a starting point for applying the KELVIN method. Figure 8.9 shows the first part of the program: Using an integrator, $\ddot{y}$ is integrated, yielding $-\dot{y}$ with an initial value of $\dot{y}(0)$.

Using a second integrator with initial value $-y(0)$ yields y, which can be multiplied by s using a coefficient potentiometer as shown in figure 8.10. Adding an inverter and a second coefficient potentiometer to derive $d\dot{y}$ and summing this term and sy results in the circuit shown in figure 8.11 yielding $-(d\dot{y} + sy)$. This can now be multiplied by $1/m$ with a coefficient potentiometer yielding the right side of equation (8.4). Feeding this signal back into the input of the first integrator closes the feedback loop. The final circuit is shown in figure 8.12.[6]

The overall setup for this program on a Telefunken RA 741 tabletop analog computer is shown in figure 8.13. Figures 8.14, 8.15, 8.16, and 8.17 show typical simulation results for various settings of the spring's stiffness and the damping coefficient with constant mass $m = 1$.[7] If the computer is running in repetitive mode and if the time-scale factors of the integrators are set to $k_0 = 10^2$ or $k_0 = 10^3$, a flicker-free image can be displayed on an oscilloscope screen. The effect of

---

[6] This setup can be further simplified by connecting the output of the leftmost integrator directly to the coefficient potentiometer d and connecting its output to a second input of the integrator yielding $-\dot{y}$, effectively saving one summer.

[7] These still frames were taken from the screen of a Nicolet 4094C digital storage oscilloscope.

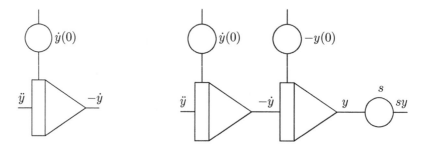

Fig. 8.9. First integration step yielding $-\dot{y}$

Fig. 8.10. Second integration step yielding y

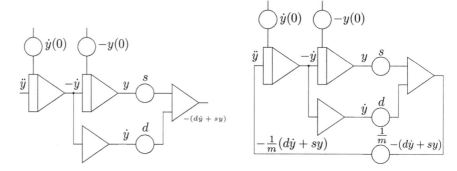

Fig. 8.11. Computing $-(d\dot{y}+sy)$

Fig. 8.12. Final computer setup for the mass-spring-damper system

changing the coefficients s, d, and m can then be seen directly while turning the potentiometer dials.[8] These runs were performed on a non-scaled model and thus yielded only qualitative results.

To derive quantitative results the following values are assumed: $m = 1.5$ kg, $s = 60\ \frac{\text{kg}}{\text{s}^2}$, $d = 3\ \frac{\text{kg}}{\text{s}}$ and $y(0) = 0.1$ m (initial displacement of the mass). Scaling the problem requires knowledge about the maximum values for all variables within the computer setup, y, $\dot{y}$ and $\ddot{y}$. Good approximations for these can be derived from an undamped harmonic oscillator with a resonant frequency of

$$\omega = \sqrt{\frac{s}{m}} = \sqrt{\frac{60\ \frac{\text{kg}}{\text{s}^2}}{1.5\ \text{kg}}} \approx 6.3\ \text{s}^{-1}.$$

---

[8] This high degree of interactivity is even today of incredible value in education as well as in research as it allows a user to rapidly get a feeling for complex dynamical systems.

8.3 Mass-spring-damper system

Fig. 8.13. Setup of the mass-spring-damper simulation

Fig. 8.14. $s = 0.2$, $d = 0.8$

Fig. 8.15. $s = 0.6$, $d = 0.8$

Fig. 8.16. $s = 0.8$, $d = 0.6$

Fig. 8.17. $s = 0.8$, $d = 1$

From $y = y(0)\sin(\omega t)$ it follows that

$$\dot{y} = y(0)\omega \cos(\omega t) \quad \text{and} \tag{8.5}$$
$$\ddot{y} = -y(0)\omega^2 \sin(\omega t). \tag{8.6}$$

Since the mass-spring-damper system in this example is a damped oscillator,

$$\max_{t>0} |y| \leq y(0)$$

holds yielding the following boundary values for $\dot{y}$ and $\ddot{y}$ based on equations (8.5) and (8.6):

$$\dot{y} < y(0)\omega \approx 0.63 \, \frac{\text{m}}{\text{s}}$$
$$\ddot{y} < y(0)\omega^2 \approx 4 \, \frac{\text{m}}{\text{s}^2}$$

Thus, a reasonable scaling for the machine variables would be $[10y]$, $[15\dot{y}]$ and $[\frac{5}{2}\ddot{y}]$ respectively. Time-scaling is not performed so that the simulation will yield results in real-time. Based on these scaled machine variables equation (8.3) becomes

$$\left[\frac{5}{2}\ddot{y}\right] = -\frac{5}{2m}\left(\frac{d}{15}[15\dot{y}] + \frac{s}{10}[10y]\right).$$

To perform a scaled simulation run the coefficient potentiometers for the problem parameters s, d and $1/m$ must be set to $[s/10]$, $[d/15]$ and $[5/2m]$.

After scaling a problem a *static check* should be performed, if the analog computer being used features this as a special mode of operation, to make sure that the patch panel setup is correct. The basic idea of such a test is simple: All integrators are switched into a mode where they act as simple summers. Then test signals, which are derived from a static solution of the problem's equations, are fed to special inputs of these integrators so that they appear at the respective outputs with inverted sign. The resulting values in the remaining circuit are then compared against this static solution, which has been prepared manually or with the help of a stored-program digital computer.

8.4 Predator and prey

While the mass-spring-damper system could be described by a single differential equation, the famous problem describing a simplified ecosystem containing predators and prey is based on two coupled differential equations. Systems like these were first studied in 1925 by ALFRED JAMES LOTKA[9] and independently in 1926 by VITO VOLTERRA.[10]

The resulting differential equations are known as LOTKA-VOLTERRA equations. Their interest was sparked by statistical data gathered by the *Hudson Bay Company* in the years between 1850 and 1900. The data sets containing the numbers of rabbits and lynxes that were captured per year showed an interesting periodicity.[11]

Consider a closed-world ecosystem with unlimited food for the prey. This ecosystem can be described by the following two coupled differential equations with l and r representing the number of lynxes and rabbits respectively:

$$\dot{r} = \alpha_1 r - \alpha_2 rl \qquad (8.7)$$
$$\dot{l} = -\beta_1 l + \beta_2 rl \qquad (8.8)$$

---

[9] 03/02/1880–12/05/1949
[10] 05/03/1860–10/11/1940
[11] See [ELTON et al. 1942] and [GILPIN 1973] for more details.

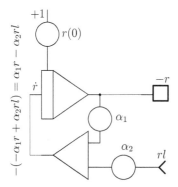
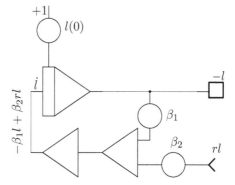

Fig. 8.18. Computing $-r$ **Fig. 8.19.** Computing $-l$

Here, α_1 represents the rate of birth for rabbits, α_2 the decimation of rabbits due to lynxes catching rabbits,[12] β_1 describes the death-rate of lynxes due to natural causes, and β_2 models the increase of the lynx population due to food supply in the form of rabbits.[13]

The two coupled differential equations (8.7) and (8.8) can be directly transformed into the two sub-circuits shown in figures 8.18 and 8.19 by using the KELVIN feedback technique.

Without the common input rl (lynxes eating rabbits), these circuits would just model an ever faster increasing rabbit population[14] and a decreasing lynx population. Noting that the two summers connected in series in figure 8.19 just act as a summer without sign-inversion readily yields the simplified circuit shown in figure 8.20 saving both summers and thus not only simplifying the overall setup but also increasing the precision of the simulation.

Figure 8.21 combines both sub-circuits by means of a multiplier yielding the shared input rl.

The overall setup of the predator-prey simulation, consisting of a Nicolet 4094C digital storage oscilloscope on the left and a Telefunken RA 741 tabletop analog computer, is shown in figure 8.22. Typical results showing an oscillatory behaviour of the ecosystem are depicted in figure 8.23. The all-important phase lag between rabbit and lynx populations is clearly visible.[15]

---

12 The *capture cross section*, so to speak.
13 The food supply for rabbits is assumed to be unlimited.
14 Natural death for rabbits has been omitted from the model.
15 A recent hydraulic analog computer for modeling predator prey dynamics is described in [DOORE et al. 2014].

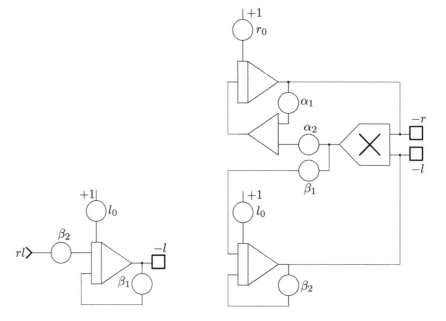

Fig. 8.20. Improved circuit for $-l$ **Fig. 8.21.** Completed predator-prey setup

Here, only a qualitative study of the simple ecosystem has been performed. Scaling a system of coupled non-linear differential equations is more difficult than in the simple case of the mass-spring-damper system, so a qualitative study is outside the scope of this example.[16]

8.5 Simulation of an epidemic

This section describes the simulation of an epidemic based on the $SEIR$ model. This model consists of four groups of individuals: Susceptible persons, i.e., persons who can be infected, exposed persons who had contact with an infected person, and recovered persons who overcame the infection. This is a typical example of a compartmental model (S, E, I, and R). The following treatment is of a qualitative nature only, so no scaling has been done.[17]

---

[16] An example for scaling such a system of coupled differential equations can be found in [SCHWARZ 1971, pp. 369 ff.].

[17] A comprehensive treatment of this model can be found in [BÄRWOLFF 2021].

 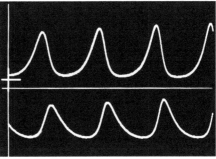

Fig. 8.22. Setup for the predator-prey simulation **Fig. 8.23.** Predator-prey simulation results

The model is described by four coupled differential equations, each describing the change of the population in one compartment over time:

$$\dot{S} = -\beta SI$$
$$\dot{E} = \beta SI - \alpha E$$
$$\dot{I} = \alpha E - \gamma I$$
$$\dot{R} = \gamma I$$

β is the transmission rate, its reciprocal is the mean time between contacts of susceptible persons with infected ones. α represents the degree with which exposed persons become actually infected. γ is the recovery rate and describes how fast infected people recover (or die).

These equations can be readily transformed into the analog computer program shown in figure 8.24. The integrator yielding S has an initial condition of -1 thus the simulation starts with the maximum number of susceptible persons. The initial condition $I(0)$ of the third integrator determines the number of persons, which are initially infected.

A typical simulation result is shown in figure 8.25. The first curve to show a peak is the one describing the compartment of exposed persons, followed by the peak of infected persons. In this case S vanishes completely at the end of the simulation. With different parameters scenarios can be found where not every individual will be infected over time.

The setup of this model on THE ANALOG THING is shown in figure 8.26. Running the analog computer in repetitive mode, a flicker-free picture can be obtained on an oscilloscope, which makes it possible to explore the effects of various settings for the coefficients β, α, and γ.

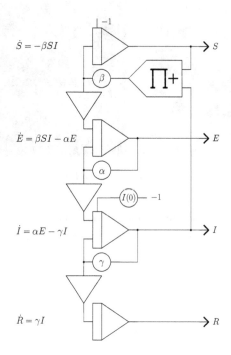

Fig. 8.24. Program for the SEIR model

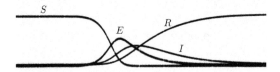

Fig. 8.25. Typical SEIR simulation result

8.6 Bouncing ball

The following section describes the simulation of a ball bouncing in a box on an analog computer with an oscilloscope as the display device. The setup follows closely an example published by Telefunken in the 1960s,[18] which was used as a

---

18 Cf. [Telefunken/1] and [PFALTZGRAFF 1969] for a much more complex but also more precise treatment of the same problem. Other vendors, such as Heathkit, also used bouncing ball simulations for marketing, although these these were usually even simpler circuits than the one described here. Such a simplified simulation can be found in [WINKLER 1961, pp. 199 ff.].

Fig. 8.26. Setup of the SEIR model on THE ANALOG THING

marketing example to show the power of their new transistorised tabletop analog computers.[19]

Figure 8.27 shows the basic parts of the simulation. A ball is thrown into a rectangular box of dimension $[-1, 1] \times [-1, 1]$ with an initial velocity $v(0)$ and a given initial position $y(0)$. It is accelerated by gravitational force until it hits the floor of the box where it rebounds fully elastically, as it also does when it hits the left or right wall. The ball loses energy over time due to air-friction.

Since the x- and y-component of the ball position are independent from each other, they can be generated independently. Figure 8.28 shows the approximate y-position of the ball with respect to time. The basis for calculating y is the acceleration $\ddot{y}$ consisting of the gravitational force and the acceleration caused by elastic collision of the ball with the bottom (or the ceiling) of the box. $\ddot{y}$ can be described by

$$\ddot{y} = -g + d\dot{y} \begin{cases} +\frac{c}{m}\left(|y| + 1\right) & \text{if } y < -1 \\ -\frac{c}{m}\left(y - 1\right) & \text{if } y > 1. \end{cases}$$

---

[19] A real-time simulation like this would have required a very expensive and large stored-program digital computer.

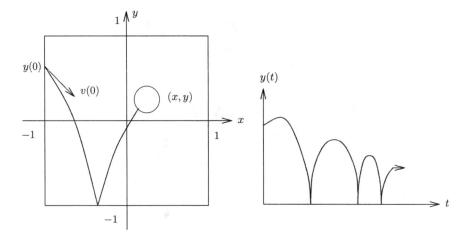

Fig. 8.27. Movement of the bouncing ball   **Fig. 8.28.** y-component of the bouncing ball

Integrating twice over $\ddot{y}$ yields the ball velocity $\dot{y}$ and position y:

$$\dot{y} = \int_0^T \ddot{y}\,dt + \dot{y}_0$$

$$y = \int_0^T \dot{y}\,dt + y_0$$

The circuit yielding y is shown in figure 8.29. The integrator on the left integrates over the sum of three terms: The gravitational force, which is set by the coefficient potentiometer labelled g, the rebound-acceleration, which comes from the inverter at the right side, and a damping signal that is proportional to the ball's velocity $\dot{y}$ with a factor d.

There are several things to be noted about this circuit: First of all, the rebound-acceleration must by very high to yield a realistic behaviour of the ball. Thus, an integrator input with a weight far exceeding the standard inputs of 1 and 10 is required. In this implementation this acceleration variable is fed directly to the summing junction through a coefficient potentiometer.[20] The acceleration signal representing the rebound of the ball has be generated whenever the ball hits either the floor or the ceiling of the box.

Normally, two comparators would be used to implement this, but the small analog computer chosen in this example did not have enough comparators, so

---

20 A better alternative would be to use a free (ungrounded) coefficient potentiometer.

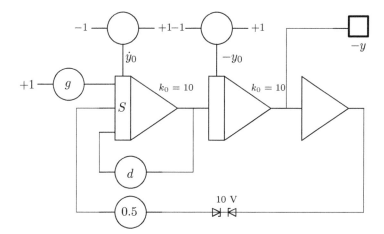

Fig. 8.29. Computing the y position of the bouncing ball

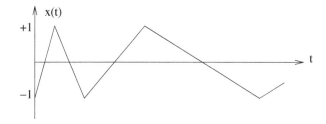

Fig. 8.30. x-portion of the ball's movement

a different approach was chosen. The two ZENER diodes in the output line of the inverter have the effect that they pass a signal only when it exceeds their respective ZENER voltage. Both ZENER diodes are selected to yield a ZENER voltage of 10 Volts in each direction thus representing the floor and ceiling of the box.[21] The initial condition inputs of both integrators are connected to free coefficient potentiometers thus allowing arbitrary initial values for the ball's y velocity and position.

Generating the x-position of the bouncing ball is even easier. It is assumed that its velocity in the x-direction is linearly decreasing, while changing the direction every time the ball hits a wall. Figure 8.30 shows the ball's x-position with respect to time while figure 8.31 shows the corresponding computer setup.

---

[21] A small error is introduced since one of the ZENER diodes always exhibits its forward voltage but this is negligible in a demonstration setup like this.

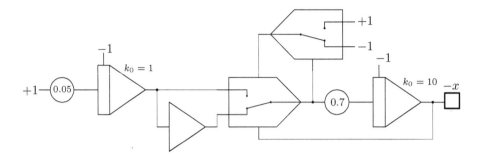

Fig. 8.31. Computing the x position of the bouncing ball

The leftmost integrator, which generated the x-velocity, starts with an initial value of -1 and integrates over a small constant like 0.05 with an integration time-scale factor $k_0 = 1$, so the output of the integrator will linearly decrease, starting at the value $+1$ representing the maximum velocity of the ball. Its output is thus

$$\dot{x}(t) = -\left(\int_0^T 0.05 \mathrm{d}t - 1\right).$$

This velocity signal is then integrated by a second integrator with a time constant ten times faster to generate the x-position. The two comparators between the integrators detect the ball hitting the left or right wall of the box. The left comparator also performs the necessary sign-reversal of the velocity signal.

Now that both the x- and y-positions of the bouncing ball are known, a sine-/cosine quadrature generator circuit is required to generate a ball shape on an oscilloscope screen. Basically this is done as described in section 8.1 but it is a bit more involved here since the generated signal pair must have a high frequency and a stabilized amplitude to display a flicker-free figure on the screen. Figure 8.32 shows the sine-/cosine generator circuit used.

The basic structure of this circuit corresponds to that shown in figure 8.2. Two integrators and an inverter form a loop effectively solving a second order differential equation like (8.1). To achieve a high value for ω the integrators need large input weights, which is achieved by connecting free potentiometers to their respective summing junction inputs.

To ensure that the amplitude of the generated sine-/cosine signal pair does not decay, a positive feedback loop is set up between the output of the inverter and one input of the right integrator. As this loop only has to provide a small positive feedback, it is attenuated by a coefficient potentiometer set to a small value like 0.02. Without additional measures this would yield an ever increasing signal amplitude, which would drive the computing elements into overload. To counteract this effect, a pair of ZENER diodes is used as a negative feedback on

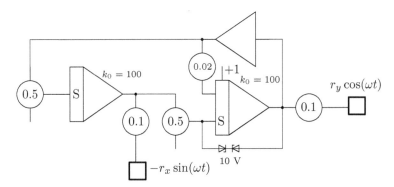

Fig. 8.32. Generating a high-frequency sine-/cosine signal pair for displaying the ball

Fig. 8.33. Generating the signals to display the ball's outline on the oscilloscope

the right integrator. These connect the integrator's output directly to its summing junction, effectively limiting its output signal to their ZENER voltage.[22]

Using two summers as shown in figure 8.33 the position signals $-x$ and $-y$ of the ball and the attenuated sine-/cosine signals are combined yielding two output signals x^* and y^*, which are connected to the x- and y-inputs of the oscilloscope used to display the simulated bouncing ball.

Figure 8.34 shows the setup of the bouncing ball simulation on a Dornier DO-80 analog computer while figure 8.35 shows a long-exposure shot of a simulation run.

8.7 Car suspension

The simulation of vibrating mechanical systems was, and still is, of high commercial value since vibrations in typical transport systems, support structures, machines, etc., can cause negative effects ranging from making a train ride un-

---

[22] A circuit like this or variations thereof are often employed when a high-frequency sine-/cosine pair is required, which does not need to be spectrally clean. "High-frequency" is a bit of a hyperbole – a frequency of a couple kHz is sufficient for a flicker-free display. Due to the combination of positive and negative feedback, harmonics are introduced, but for driving a display their amplitude is small enough not to cause any visible distortions.

 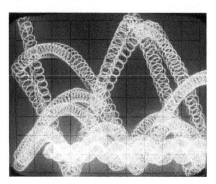

Fig. 8.34. Setup of the bouncing ball simulation (Dornier 80)

Fig. 8.35. Time exposure shot of a bouncing ball simulation run (photo taken by TORE SINDING BEKKEDAL, reprinted with permission)

comfortable to dangerous situations when resonant frequencies are encountered and even rigid structures collapse.

In the late 1950s the simulation and analysis of the dynamic behaviour of automobiles and railway vehicle gained a lot of interest and several manufacturers installed large analog computing facilities. The rationale behind this is made clear by the following quotation from ROBERT H. KOHR:[23]

> "However, the general trend toward heavier cars with softer tires and the increasing adoption of power steering and air suspensions calls for a complete dynamic analysis of the automobile with a view to gaining a basic understanding of the automobile's behaviour on the road."

Realistic simulations of a car suspension system require up to seven degrees of freedom[24] and many function generators to model the non-linear behaviour of wheels, suspension springs, and other mechanical parts.[25] To make the analog model as accurate as possible, actual road profiles were traced and stored on analog tape units to be used as inputs for the simulations. Using multiple read heads, the time-delayed excitation of the front and back wheels could be easily implemented.

Simulations like these were also standard for the development of railway vehicles as this citation shows:[26]

---

23 Vehicle Dynamics Section, Engineering Mechanics Department of the *General Motors* Research Staff at Warren, Michigan, cf. [MCLEOD et al. 1958/6, p. 1994].
24 Cf. [MCLEOD et al. 1958/6, p. 1994].
25 A thorough introduction to the mathematics of vibrating multi-mass systems can be found in [MACDUFF et al. 1958, pp. 193 ff.] and [Telefunken/2].
26 See [HELLER et al. 1976, p. 2].

Fig. 8.36. Analog computer installed at the Pullman-Standard Car Manufacturing Company in the early 1950s (see [ROEDEL 1955, p. 42])

> "*A railroad freight vehicle is a complex dynamic system consisting of numerous interrelated physical components [...] Comprehensive models for such a system will generally consist of a large number on non-linear simultaneous differential equations with complicated functional relationships between the variables to be integrated. The computer implementation of these models will require a significant amount of computer resource to run.*"

Due to the complex structure of the bogies of such a vehicle typical simulations featured between 17 and 23 degrees-of-freedom.[27] A 5 degree-of-freedom simulation of a single bogie is described in [MALSTROM et al. 1977] while [ROEDEL 1955/2] covers the simulation of the dynamic behaviour of railway vehicles in general. Figure 8.36 shows the analog computer installation that was in use at the Pullman-Standard Car Manufacturing Company in the early 1950s. This installation was mainly used to simulate complete railway vehicles as well as individual bogies.

A much simpler setup was demonstrated in the early 1960s at an industry exhibition in Germany (Hanover fair). This setup, which employed a small Telefunken RAT 700 tabletop analog computer, turned out to be an eye-catcher and generated a lot of interest in these small analog computers.[28]

---

[27] Cf. [HELLER et al. 1976, p. 39].
[28] This particular setup is described in [BEHRENDT 1965].

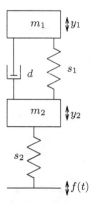

Fig. 8.37. Two-mass system

This setup is recreated in the following example. At the heart of the simulation is a simple coupled two-mass system with linear springs and a damper, as shown in figure 8.37.[29] This mechanical system is described by the following two coupled differential equations

$$0 = m_1\ddot{y}_1 + d(\dot{y}_1 - \dot{y}_2) + s_1(y_1 - y_2)$$
$$0 = m_2\ddot{y}_2 + d(\dot{y}_2 - \dot{y}_1) + s_1(y_2 - y_1) + s_2(y_2 - y_3),$$

which yield

$$\ddot{y}_1 = -\left(\frac{d}{m_1}(\dot{y}_1 - \dot{y}_2) + \frac{s_1}{m_1}(y_1 - y_2)\right) \quad \text{and} \tag{8.9}$$

$$-\ddot{y}_2 = \frac{d}{m_2}(\dot{y}_2 - \dot{y}_1) + \frac{s_1}{m_1}(y_2 - y_1) + \frac{s_2}{m_2}(y_2 - f(t)) \tag{8.10}$$

as the starting point for applying the KELVIN method to derive a computer setup.

Figure 8.38 shows the circuit equivalent to equation (8.9). This circuit is basically a damped oscillator consisting of two subcircuits. The oscillator itself consists of two integrators in series with a sign-inverting summer in the feedback path. A second feedback path containing two summers in series implements the damping effect $e^{-\alpha t}$ with α being defined implicitly by d/m_1.

Equation (8.10) is implemented by the circuit shown in figure 8.39. It, too, is basically a damped oscillator. Both circuits are coupled through the variables y_2, $\dot{y}_2$, $-\dot{y}_1$ and $-s_1(y_2 - y_1)/m_1$. For an engineering simulation these two coupled circuits are already sufficient, but to replicate the historic exhibition setup a nice real-time display of a car bouncing up and down is required. The car is shown

---

29 A more detailed description of such a two-mass system, including scaling, can be found in [GILOI et al., pp. 48 ff.], [EAI TR-10, pp. 44 ff.] and [CARLSON et al. 1967, p. 91].

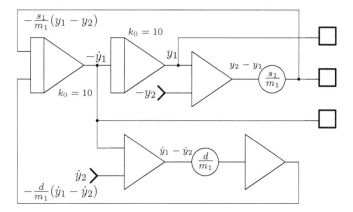

Fig. 8.38. Circuit corresponding to equation (8.9)

from the side, with the car's body representing m_1 and the wheels, which are not modelled separately, representing m_2.

A high-frequency[30] sine-/cosine signal pair is an ideal basis for displaying closed figures on an oscilloscope screen. Consequently, the additional circuit shown in figure 8.32 is required for this simulation. In addition to this, an electronic switch is necessary so that three individual figures (both wheels and the car frame) can be displayed in rapid succession on the oscilloscope. The wheels can be directly generated by feeding the x- and y-deflection channel of the oscilloscope with the sine-/cosine-pair plus the wheel displacement y_2. The car frame is also based on this signal pair with the cosine part suitably modified by means of a function generator. Adding its corresponding height signal y_1 controls the y position at which it is displayed.

The overall program setup on a Telefunken RA 770 precision analog computer is shown in figure 8.41. In addition to the circuits described above, it also contains the three pairs of electronic switches to rapidly switch the oscilloscope's inputs between the three x/y-outputs generated by the simulation. Additionally, a sweep generator like the one shown in figure 8.4 has been implemented to explore the dynamic response of the two-mass vibrating system to various excitation frequencies and amplitudes. Also, a second excitation source is available in this setup since this analog computer is equipped with a noise generator. A screen shot of the display generated by this setup is shown in figure 8.42.

---

[30] Several kHz.

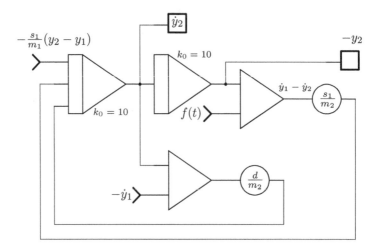

Fig. 8.39. Circuit corresponding to equation (8.10)

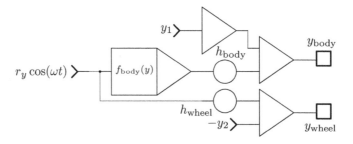

Fig. 8.40. Generating the signals for the graphic representation of the two-mass system

Fig. 8.41. Setup of the car suspension simulation (Telefunken RA 770)

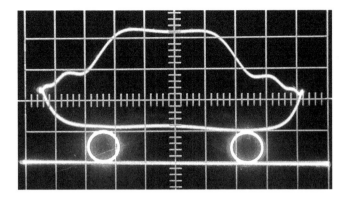

Fig. 8.42. Display of the car suspension simulation

8.8 Lorenz attractor

A well-known chaotic system is the LORENZ *attractor*, discovered and published by EDWARD N. LORENZ.[31] in 1963.[32] He devised a simplified model for atmospheric convection, which he analysed on a simple digital computer, a Royal McBee *LGP-30*, and it turned out that this system exhibits a behaviour nowadays called *chaotic*.[33] An analog computer is ideally suited to implement a chaotic system like this, based on the defining differential equations:[34]

$$\dot{x} = \sigma(y - x) \tag{8.11}$$
$$\dot{y} = x(\rho - z) - y \tag{8.12}$$
$$\dot{z} = xy - \beta y \tag{8.13}$$

Applying the KELVIN method to each of these equations yields the three circuits shown in figure 8.43.

Obviously, these circuits can be further simplified, saving some computing elements, thus improving the precision of the computation. The circuit shown on top in figure 8.43 contains two summers in series, which can be eliminated by using

---

[31] 05/23/1917–04/16/2008
[32] See [LORENZ 1963].
[33] See [JÖNCK et al. 2003] for a comprehensive description of the system. Recent work on this is described by [TUCKER 2002]. Here a proof is given that the attractor is robust, "*it persists under small perturbations of the coefficients in the underlying differential equations*". It is also proven that the Lorenz equations "*support a strange attractor*", which was conjectured by Lorenz as early as 1963.
[34] [SPROTT 2016] described a wealth of chaotic systems, most of which can be easily implemented on analog computers. [ARGYRIS et al. 1995] give a good overview of chaotic systems and their mathematical treatment. [449] also contains a wealth of information about such systems.

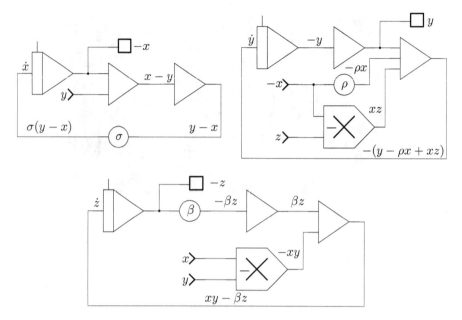

Fig. 8.43. Straightforward setup for equations (8.11), (8.12) and (8.13)

multiple inputs of the integrator. The resulting circuit is shown in the upper left of figure 8.44. A further change has been implemented as well: Instead of generating $-x$ based on y, the signs have been reversed since $-y$ is available from another circuit.

The circuit shown in the upper right of figure 8.43 can be similarly simplified: The two summers in series are eliminated using multiple inputs of the integrator. The input x is changed to $-x$, which is available from the optimised circuit. Accordingly, this circuit will yield y feeding the preceding circuit. The resulting circuit is shown on the upper right of figure 8.44.

Applying the same rationale to the lower circuit of figure 8.43 yields the corresponding circuit of figure 8.44, which saves two summers. It should be noted that the sign-inverting multipliers are most easily set up by using quarter-square multipliers with the polarity of one of the input-signal pairs reversed.

Figure 8.45 shows the setup of the program generating a picture of the LORENZ attractor. In this case, a Telefunken RA 770 precision analog computer has been used. With parameters

$$\sigma = 0.357$$
$$\rho = 0.1$$
$$\beta = 0.374$$

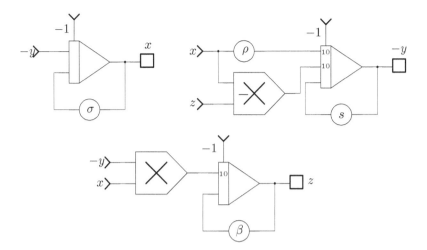

Fig. 8.44. Simplified setup for equations (8.11), (8.12) and (8.13)

Fig. 8.45. Setup for the Lorenz attractor (Telefunken RA 770)

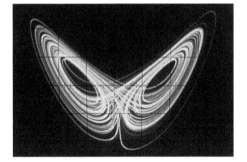

Fig. 8.46. Screen shot showing a Lorenz attractor

a *phase space*[35] plot like the one shown in figure 8.46 can be obtained.[36] This gives detailed information about the overall behaviour of the system.

---

35 The phase space of a dynamic system contains all possible states of the system, each state being represented by one point in the space. A visualisation of the phase space shows the overall behaviour of the underlying dynamic system.
36 A slightly different implementation is described in [ULMANN 2016].

8.9 Mathieu's equation

One of the author's favourite equations is MATHIEU's differential equation

$$\ddot{y} + (a - 2q\cos(2t))\,y = 0, \tag{8.14}$$

with initial conditions

$$y(0) = 1 \text{ and} \tag{8.15}$$
$$\dot{y}(0) = 0,$$

which was devised by ÉMILE LÉONARD MATHIEU[37] during his theoretical studies of vibrating surfaces such as drum skins.[38] Nowadays, this equation finds applications in quantum mechanics, optical systems and many other areas.[39]

This equation is a good example of the scaling process in analog computer programming.[40] Without loss of generality,

$$a := 2q \text{ and}$$
$$x := 1 - \cos(2t) \tag{8.16}$$

are defined, thus transforming (8.14) into

$$\ddot{y} + axy = 0. \tag{8.17}$$

The first task is now to generate the function $x(t)$, which is done by devising an auxiliary differential equation yielding $x(t)$ as a solution. Therefore, (8.16) is differentiated twice yielding

$$\dot{x} = 2\sin(2t) \text{ and} \tag{8.18}$$
$$\ddot{x} = 4\cos(2t). \tag{8.19}$$

The resulting differential equation is then

$$\ddot{x} + 4x = 4. \tag{8.20}$$

Using two integrations the unscaled program would basically look like this:

$$\ddot{x} = 4 - 4x, \tag{8.21}$$

---

[37] 05/15/1835–10/19/1890
[38] Solutions of this differential equation are called MATHIEU *functions*. [MCLACHLAN 1947] contains a comprehensive analytical treatment of this equation. [ARSCOTT 1964, pp. 26 ff.] is also read worthy in this respect.
[39] See [RUBY 1996]. [RANDERY 1964] shows the application of MATHIEU's equation to the simulation of a parametron.
[40] This section basically follows [EAI 7.7.4a 1964].

$$\dot{x} = \int \ddot{x}\,dt, \text{ and} \qquad (8.22)$$

$$x = \int \dot{x}\,dt. \qquad (8.23)$$

According to (8.16) $0 \leq x \leq 2$, so (8.23) must be scaled with a factor $\lambda_x = \frac{1}{2}$ to ensure that x does not exceed the machine interval.[41] This yields the scaled equation[42]

$$\hat{x} = \lambda_x \int \dot{x}\,dt = \frac{1}{2}\int \dot{x}\,dt. \qquad (8.24)$$

This scaling factor must now be compensated for in (8.21):

$$\ddot{x} = 4 - \frac{1}{\lambda_x}4\hat{x} = 4 - 8\hat{x}. \qquad (8.25)$$

$-4 \leq \ddot{x} \leq 4$, according to (8.19), so (8.25) must be scaled with a factor $\lambda_{\ddot{x}} = \frac{1}{4}$ yielding

$$\hat{\ddot{x}} = \lambda_{\ddot{x}}4 - \lambda_{\ddot{x}}8\hat{x} = 1 - 2\hat{x},$$

which must be compensated for in (8.22):

$$\dot{x} = \frac{1}{\lambda_{\ddot{x}}} \int \hat{\ddot{x}}\,dt = 4\int \hat{\ddot{x}}\,dt$$

According to (8.18) this is bounded by $-2 \leq \dot{x} \leq 2$ requiring a third scale factor $\lambda_{\dot{x}} = \frac{1}{2}$ yielding

$$\hat{\dot{x}} = \lambda_{\dot{x}}4 \int \hat{\ddot{x}}\,dt = 2\int \hat{\ddot{x}}\,dt.$$

The factor $\lambda_{...x}$ cancels out the factor λ_x in equation (8.24). This results in the following set of scaled equations to implement the auxiliary differential equation:

$$\hat{\ddot{x}} = 1 - 2\hat{x},$$

$$\hat{\dot{x}} = 2\int \hat{\ddot{x}}\,dt, \text{ and}$$

$$\hat{x} = \int \hat{\dot{x}}\,dt.$$

This ensures that x, $\dot{x}$, and $\ddot{x}$ all stay within the machine interval $[-1, 1]$. The resulting analog computer setup generating x is shown in figure 8.47.

Now equation (8.17) must be implemented. Since $0 \leq x \leq 2^{43}$ this can be rewritten as

$$\ddot{y} + 2ay = 0 \qquad (8.26)$$

---

[41] One might think about shifting x down by subtracting 1 but this would complicate the remaining parts of the program.
[42] A hat over a variable typically denotes a scaled machine variable.
[43] The fact that $0 \leq \hat{x} \leq 1$ will be taken care of later.

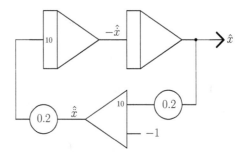

Fig. 8.47. Scaled setup for equation (8.20)

for scaling. This equation describes a harmonic oscillator, so it has a solution

$$y = y(0) \cos(\omega t) \tag{8.27}$$

with $y(0) = 1$ according to (8.15). For this harmonic oscillator $\omega^2 = 2a$. Differentiating (8.27) yields

$$\dot{y} = -y(0)\omega \sin(\omega t). \tag{8.28}$$

Based on (8.26) a basic analog computer program could be set up according to

$$\dot{y} = \int \ddot{y}\, dt,$$

$$y = \int \dot{y}\, dt, \text{ and} \tag{8.29}$$

$$\ddot{y} = -2ay. \tag{8.30}$$

Since MATHIEU's equation tends to instability, a scaling factor $\lambda_y = \frac{1}{5}$ is introduced into equation (8.29), which must be compensated for in (8.30):

$$\dot{y} = \int \ddot{y}\, dt \tag{8.31}$$

$$\hat{y} = \lambda_y \int \dot{y}\, dt = \frac{1}{5} \int \dot{y}\, dt \tag{8.32}$$

$$\ddot{y} = -\frac{2}{\lambda_y} ay = -10ax$$

$0 \le |\dot{y}| \le y(0)\omega$ according to (8.28). If $0 \le a \le 10$ then $\omega = \sqrt{20} \approx 5$. This, in conjunction with the safety scaling factor λ_y, which was chosen to at least partially counteract the blow-up tendencies of the equation, yields to the scaling factor $\lambda_{\dot{y}} = \frac{1}{25}$, which must be applied to (8.31) and compensated for in (8.32):

$$\hat{\dot{y}} = \lambda_{\dot{y}} \int \ddot{y}\, dt = \frac{1}{25} \int \ddot{y}\, dt$$

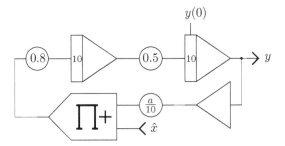

Fig. 8.48. Second half of the implementation of MATHIEU's equation

Fig. 8.49. Setup of MATHIEU's equation on THE ANALOG THING

$$\hat{y} = \frac{\lambda_y}{\lambda_{\dot{y}}} \int \hat{\dot{y}}\, dt = t \int \hat{\dot{y}}\, d5$$
$$\ddot{y} = -10a\hat{y} \tag{8.33}$$

The scale factors can now be regrouped, i. e., the factor 10 in front of a in (8.33) will be moved into $\lambda_{\dot{y}}$. Scaling a down by another factor of 10 to simplify setting of this parameter introduces another factor of 10, which can be implemented by using an input with weight 10 at the integrator yielding $-\dot{y}$. Finally, including a multiplier in the loop of this program, yielding $\frac{a}{10}\hat{x}y$ introduces another factor of 2 since $0 \leq \hat{x} \leq 1$ instead of $0 \leq x \leq 2$. The resulting program is shown in figure 8.48. The overall implementation on THE ANALOG THING is shown in figure 8.49.

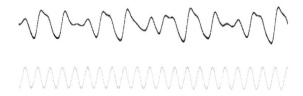

Fig. 8.50. Typical behaviour of the MATHIEU equation

A typical solution for the MATHIEU equation is shown in figure 8.50. This picture was taken with a digital oscilloscope and shows the first few milliseconds of the solution.

8.10 Projection of rotating bodies

The following example shows how an analog computer operating in repetitive mode can be used to display the projection of three-dimensional rotating figures.[44] The following example, which generates a two-dimensional display of a rotating three-dimensional spiral, is based on [Telefunken/3] and [Hitachi 1967].

At first, the coordinates of a three-dimensional spiral must be generated. This is done by the modified sine-/cosine-generator shown in figure 8.51 yielding

$$u(t) = \cos(\omega_{\text{rep}}t)e^{-\alpha k_0 t} \text{ and}$$
$$v(t) = -\sin(\omega_{\text{rep}}t)e^{-\alpha k_0 t}.$$

The attenuating term $e^{-\alpha k_0 t}$ results from the negative feedback path of the leftmost integrator, containing the coefficient potentiometer labelled α. $u(t)$ and $v(t)$ only describe a two-dimensional spiral, so a third function $w(t)$ is required for a three-dimensional figure.

This function is defined as

$$w(t) = -1 + \int_{t=0}^{t_{\text{rep}}} \beta k_0 \, dt$$

with t_{rep} denoting the operation time during one repetition cycle of the analog computer. The corresponding setup is shown in figure 8.52. It is to be noted that the three integrators shown in figures 8.51 and 8.52 are marked with a black bar. This denotes that these integrators are part of an integrator-group that is controlled by an external clock generator.

---

44 [MACKAY 1962, p. 137] describes the projection of a four dimensional hypercube on an oscilloscope by means of an analog computer.

8.10 Projection of rotating bodies

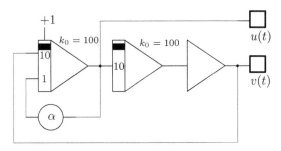

Fig. 8.51. Generating $u(t)$ and $v(t)$ for the spiral

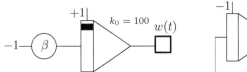

Fig. 8.52. Generating $w(t)$ for the spiral

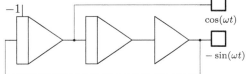

Fig. 8.53. Generating a sine-/cosine signal pair for the rotation of the spiral

To generate a display of a slowly rotating three-dimensional spiral, two time-scales are necessary: A very short time-scale for the integrator group generating the spiral coordinates $u(t)$, $v(t)$, and $w(t)$, and a much slower time-scale for a second integrator group yielding a sine-/cosine-pair for the rotation of the figure. The three integrators denoted by the black bar have an effective $k_0 = 10^3$ due to $k_0 = 10^2$ being set at the integrators and using inputs with weight 10.

The second integrator group is run in continuous mode and is shown in figure 8.53. It is a common sine-/cosine signal generator. Its output signals are fed to the circuit shown in figure 8.54, which performs the actual rotation and projection of the three dimensional figure. It implements the functions

$$Y = u(t) \quad \text{and}$$
$$X = w(t)\sin(\omega t) - v(t)\cos(\omega t),$$

which are then connected to the x- and y-inputs of an x, y-display.

Control of the first group of integrators is typically done by means of two control signals OP and IC controlling the three modes *operate*, *initial condition*, and *halt* of the integrators. Figure 8.55 shows these two signals during repetitive operation of the analog computer. A short period required to reset the integrators to their respective initial conditions. This is followed by a longer period during which the actual computation takes place. These two modes of operation are repeated rapidly to generate a flicker-free display of the rotating spiral.

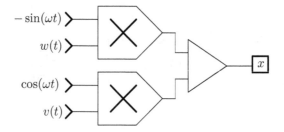

Fig. 8.54. Rotation and projection

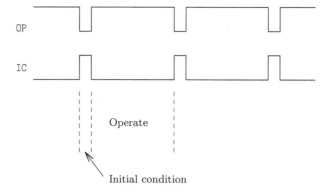

Fig. 8.55. Control signals for the integrators operating in repetitive mode

Figure 8.56 shows a suitable setup of the digital elements of an EAI 580 analog computer to generate these control signals.[45] A snapshot of the display of the rotating three-dimensional spiral is shown in figure 8.57.

8.11 Conformal mapping

The last and most complex example covered in this chapter is based on [Telefunken/4, p. 123] and [SYDOW 1964] and implements a *conformal mapping*, which is an angle-preserving mapping. Such mappings are described either explicitly by analytic functions of a complex variable such as

$$w = f(z) = u(x,y) + iv(x,y) \quad \text{where } z = x + iy$$

---

[45] The gate labelled *AND* is unusual as it is not used as an and-gate but as a signal driver with two outputs, one inverted and one non-inverted.

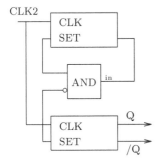 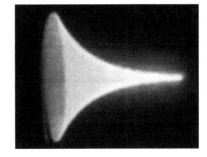

Fig. 8.56. Generation of the control signals for the integrators operating in repetitive mode

Fig. 8.57. Oscilloscope screen capture of a rotating spiral

or implicitly by the solution of suitable ordinary differential equations.[46] Implementing a conformal mapping on an analog computer is quite straightforward. The complex variable is split into its real and imaginary parts, which are handled explicitly as a two-dimensional vector.

A particularly interesting conformal mapping is

$$w(z) = (z - z_0) + \frac{\lambda^2}{z - z_0}, \qquad (8.34)$$

which describes a JOUKOWSKY[47] *airfoil*.[48] Applying this mapping to a unit circle yields an airfoil that can then be used to analyse the flow of a medium such as air around the structure. Implementing equation (8.34) on an analog computer is based on the following two equations, which describe the real and imaginary part separately:

$$u(x(t), y(t)) = (x(t) - x_0(t)) + \frac{\lambda^2(x(t) - x_0(t))}{(x(t) - x_0(t))^2 + (y(t) - y_0(t))^2} \qquad (8.35)$$

$$v(x(t), y(t)) = (y(t) - y_0(t)) - \frac{\lambda^2(y(t) - y_0(t))}{(x(t) - x_0(t))^2 + (y(t) - y_0(t))^2} \qquad (8.36)$$

The mechanization of equations (8.35) and (8.36) is shown in figure 8.58. This function is required twice to map a unit-circle into an airfoil and to map the path

---

[46] Cf. [HEINHOLD 1959, pp. 46 ff.] for more information on this technique.
[47] NIKOLAY YEGOROVICH JOUKOWSKY, 01/17/1847–03/17/1921
[48] A detailed description of this type of airfoil can be found in [THWAITES ed. 1987, pp. 112 ff.] and [ECK 1954, pp. 237 f.]. Applying suitable changes to equation (8.34) more complex and realistic airfoils can be described. [ASHLEY et al. 1985, pp. 52 ff.] described two such enhanced mappings based on the MISES (RICHARD VON MISES, 04/19/1883–07/14/1953) the KÁRMÁN-TREFFTZ transformation (THEODORE VON KÁRMÁN, 05/11/1881–05/07/1963, ERICH TREFFTZ, 02/21/1888–01/21/1937) and the THEODORSEN mapping (THEODORE THEODORSEN, 01/08/1897–11/05/1978). See [WARSCHAWSKI 1945] for details. Numerical approaches for the generation of such airfoils and the simulation of particle flows are described in [ZINGG 1989].

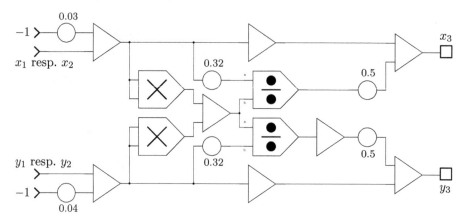

Fig. 8.58. Transforming a unit circle into a Joukowsky air foil using a conformal mapping

of a particle flowing around a rotating cylinder into its corresponding path around this airfoil. Since the setup is quite convoluted, the circuit shown is applied to both tasks in an alternating fashion using analog switches.

The path of a particle flowing around a rotating cylinder is described by the complex velocity potential

$$f(z) = v_0 \left(e^{i\delta} z + \frac{e^{-i\delta} r^2}{z} \right) - i \frac{\Gamma}{2\pi} \ln(z),$$

which is realized by the circuit shown in figure 8.59. The upper leftmost integrator generates the time varying x-coordinate of a particle, the path of which is determined by two parameters: The angle of attack, which is controlled by the term $\tan(\delta)$, and the y-position of the particle with respect to the rotating cylinder. y is generated by the lower leftmost integrator, which is controlled separately from the remaining integrators of the computer setup.[49]

The overall setup of this program on an Telefunken RA 770 precision analog computer is shown in figure 8.60. Figure 8.61 shows a single flow line around a JOUKOVKSY airfoil on top and a group of flow lines spaced equidistantly in the bottom half.

---

[49] This is, as before, denoted by a black bar at the top of the integrator symbol. The simple digital control circuit is set up on the digital control subsystem of the RA 770 analog computer and is not shown here. This control system also generates the signals used to switch the circuit for the actual conformal mapping between both of the inputs, in an alternating fashion.

8.11 Conformal mapping — 199

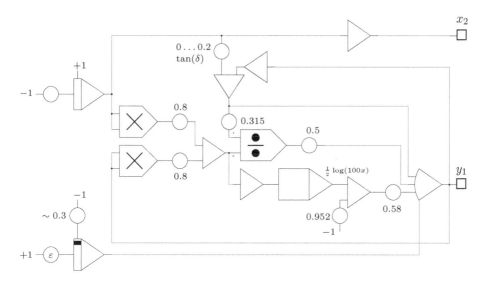

Fig. 8.59. Generating the flow lines

Fig. 8.60. Implementation of the conformal mapping and the generation of flow lines

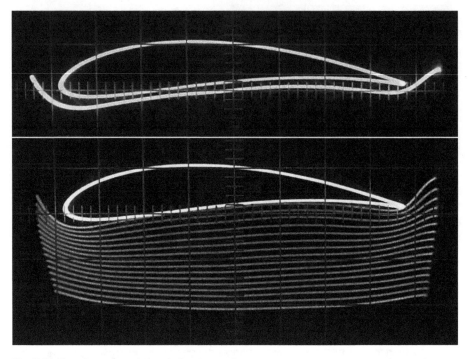

Fig. 8.61. Joukowsky air foil with flow lines

9 Hybrid computers

In the mid-1950s it became clear that analog computers had some drawbacks that could not easily be overcome by analog electronic means. One of the main challenges was the cumbersome process of setting up function generators and the generation of functions of more than one variable – a task often required in aerospace applications or whenever functions are defined by measurement data and not analytically. The classic stored-program digital computer conversely excels at tasks like these. Thus, the idea of coupling analog computers with digital computers, forming *hybrid computers*, was born.

9.1 Systems

An early account of this rationale is given in [WALTMAN 2000, p. 69] describing the application of analog computers for the simulation of the *X-15* aircraft:

> "*The thought of building up another set of function generators like those already in use was probably considered, but not by me or any other X-15 simulation programmers. We had had enough of those fuses and dinky pots. The idea of using a digital computer to do this job was unanimously and immediately accepted. No discussion was needed. We were going hybrid.*"

To couple an analog computer with a digital computer, *Analog-Digital-Converters* (*ADCs*) and *Digital-Analog-Converters* (*DACs*) are required. These devices translate analog voltages into digital signals and vice versa. In addition to that, typical hybrid computers allow control of the integrators by the digital computer. They also typically feature digital interface lines, which can be used to trigger interrupts on the digital system by comparators on the analog side and to control electronic or relay switches on the analog system from the digital computer. Figure 9.1 shows the structure of an EAI 690 hybrid computer system.

One of the first hybrid controllers was the *ADDAVERTER* shown in figure 9.2. This system was developed and built by *Space Technology Laboratories*[1] in 1956.[2] The system shown features 15 ADC- and 10 DAC-channels[3] with a precision of ±0.1%, matching well the precision of large analog computers of that time.

This development was the result of previous research efforts by Ramo-Woolridge Corporation and *Convair Astronautics*. Both companies were working

---

[1] A subsidiary of *Ramo-Woolridge Corporation*.
[2] See [BAUER et al. 1956].
[3] A fully expanded system had up to 15 DAC channels.

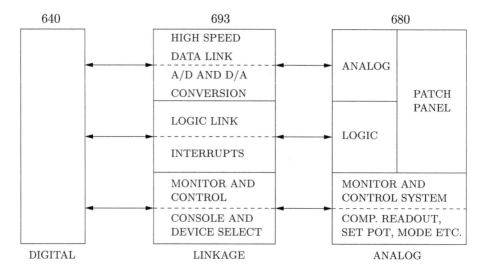

Fig. 9.1. Block diagram of the EAI 690 hybrid computer system consisting of an EAI 640 digital computer, the EAI 693 linkage system and an EAI 680 analog computer (cf. [EAI 690])

on simulations of intercontinental ballistic missiles, which involved very large analog computer systems.[4] The requirements of these simulations are summarized by [MCLEOD et al. 1957, p. 1127] as follows:

> "It was imperative that the simulation be done in real time to allow inclusion of weapon system hardware, and to conserve operating time. This ruled out an all-digital simulation because the large number of individual computations could not be made fast enough, and because analog equipment would be necessary to connect to some of the weapon system components which were to be included. An all-analog simulation was ruled out by accuracy requirements, particularly with respect to the navigational problem. Clearly a combined simulation was necessary to fulfill the requirements [...]"

The system that was finally put into operation at Convair Astronautics consisted of an IBM 704 digital computer, an ADDAVERTER, and a large EAI analog computer. The overall cost for this system was 2.3 million US$ for the IBM 701, 200,000 US$ for the ADDAVERTER and 1.6 million US$ for the analog computer[5] – very substantial sums at that time.[6] A similar system is described in [BURNS et al. 1961] where an IBM 704 is coupled with a REAC system.

---

4 Cf. [BEKEY et al. 1968, p. 154].
5 See [MCLEOD et al. 1957, p. 1130].
6 Adjusted for inflation, 4.1 million USD in 1957 is equivalent to about 42 million USD in 2022.

Fig. 9.2. Space Technology Laboratories ADDAVERTER (see [MCLEOD et al. 1957, p. 1129])

Although this system was never used for the proposed task, which had been solved by conventional means in the meantime.[7] Nevertheless, it served as a demonstrator of the efficiency and power of hybrid computers.[8] Subsequently the market demanded more and more hybrid systems, which was aided by the development of fully transistorised precision analog computers using a machine unit of ± 100 V.[9] One of the first such analog computers was developed by *Comcor* in 1964 and offered the same precision as previous vacuum tube based machines but at lower cost, lower power consumption, and longer maintenance intervals.[10]

---

7 [BEKEY et al. 1968, pp. 154 f.] notes that "*[i]n the time that elapsed between the original specification of the hybrid system and the delivery and acceptance of the conversion equipment, it was established that the guidance and control problems associated with missile flight are not closely coupled and can be studied separately. Consequently the basic problem for which the hybrid computing system was designed, vanished.*"
8 A collection of typical problems that could be solved by hybrid computers is given in [BENHAM 1970].
9 These large machine units were commonplace in aerospace technology and it was an imperative for any analog computer being coupled with simulator and flight hardware to support this voltage range.
10 See [BEKEY et al. 1968, p. 155].

204 — 9 Hybrid computers

Fig. 9.3. Hybrid computer installation at the Department for Electrical Engineering of the Naval Postgraduate School in the late 1960s (with permission of ROBERT LIMES)

A typical 1960s hybrid system is shown in figure 9.3. This system consists of an SDS-9300 digital computer[11] visible on the left and a Comcor *CI-5000* analog computer on the right with a multi-channel recorder in front of it.

Suitable digital computers for a hybrid computer setup must be fast enough to keep up with the analog system in a simulation.[12] Particularly, simulations of nuclear reactors showed that the analog computer could easily outperform the digital system and were consequently slowed down considerably in a hybrid setup. In some cases a hybrid computer approach was only about three to ten times faster than a pure digital solution.[13]

Apart from this obvious idea of coupling analog and stored-program digital computers, other schemes were devised as well. In [KARPLUS et al. 1972], an analog co-processor consisting of a complex resistor-network is described, aiding a digital computer in the solution of parabolic partial differential equations.

Another interesting idea is the implementation of hybrid number systems in which values are represented by a combination of continuous signals like voltages and sequences of bits, thus forming a generalized form of floating point numbers consisting of a mantissa part and an exponent. Here the analog part of a num-

---

11 See [Scientific Data Systems/2].
12 This is even today not an easy task, mostly due to high interrupt latencies on many modern digital computers.
13 See [FRISCH et al. 1969, p. 36].

ber interpolates between two values that can be described exactly by a binary sequence.[14]

In 2005 GLENN EDWARD RUSSELL COWAN and NING GUO in 2016 developed two reconfigurable analog computers in the form of $VLSI$[15] integrated circuits.[16] Although no products resulted from these academic developments, it was demonstrated that analog computers can still be used as co-processors to speed up digital computers considerably for certain types of application.[17]

9.2 Programming

Programming the analog half of a hybrid computer is not different from classic analog computer programming. The only difference are some additional input/output and computing elements such as ADCs, DACs, and function generators. Programming the digital half of such a setup is more challenging, as it has to work closely coupled with the very fast analog computer. This typically demands (very) low interrupt latencies, reentrant code, etc. As requirements like these are typical for systems used for process automation tasks, programming paradigms from this area can at least partially be applied to hybrid computers.

Basically, two modes of operation must be distinguished in a hybrid computer system:

Alternating operation: In this mode the analog and digital subsystems work in an alternating fashion. Typically, the digital computers sets coefficients on the analog computer, which is then run for some time to yield a solution. At the end of or even during this analog computer run data is collected by the digital computer, which can then be used to derive an improved parameter set for the analog computer, etc. From a programming perspective, this mode of operation is quite simple since timing is not critical.[18] Furthermore, the digital computer has full control of the analog computer.

Simultaneous operation: This mode of operation is by far more complicated. Here both computers operate in a closely coupled fashion, and the digital computer must respond to interrupts generated by the analog computer, it

---

[14] More information about this can be found in [GILOI 1963, pp. 268 f.] and [SKRAMSTAD 1959].
[15] *Very Large Scale Integrated circuit*
[16] See [COWAN 2005] and [GUO 2017].
[17] See also section 14.
[18] Nevertheless, setting coefficients and reading data from the analog computer by means of the digital computer can take a considerable amount of time, thus slowing down the overall time to solution.

must generate functions of more than one variable, it must read values in real time, perform calculations and generate input signals for the analog system.

Historically two main approaches for programming the digital part of a hybrid computer have been pursued. The first is based on extending traditional programming languages such as *FORTRAN* or *ALGOL* with special library calls and sometimes additional language features to support the control of the various interfacing devices, as well as the implementation of fast interrupt routines.[19]

A good overview of the various software subsystems included in the EAI 8900 hybrid computer system is given in [BEKEY et al. 1968, p. 181]. Apart from an extended FORTRAN dialect, a special purpose language, *HYTRAN*, was also available as well as a powerful macro assembler, which was typically used to implement highly time-critical routines. These systems also had the advantage that specialised diagnostic routines were available to automatically check the digital computer as well as the analog computer. Typical tests included *rate checks* and *static checks*.[20] A rate check tests the time-scale factors of the integrators by applying a known constant input signal for a precisely determined amount of time.

A complex, yet instructive example of the application of a hybrid computer system to an optimisation task can be found in [WITSENHAUSEN 1962]. This paper gives a thorough description not only of the analog computer setup but also describes the necessary control routines written for the digital processor.

9.3 Example

A modern hybrid controller implementation is shown in figure 9.4. This particular device allows an Analog Paradigm *Model-1* analog computer to be coupled with a digital computer via a USB[21] connection. At the heart of the hybrid controller is a variant of an *Arduino MEGA-2650* controller, which handles the communication with the digital computer as well as control of the analog computer. This module also contains eight *digital potentiometers* with 10 bit resolution, eight digital input, and eight digital output lines.

An interesting application of a hybrid computer based on this module is described in [HOLZER et al. 2021]. The analog part of the setup is programmed to simulate the behaviour of an inverted pendulum mounted on a cart, which can be moved along the x-axis under control of the digital computer. The digital

---

19 A description of an extended ALGOL system can be found in [HERSCHEL 1966] while a more general description of this approach is given in [FEILMEIER 1974, pp. 133 ff.].
20 See section 5.5.
21 *Universal Serial Bus*

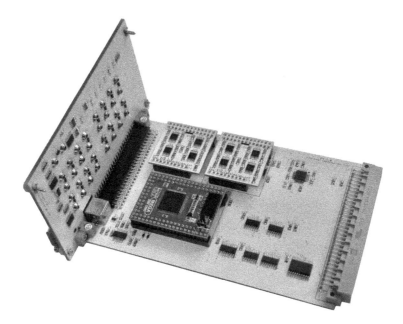

Fig. 9.4. Modern USB-based hybrid controller for an Analog Paradigm Model-1 analog computer

computer runs a reinforcement learning system, which is trained to balance the inverted pendulum by moving the cart in little steps.

The analog computer program simulating the behaviour of the inverted pendulum is shown in figure 9.5. Its derivation, based on the EULER-LAGRANGE-equations, is described in detail in [HOLZER et al. 2021]. Input to this program is $\ddot{x}$, the cart's acceleration in x-direction. Using two digital output lines D0 and D1, the digital computer can push the cart to the left or to the right in small increments using the circuit shown in figure 9.6.[22] The output signals x and y can be used to display the inverted pendulum on an x, y-display.

The reinforcement learning system implemented on the attached digital computer uses a *Q-learning*[23] approach to learn how to balance the inverted pendulum. At the start of a simulation/training run the cart is positioned in the middle of the simulated x-space with the pendulum in an upright position. The digital computer continuously reads the pendulums angle φ from the analog simulation and tries to keep the pendulum in this position. Whenever the pendulums angle

---

[22] The manual input was used during this study to actively unbalance the pendulum and observe the reaction of the trained control system to this perturbation.
[23] See [SUTTON et al. 2018] or [FRANÇOIS-LAVET et al. 2018] for more details on Q-learning.

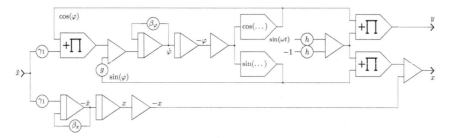

Fig. 9.5. Analog computer program simulating an inverted pendulum mounted on a moving cart with one degree of freedom

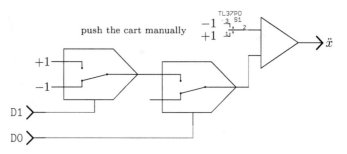

Fig. 9.6. Control circuit for the cart's motion

exceeds a predetermined value or when the cart hits the left or right boundaries, the simulation is restarted.

After a few dozen runs, the digital computer showed first signs of actively balancing the pendulum, which evolved into near perfect control after several hundred runs. After this, even manual interventions, i.e., perturbing the cart by applying little pushes to the left or right, were easily corrected by the digital controller.[24]

---

[24] This experiment is also described in https://www.youtube.com/watch?v=jDGLh8YWvNE (retrieved 06/06/2022).

10 Digital differential analysers

The essence of an analog computer is that it forms an analog of the problem to be solved. Although most analog computers are based on analog electronic implementations of their respective computing elements working in continous time and using continuous voltages or currents to represent values, there is no need to restrict analog computers to this particular type of implementation.

A *digital differential analyser* (*DDA*) is basically an analog computer but one working in discrete time with values represented as serial or parallel bitstreams.[1] The earliest developments of DDAs date back to the late 1940s when *Northrop* started the development of a cruise-missile like system.[2]

Representing values as bit sequences makes it possible to achieve much higher precision than with a traditional analog computer, and it is even possible to represent values as floating point numbers, thus eliminating the need for scaling altogether, if the complexity of the implementation is acceptable. Using logic gates as the foundation also eliminates any problems caused by drift or ageing of components. Nevertheless, these advantages come at a price: Computation is no longer continuous, neither with respect to the value representation nor regarding time. Furthermore, the energy efficiency of such machines is not as high as that of machines using analog electronic computing elements.[3]

Basically there are two different approaches to implement a DDA. A straightforward *parallel* DDA[4] consists of a whole complement of typical analog computing elements, which can be interconnected by means of a traditional patch panel or the like. An example for this class of machines is *TRICE*, which is described in more detail in section 10.4.4. Machines like this exhibit the same amount of fine-grained parallelism as a traditional analog computer, thus easily outperforming classic stored-program digital computers. Nevertheless, this high computational power comes at a very high cost since the number of computing elements is dictated by the complexity of the problems to be solved, with large problems requiring a vast number of computing elements.

The second basic approach is based on a time-multiplexed use of a few – in some cases only one – central computing elements. Like a simple stored-program

---

[1] [MICHELS 1954, p. 2] defines a DDA as "*an electronic computer which solves differential equations by numerical integration.*". This definition is debatable as it would also hold true for a classic stored-program digital computer and does not take the central feature of setting up an analog to solve a problem into account.
[2] See section 10.4.1.
[3] It can be shown that analog computation is more energy efficient due to physical reasons for a *SNR* (*signal to noise ratio*) of up to 60 dB (see [SARPESHKAR 1998, p. 1615]).
[4] Also known as *simultaneous* DDA, see [OWEN et al. 1960, p. 6].

https://doi.org/10.1515/9783110787740-010

digital computer this *sequential* DDA has one or a few central *arithmetic/logic units (ALUs)*, which are fed with data under control of a sequencing unit that in turn is controlled by a machine-readable description of the connections between the virtual computing elements. This second approach is much cheaper to implement due to its reduced complexity and the lower number of computing elements required. In addition to that, it scales quite well since large programs just require more time to solution whereas they could not be implemented on a parallel DDA if it did not contain enough computing elements. The major disadvantage of sequential DDAs is their relatively low computational power.

10.1 Basic computing elements

The following section describes some basic DDA computing elements regardless how they are implemented.[5]

10.1.1 Integrators

As with an analog computer, the central element of a DDA is the integrator, which is most easily implemented based on an *accumulator*. Although simple accumulation of values will not be sufficient for more demanding computations, this basic approach led to the term *incremental computer* for a DDA.[6] An accumulator-based integrator has the distinct advantage that not just time but every variable can serve as the free variable of integration so that integrals like

$$\int_{y_0}^{y_1} f(x) \mathrm{d}x$$

can be treated directly.

The basic structure of a simple DDA integrator is shown in figure 10.1. In contrast to an analog electronic integrator there are two inputs $(\Delta Y)_i$ and $(\Delta X)_i$. These variables are often represented not as absolute values but as incremental values, which simplifies the implementation of the DDA and reduces the number of signal lines required to interconnect computing elements. These incremental

---

5 For further reading see [SHILEIKO 1964], [BYWATER 1973], [FORBES 1957, pp. 215 ff.], [WINKLER 1961], [BECK et al. 1958], [KLEIN et al. 1957, p. 1105], [GOLDMAN 1965], and [JACKSON 1960, pp. 578 ff.].
6 See [MCLEOD et al. 1958/4, p. 1223]

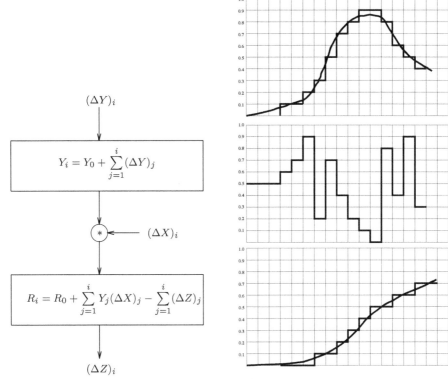

Fig. 10.1. Basic integrator of a DDA (cf. [Bendix 1954, p. 1])

Fig. 10.2. Operation of a DDA integrator (cf. [Michels 1954, p. 4])

values are often restricted to the set $\{-1; 0; 1\}$.[7] $(\Delta Y)_i$ represents the change of the integrand, while $(\Delta X)_i$ is the change of the variable of integration and corresponds to dx.

At the heart of an integrator like this are two accumulators: One accumulates the series of $(\Delta Y)_i$ values at its input and yields

$$Y_i = Y_0 + \sum_{j=1}^{i}(\Delta Y)_j$$

where Y_0 is an initial value. The resulting Y_i is then multiplied by the incremental input $(\Delta X)_i$. Since these incremental values are restricted to $(\Delta X)_i \in \{-1; 0; 1\}$

---

[7] Since two signal lines are necessary to transmit such values, some DDA implementations restrict the domain of these incremental values to $\{-1; 1\}$, thus simplifying the hardware implementation but requiring constant values to be represented by an alternating sequence of -1 and 1 increments. See [Michels 1954, p. 19].

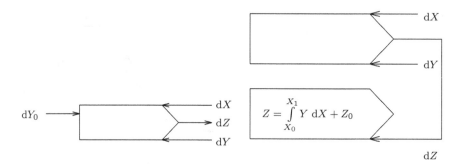

Fig. 10.3. Symbol of a DDA integrator

Fig. 10.4. Explicit computation of a simple integral (see [Bendix 1954, p. 7])

this multiplication step is reduced to either addition or subtraction. The result is then accumulated in the second accumulator yielding

$$R_i = R_0 + \sum_{j=1}^{i} Y_j (\Delta X)_j - \sum_{j=1}^{i} (\Delta Z)_i.$$

The output of this integrator is not R_i but instead an incremental value, $(\Delta Z)_i$ representing the over- or underflow of this second accumulator stage. $(\Delta Z)_i$ can now be used as an input signal for other computing elements of the DDA. The behaviour of such an integrator is depicted in figure 10.2: The input function and its representation as a sequence of increments and decrements $(\Delta Y)_i$ is shown in the upper graph. The output of the second accumulator is shown in the middle graph. Accumulator overflows generate a corresponding incremental output signal $(\Delta Z)_i$ representing the result of the integration, which is shown in the bottom graph.

Although the use of incremental values simplifies the implementation of a DDA and thus saves cost, this representation limits the maximum rate of change of variable values in a computation. To overcome this restriction, additional time-scaling of the computer setup may be required, which may result either in non-real-time operation or the very high clock rates of the DDA.

The symbol of a DDA integrator is shown in figure 10.3. A typical setup of an DDA yielding the integral over a function is shown in figure 10.4. The first integrator is fed with the incremental input values dX and dY yielding an incremental output signal dZ, which is accumulated in a second integrator resulting in the actual value of the integration. The missing second incremental input of this integrator is assumed to be $+1$.[8,9]

Integrators in a DDA have an even more central role than those of a traditional analog computer. While the latter uses free or open operational amplifiers to create implicit functions by means of a function generator in the feedback path, DDAs use specially configured integrators for this purpose. These are called *Servos*.[10]

10.1.2 Servos

As its name implies a *Servo* is a computing element that is typically used to minimise some error term in a feedback loop arrangement (this is then called a *servo loop*), which can be used to generate implicit functions. Typical servos are based on DDA integrators but feature only one incremental input $(\Delta Y)_i$ and one accumulator. The output $(\Delta Z)_i$ of this element is defined as

$$(\Delta Z)_i = \begin{cases} +1, & \text{if } Y_i > 0, \\ 0, & \text{if } Y_i = 0, \\ -1, & \text{if } Y_i < 0. \end{cases} \quad (10.1)$$

Figure 10.5 shows the symbol of a servo element. Some implementations allow for an additional initial value input, which has been omitted here. Using such an initial value, the threshold level of the servo can be set to an arbitrary value $\neq 0$ making it possible to use the servo as a generalized decision element.

Figure 10.6 shows a simple application example of a servo element: Using a function generator, an implicit function based on the condition $F(U,V) = 0$ is generated.[11] The input of the servo is the incremental output signal $dF(U,V)$ of the function generator circuit. The output signal of the servo, defined by (10.1), is in turn used as one input to the function generator while its second input is fed by dV from some external source. The servo tries to drive the output of the function generator to 0 thus yielding the desired implicit function.

---

8 A comprehensive study of this type of integrator can be found in [ZOBERBIER 1968].
9 This second integrator is not normally required when all values within a DDA setup are represented by incremental values.
10 See [Bendix 1954, pp. 16 ff.].
11 Whenever possible $F(U,V)$ should be generated by solving a suitable differential equation.

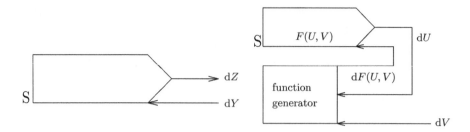

Fig. 10.5. Symbol of a DDA servo element

Fig. 10.6. Application of a servo element (see [Bendix 1954, p. 18])

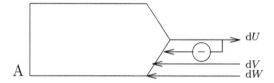

Fig. 10.7. Symbol of a DDA summer

10.1.3 Summers

Typical incremental DDAs, like the *Bendix D-12* described in section 10.4.2 reduce the addition of two incremental values $dU = dV + dW$ to solving the equation

$$V + W - U = 0. \tag{10.2}$$

A summer can thus be realized by using a modified servo element featuring more than one incremental input. Figure 10.7 shows the symbol of a DDA summer having two inputs dW, dV, an output dU, which also feeds a third input after being sign reversed. According to equation (10.2) dU represents the desired sum.

10.1.4 Additional elements

In addition to integrators, servos, and summers, typical incremental DDAs often feature circuit elements like *output multipliers*, which multiply incremental values by fixed rates thus acting as coefficient units,[12] function generators, which are often based on table-lookup techniques, etc.

Since integration can be performed with respect to any variable and not just time t, multiplication in a DDA is most often based on two integrations as de-

---

[12] See [Bendix 1954, pp. 21 ff.].

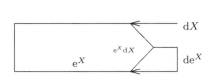

Fig. 10.8. Computing e^x with a DDA (cf. [Bendix 1954, p. 9])

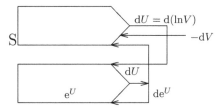

Fig. 10.9. Implicitly generating a logarithm function (see [Bendix 1954, p. 19])

scribed by equation (2.1) in section 2.5.4. An example of this is given in the following section.

10.2 Programming examples

Figure 10.8 shows a DDA-setup to generate an exponential function, which is based on a single integrator with its output fed back to its incremental dY-input. Its incremental dX-input is fed with the incremental values representing the desired exponent value.

This function can be used in conjunction with a servo element to implicitly generate a logarithm function as shown in figure 10.9. The servo at the top has two inputs: The incremental input value $-dV$ determining the argument of the logarithm function and a second input, which is fed with the output of an exponential function generator as described above. At the output of the servo a signal representing $d(\ln V)$ is available. This is due to the fact that the servo tries to ensure $F(U, V) = e^U - V = 0$ holds, so $e^U = V$ yielding $dU = d \ln V$.

Generating trigonometric functions can be done equally simply by solving an auxiliary differential equation of the form $\ddot{y} + \omega^2 y = 0$ as shown in figure 10.10. In contrast to a traditional analog computer implementation the sign reversal operation does not require an additional inverter since it can be done directly using incremental values.

Figure 10.11 shows a typical DDA setup to perform the multiplication of two incremental variables dV and dU. As already described in section 2.5.4, the product rule can be used to perform a multiplication of two values if integration is not limited to time being the only free variable. The two integrators shown on top of the figure perform the actual multiplication of the incremental input values yielding an incremental output that is fed into a third integrator yielding the desired product.

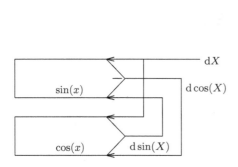 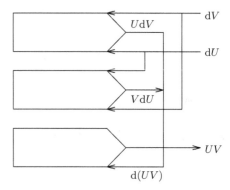

Fig. 10.10. Generating a sine-/cosine signal pair (cf. [Bendix 1954, p. 10])

Fig. 10.11. Multiplying two incremental values using two coupled integrators (see [Bendix 1954, pp. 10 f.])

Additional programming examples of varying complexity for DDAs can be found in [Bendix 1954], [Palevsky 1962] and [Forbes 1972], the latter focusing on the solution partial differential equations by means of DDAs.

10.3 Problems

The advantages of DDAs, principally their ability to use any variable as the free variable of integration, their immunity to drift and aging effects of components as well as the advantage of using basically arbitrarily long bit sequences to achieve any degree of precision required, come at a price.

A simple bit-serial incremental DDA working with values -1 and $+1$, and featuring a word-length of n bit exhibits a basic time-constant of

$$t = \frac{2n}{f}$$

with f denoting the systems clock frequency. This limits the real-time capabilities of such a basic DDA considerably.[13] Making n smaller would yield a faster response time but would also diminish the precision of the machine, so increasing f is often the easiest way to speed up an incremental DDA without sacrificing precision.

---

[13] See [Campeau 1969, p. 711].

Another approach is to allow incremental values of a larger domain – ideally floating point numbers could be used throughout a DDA.[14]

One source of error is the integration scheme used in a particular DDA. Using just simple accumulators implementing the rectangle rule is not sufficient for more demanding applications. Other techniques, implementing at least a trapezoidal rule, yield better results but complicate the hardware implementation considerably. In addition to that, a DDA cannot speed up a calculation as easily as a traditional analog computer where the integrator capacitors can be changed to speed up or slow down a calculation.

The limited precision and speed of incremental DDAs makes them unsuited to many applications as shown by FRED LESH in [MCLEOD et al. 1958/3, pp. 488 f.]. In this study a problem involving the computation of rocket trajectories is treated on a DDA as well as on a traditional analog computer. It turned out that the analog computer took only 30 seconds for a solution while the DDA was running for 3 minutes, 30 minutes or up to 5 hours depending on the integration step size. While the analog computer yielded results with an error of only about 2%, compared to the exact solution, it took the DDA 30 minutes to product a solution with an error of 20%.

10.4 Systems

Over the years many different DDAs have been implemented. All of these are of the incremental type, with many of them being based on sequential operation. Most of these systems are programmed by specifying the connections between the computing elements in binary form on some storage system with *CORSAIR*, which uses a traditional patch panel, being an exception. Only one relevant system, *TRICE*, was of the parallel type, using many individual computing elements and a patch panel.

10.4.1 MADDIDA

Shortly after the end of World War II, in March 1946, *Northrop* started the development of a subsonic cruise missile designated *MX-775A*, which was later desig-

---

[14] A detailed analysis of DDA specific questions regarding the precision of computations can be found in [KELLA 1967] and [KELLA et al. 1968]. Optimal ratios of the length of the integrator accumulator registers and incremental values are investigated in [MCGHEE et al. 1970].

nated *SSM-A-3 Snark*.[15] This system should be able to hit a target at a distance of up to 5,000 miles with a precision of 200 yards – much better than the German vengeance weapons V1 and V2.[16] To achieve this level of accuracy, a new guidance system, based on astronavigaion, was developed. Using a built-in telescope, azimuth and elevation angles for sighting predetermined stars were determined to get a navigational fix. This system was then coupled with an inertial guidance system.[17]

While HELMUT HOELZER used an analog electronic analog computer in the guidance system of the A4 rocket, the long time-of-flight and the required high precision of Snark required a digital approach to minimise computational errors. FLOYD STEELE[18] developed the idea of implementing an analog computer using only digital elements, which was called *DIDA*, short for *DIgital Differential Analyser*.

DONALD E. ECKDAHL,[19] RICHARD SPRAGUE,[20] and FLOYD STEELE[21] headed a team that developed a laboratory prototype of a DIDA, which was called *MAD-DIDA*.[22] This machine, actually built by *Hewlett Packard*,[23] was based on a magnetic drum like those pioneered in sophisticated radar systems of World War II. The tracks of this drum housed bit-serial accumulators, one for each of the integrators of the machine, as well as data representing the interconnection of the computing elements. The actual computations were performed on a single, central ALU that was time-shared between the various computing elements used in any given setup. This remarkable prototype, which only used 53 vacuum tubes and 904 diodes and was built in "*less than 600 man hours*",[24] is shown in figure 10.12.[25]

Clearly visible on the left is the small magnetic drum with its read/write heads and the amplifier tubes hanging down. The heart of the machine, the digital logic elements implementing the bit-serial ALU with its surrounding control logic is visible on the right in its raised service position.

---

15 See [WERRELL 1985, pp. 82 ff.]. The system was named after LEWIS CARROLS's (01/27/1932–01/14/1898) *The Hunting of the Snark*. The follow-up cruise missile was aptly named *Boojum* following the last line of this poem.
16 The V2 was admittedly no cruise missile. Its engineering version was known as *A4*.
17 See [CERUZZI 1989, pp. 22 ff.].
18 06/28/1918–09/23/1995
19 04/29/1924–07/23/2001
20 08/27/1921–01/27/1996
21 See [CERUZZI 1989, p. 25].
22 Short for *MAgnetic Drum DIfferential Analyser*. Due to reliability problems of this prototype, it was often pronounced as *mad Ida*.
23 See [CERUZZI 1989, p. 25].
24 See [DONAN 1950, p. 1].
25 See also [Popular Mechanics 1950].

Fig. 10.12. MADDIDA prototype (with permission of the *Computer History Museum*, DAG SPICER)

Fig. 10.13. Two MADDIDA systems in use at the Navy Electronics Laboratory (NEL, file number E1278)

In 1950 MADDIDA was demonstrated to JOHN VON NEUMANN,[26] who was excited to see the system computing BESSEL functions and characterized the system as "*a most remarkable and promising instrument*".[27] Although no actual guidance system for Snark resulted from the work on MADDIDA,[28] it became the ancestor of most DDAs to follow. A direct successor was built in a small production batch and used mostly for aerospace research. Figure 10.13 shows two of these systems.[29]

10.4.2 Bendix D-12

In the early 1950s the *Bendix Corporation* developed and marketed the *D-12*, a DDA quite similar to MADDIDA. Figure 10.14 shows the DDA on the right with

---

[26] 12/28/1930–02/08/1957
[27] Cf. [ECKDAHL et al. 2003].
[28] Snark finally employed an analog electronic analog computer for its guidance, see [CERUZZI 1989, p. 25]. Nevertheless, a DDA was developed for the *Polaris* guidance system, see section 13.15.8.
[29] A thorough description of these production systems can be found in [Northrop 1950].

Fig. 10.14. Bendix D-12 DDA (cf. [Bendix, p. 2])

Fig. 10.15. Bendix D-12 coupled with a Bendix G-15D stored-program computer (see [KLEIN et al. 1957, p. 1105])

its operator desk on the left. On this desk, from left to right, are the control console, an xy-plotter and a console typewriter.[30]

The D-12, operating on a serial stream of decimal values represented by four bits each,[31] is based on a magnetic drum holding the integrator registers as well as the necessary information about signal sources and destinations. The system can work in two modes with either 30 integrators operating at 200 iteration steps per second or with 60 integrators running at half this speed.[32] Apart from the simple integration scheme described in section 10.1.1, the D-12 integrators can also work based on

$$\frac{(\Delta Y)_i + (\Delta Y)_{i+1}}{2}$$

yielding better results than those obtained by the rectangle rule.[33]

While MADDIDA was basically a stand-alone system, the D-12 could be coupled directly to a Bendix *G-15D* stored-program digital computer as shown in figure 10.15. On the left the G-15D stored-program digital computer can be seen with the DDA on its right. The enclosure housing the DDA in this configuration is considerably smaller than that of the stand-alone system since the G-15D and the D-12 share a common magnetic drum, which is part of the G-15D stored-program computer, thus saving a considerable amount of hardware on the DDA part of the machine.

---

[30] See [EVANS et al. 1966] for an in-depth description of the internal structure of the D-12.
[31] Excess-three coded decimals.
[32] This is a good example of the ability of a sequential DDA to trade speed against problem complexity.
[33] See [Bendix 1954, pp. 11 ff.].

A typical program for the D-12 is shown in figure 10.16. The four integrators, labelled (4) to (8), solve the VAN DER POL differential equation[34]

$$\ddot{x} - k(x^2 - 1)\dot{x} + x = 0.$$

Integrator (9) controls the printout of the values generated by the integrators (3), (4), and (5) on the console typewriter. The integrators (0) and (1) generate the value-pair $(x(t), t)$ that is used to control plotter #1. A phase-space plot is generated on plotter #2 based on the signal pair $(\dot{x}(t), x(t))$, which is generated by the integrators (18) and (19). The remaining three integrators (10), (11), and (12) perform some post-processing of the signals required for the phase-space plot.

This DDA setup is implemented by the program listing shown in figure 10.17. Each line corresponds to one computing element, the number of which is specified in columns 2 and 3. The initial values used for integrators is specified by the digits in columns 6 to 14.[35] Columns 16 to 20 specify integrator control:

Column 16: This column controls the mode of operation for the integrator. The values "1" to "4" select integrator-mode with one out of four different integration schemes, while a value of "5" switches the integrator into summing mode and "6" specifies a servo.

Column 17: Possible values are "1" or "2", selecting normal or automatic reset mode of the integrator.

Column 18: Integrators flagged with a value of "1" in this column are running in normal (silent) mode while a value of "2" causes the contents of the second accumulator register to be printed on the console printer after each iteration.

Column 19: This can be used to select an automatic sign-reversal for the incremental output signal of an integrator. "1" specifies normal operation while "-" causes a change of sign.

Column 20: Using this column a constant output multiplication factor of 1, 2 or 5 can be selected. The special value "6" selects the corresponding integrator as control circuit for the typewriter.[36]

Columns 22 and 23 specify the source of the dX input while columns 25 and 26 select the source of the initial value of the integrator. Up to 8 addresses of source

---

[34] This equation describes an oscillator with amplitude stabilisation due to the term $x^2 - 1$. For $x < 1$ this becomes negative, thus increasing the oscillator's amplitude and decreasing it for $x > 1$. See [VAN DER POL et al. 1928] for more information on this topic.

[35] All values are represented as normalized fixed-point values with one decimal digit before and 5 after the implicit decimal point.

[36] See bottom left of figure 10.16.

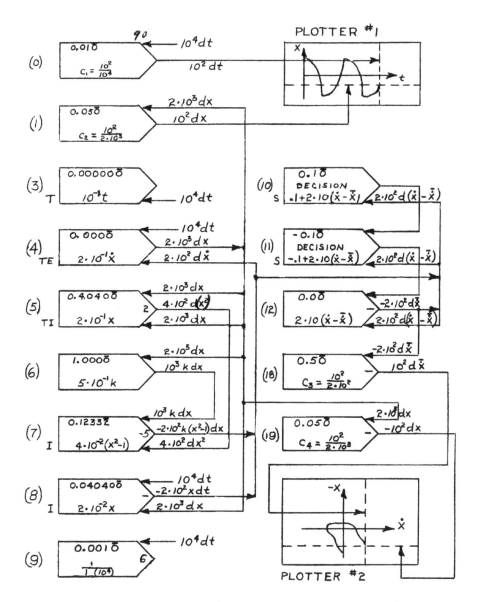

Fig. 10.16. DDA setup for solving $\ddot{x} + k(x^2+1)\dot{x} + x = 0$ (see [Bendix 1954, p. 65])

```
           1111111111222222222233333333334444444445
  12345678901234567890123456789012345678901234567890
  --------------------------------------------------
     00  0010        11111 090
     01  0050        11111 04
     03  0000000     11211          90
     04  00000       31211 90       07 08
     05  040400      21212 04       04
     06  10000       11111 04
     07  012332      211-5 06       05
     08  0040400     211-1 90       05
     09  00010       11116 90
     10  010         91111          07 08 12
     11  -000        91111 10       07 08 12
     12  000         111-1 11       07 08 12
     18  050         111-1 12
     19  0050        111-1 04
```

Fig. 10.17. D-12 example program (cf. [Bendix 1954, p. 69])

elements for the dY inputs can be selected by values in the columns starting at 28.[37]

The *UNIVAC incremental computer*, see [Remington 1956], a system developed for the *Operational Flight Trainer* described by [GRAY 1958], and *STAR-DAC*, see [MILAN-KAMSKI 1969] are similar sequential incremental DDAs, to name just a few.

10.4.3 CORSAIR

An interesting variant of such a sequential incremental DDA is the *CORSAIR* system,[38] developed at the *Royal Aircraft Establishment*, Farnborough, starting in the late 1950s. This system is remarkable in several respects: It is fully transistorised, uses core memory[39] and is programmed by a patch panel similar to those found on traditional analog computers. The name CORSAIR consists of three parts: "COR" representing the core memory, "S" for the surface barrier transistors used

---

[37] Each address occupies two columns and successive input addresses are separated by a space character.
[38] See [OWEN et al. 1960] for details on this system.
[39] See [ULMANN 2014, pp. 40 ff.] for more details on core memory and its development.

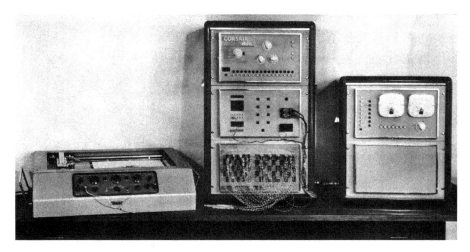

Fig. 10.18. Typical CORSAIR installation (see [OWEN et al. 1960, p. 33])

in construction, and "AIR" for its aircraft association. Figure 10.18 shows the overall system. On the left is a plotter, the DDA is in the middle of the picture with its power supply on the right.

The system features a 500 kHz bit rate resulting in 500 integrations steps per second.[40] The core memory consists of a 50×32 matrix of tiny ferrite cores, capable of storing fifty value pairs, each associated with one computing element. Each connection made on the central patch panel consists of two wires, one for each of the two incremental bit-serial signals.[41]

10.4.4 TRICE

A radically different and far more complex and costly implementation of a DDA was developed by *Packard Bell* in 1958. In contrast to the systems described above, which operate sequentially, this machine, named *TRICE*, was the first parallel DDA. This required a fully transistorised implementation allowing a high packing density of the circuitry while minimising energy consumption and thus simplifying the problem of cooling the densely packed electronics.[42]

Figure 10.19 shows a fully expanded TRICE system. From left to right it consists of four racks housing the various computing elements, a wide rack containing

---

[40] Two and a half times faster than the much larger Bendix D-12.
[41] [BEECHAM et al. 1965] describes a practical application of the CORSAIR system for the kinematic simulation and wind tunnel control in a wind tunnel based flight dynamics simulator.
[42] See [AMELING 1963/2], [RECHBERGER 1959], and [Packard Bell].

Fig. 10.19. Fully expanded TRICE system (cf. [AMELING 1963/2, p. 30])

the central patch panel and the operator console, a rack with a *PB-250* bit-serial stored-program digital computer, which controls the overall system, loads initial values into integrators, etc. Next to this is a rack containing a paper-tape reader/puncher and various ADCs and DACs allowing TRICE to be coupled with a traditional analog computer or to feed it with data from analog signal sources. The two racks on the right contain additional computing elements.

TRICE ran at a clock rate of 3 MHz, which was quite remarkable for such a large and early transistorised system. Its word length is 30 bits[43] while values are represented in binary and not in excess-three code decimal form as in the D-12, allowing a maximum iteration frequency of 10^5 s^{-1}. Due to the completely parallel operation of TRICE, repetitive operation is possible with a frequency of up to 100 Hz thus challenging contemporary analog computers. TRICE has been described as the *"most advanced DDA that was ever built."*[44]

TRICE contains summers, integrators with only one dX and dY input,[45] special summing-integrators, constant value multipliers, variable multipliers, servos, and a variety of ADCs and DACs. A typical TRICE module, a multiplier, is shown

---

[43] The internal registers of the basic computing elements are implemented as magnetostrictive delay lines, so word length and clock frequency are fixed.
[44] See [GILOI 1975, p. 23].
[45] If more inputs are required for an integrator, a summer in front of the integrator is necessary.

Fig. 10.20. TRICE multiplier (cf. [AMELING 1963/2, p. 30])

in figure 10.20. This multiplier consists of two dedicated integrators implementing the product rule technique described earlier.[46]

A typical application of TRICE is described in [AMELING 1963/2, pp. 40 f.]: A complex flight-dynamics problem is solved on a TRICE system coupled to a large analog computer. While TRICE performs those parts of the simulation requiring utmost precision such as trajectory calculations and the like, the analog computer implements operations requiring extreme high speed but won't introduce errors into operations such as coordinate transformations, etc. The TRICE system used in this particular simulation contained 42 integrators, 11 multipliers, 5 constant value multipliers, 5 servos, 6 ADCs and 13 DACs, while the analog computer featured 200 operational amplifiers, 14 multipliers, 5 function generators and 20 servos for coordinate transformation and rotation.

Due to the large amount of hardware necessary to implement a system like TRICE, only a few of these systems were eventually built. According to TOM GRAVES three TRICE installations were used at NASA's flight simulation labo-

---

[46] See [JACKSON 1960, pp. 585 f.].

ratory in Houston. While much simpler systems such as the Bendix D-12 were not successful in the long run due to their inherently slow speed of calulcation, TRICE, too, wasn't a market success due to its complexity and associated cost.

The legacy of DDAs could be found long after the demise of these marvelous machines in line drawing algorithms for digital computers and in CNC[47] machines, where DDA techniques were used to perform surface and line interpolation for controlling cutter paths.

---

[47] *Computerised Numerical Control*

11 Stochastic computing

In a series of lectures held at the California Institute of Technology (Caltech) from January 4th to January 15th 1952, JOHN VON NEUMANN wondered how one could build "*reliable organisms from unreliable components*". He advocated treating (unavoidable) errors in computer elements by thermodynamical methods instead of the classic techniques typically used in digital computers. [MANOHAR 2015, p. 119] describes this approach aptly as "*embracing, not competing the device physics*".

One idea to introduce more robustness in digital circuitry was the *multiple line trick*.[1] Here, instead of a single line transferring a bit, n lines in parallel are grouped, each transmitting the same value of 0 or 1. If at least Δn of these lines are set to 1 the overall value is assumed to be 1 by the receiver, 0 otherwise. Despite the obvious shortcomings of such an arrangement with respect to its complexity, it could easily tolerate the failure of multiple lines in a group without disabling the overall digital circuit it is part of.[2]

Later in this lecture series VON NEUMANN discussed different ways of representing and processing information in a fault tolerant way by taking biological neural networks as an antetype. He notes that[3]

> "*message pulse trains seem to convey meaning by certain analogic traits (within the pulse notation – i.e., this seems to be a mixed, part digital, part analog system), like the time density of the pulses in one line, correlations of the pulse time series between different lines in a bundle, etc.*"

Obviously, small errors such as transients or glitches in a system employing this type of value representation would only superficially impair the operation of the overall system.[4]

In 1967 BRIAN R. GAINES[5] elaborated on the idea of using a stochastic bit sequence for representing values, for which the term *stochastic computing* was coined.[6] The *unipolar stochastic number* is a simple example for such a value representation. Given a machine unit $m \in \mathbb{R}^+$ the two states of a bit can be interpreted as 0 and m respectively. Accordingly, a continuous variable $x \in [0, m]$

---

[1] Cf. [VON NEUMANN 1956, pp. 63 f.].
[2] This idea eventually led to modern majority voting systems used in fault tolerant digital systems.
[3] Cf. [VON NEUMANN 1956, p. 91].
[4] See also [VON NEUMANN 1956, pp. 95].
[5] 1938–
[6] See [GAINES 1967]. A comprehensive survey of stochastic computing can be found in [ALAGHI et al. 2013], while [GAUDET et al. 2019] and [GAINES 2019] give an overview of its historic roots.

https://doi.org/10.1515/9783110787740-011

can then be represented as the arithmetic mean of a sequence of a stochastic bitstream as shown in the following two examples:[7]

$$01010101\ldots \mathrel{\widehat{=}} \frac{m}{2}$$
$$001001001\ldots \mathrel{\widehat{=}} \frac{m}{3} \text{ etc.}$$

Ideally, the bit sequence would not show any regularity but instead be of (pseudo-)random[8] nature with the probability of a 1 in the bitstream being

$$P_1(x) = \frac{x}{m}.$$

This idea can be easily extended to *bipolar stochastic numbers* with defining

$$P_1(x) = \frac{1}{2}\left(1 + \frac{x}{m}\right).$$

An alternating sequence of 0 and 1 would then represent $x = 0$, all zeros would correspond to $x = -m$ and all ones to $x = m$.[9]

Based on such a number representation scheme, basic operations can then be implemented using logic gates to build an actual computer.[10] Maybe the most fascinating operations is multiplication, which can be implemented by a single two-input *AND gate*,[11] which combines the two stochastic bitstreams representing the numbers to be multiplied. [POPPELBAUM 1979, p. 1] notes

> "*To multiply two numbers with 7 bit (i.e., 1% accuracy a 2000 transistor microprocessor needs about 10 clock-periods. A stochastic computer gives the same accuracy (within one standard deviation) in 10^4 clock periods – but it only uses a 2 transistor AND for the operation.*"[12]

It is interesting to note that although the multiplication of two independent stochastic bitstreams is simple, squaring a value is a bit more complicated since

---

[7] See [WINSTEAD 2019, pp. 41 ff.] for more details on stochastic numbers in general. This work also contains a very readable introduction to stochastic computing.

[8] From an implementation perspective, stochastically independent pseudorandom bit sequences are very desirable as they can be generated without having to worry about the stability of its characteristics. [HOLT et al. 1995] describes an interesting approach to such pseudorandom generators suitable for stochastic computers.

[9] There are more ways to interpret a stochastic number such as *likelihood ratio stochastic numbers* and *log likelihood ratio stochastic numbers*, see [WINSTEAD 2019, pp. 42 f].

[10] See [POPPELBAUM 1979, p. 1].

[11] A *logic gate* implements a certain basic boolean operation such as *AND* (the output of such a gate is *true* if and only if all of its inputs are true), *OR* (its output is true if at least one of its inputs is true, so this is aptly called an *inclusive OR*), negation (the output is true of the input is *false* and vice versa), etc.

[12] Division is considerably more involved, see [POPPELBAUM 1979, p. 24], but still much simpler than in a classic digital computer.

an AND gate fed with the same bitstream applied to both of its inputs just yields
an identical bitstream at its output. To implement a squaring function one input
of the AND gate is fed with a delayed version of the bitstream representing the
value to be squared. This delay can be as short as one bit-time and implemented
as a simple flip-flop circuit.[13]

Addition and subtraction are a bit more involved. Two bitstreams are added
by yielding an output of 1 if a 1 is received at either input and storing a 1 when
both inputs happen to by 1 until a free *slot* is available in the output sequence
(i. e., until the next 0 in the output bitstream). Subtraction can be implemented
similarly by subtracting a subtrahend 1 from the next 1 in the minuend stochastic
bitstream.

Even an integrator is not that difficult to implement, as it just requires an
up/down counter and some means to generate a stochastic output bitstream with
$P_1(x)$ corresponding to its current counter value; this can be implemented using
a comparator and a noise source.[14]

Apart from the very simple implementation of basic operations, stochastic
computers have the characteristic trait that a rough approximation of some solution for a problem can be quickly obtained while computing times rapidly increase
when higher precision is required, a behaviour called *progressive precision*.[15] In
short, a stochastic computer features an extremely simple hardware implementation but at the price of low computational bandwidth and low precision.[16]

Figure 11.1 shows the *RASCEL*[17] computer, developed by JOHN WILLIAM
ESCH in the late 1960s.[18] This system represents values as fractions with numera-
tor and denominator consisting of independent stochastic bitstreams. It contains
a number of computing elements arranged in the tree-like structure visible on top
of the computer. Each such element can be configured to perform either squaring, addition, subtraction, multiplication, or division. RASCEL could yield results
spanning three to four orders of magnitude with an accuracy of 1% and a slewrate
of 100 ms.

---

13 See [WINSTEAD 2019, pp. 55 f.].
14 See [GAINES 1967, pp. 152 f.] for more details.
15 See [ALAGHI et al. 2013, p. 92:9].
16 See [POPPELBAUM 1968, p. 22]. [POPPELBAUM 1979] also describes technique called *burst processing*, which lies between a classic binary value representation and stochastic computing requiring only about 10% of the circuitry of a weighted binary value representation.
More detailed information on the theory and basics of stochastic computing can be found in
[GAINES 1967], [ESCH 1969], and [MASSEN 1977]. [RIEDEL et al. 2019, pp. 104 ff.] analyses what
can be computed by stochastic computers in general. Example circuits for stochastic computers
can be found in [LANGHELD 1979/1], [LANGHELD 1979/2] and [POPPELBAUM et al. 1967].
17 Short for *Regular Array of Stochastic Computing Element Logic*.
18 See [ESCH 1969].

Fig. 11.1. The RASCEL system in 1969 (JOHN WILLIAM ESCH on the left), see [ESCH 1969, p. 44]

The main advantage of stochastic computers is the simplicity of the individual computing elements, which are not only quite tolerant to jitter or skew in the various bitstreams bit are also robust with respect to noise. Unfortunately, this is more than outweighed by several disadvantages:[19] First of all the precision is very limited. 2^n different states of a variable could be represented in standard binary form by n bits. A stochastic computer would require a bitstream of 2^{2n} bits to represent a value with the same precision, which not only becomes quickly impractical but also very slow given the serial nature of such a bitstream.[20] Another disadvantage is the fact that it is not easy to generate a multitude of truly independent stochastic bitstreams, which is essential for a stochastic computer.[21]

This is reflected by the fact that the first, and at the same time last, symposium on stochastic computing took place in 1978 in Toulouse, France. Nevertheless, stochastic computing might experience a revival in the 21st century as it is suitable for implementing artificial neural networks.[22]

---

[19] See [MANOHAR 2015].
[20] See [RIEDEL 2019, p. 122].
[21] [HSIAO et al. 2019] describes various techniques for generating stochastic bitstreams, [JIA et al. 2019, pp. 166 ff.] describes an interesting approach to generating such bitstreams based on magnetic tunnel junctions. [RIEDEL 2019] details on deterministic approaches to stochastic computing.
[22] See [WINSTEAD 2019, pp. 69 ff.] and [ONIZAWA et al. 2019].

12 Simulation of analog computers

Clearly everyone simulating an analog computer on a stored program digital computer is in a state of sin[1] as the main advantages of an analog computer (its extremely high degree of parallelism, speed, and energy efficiency) are forsaken by simulating its computing elements in a sequential fashion on a digital computer.

Nevertheless, simulating an analog computer on a digital computer can be justified in certain cases – such as education, the scaling of differential equations, or the treatment of problems involving variables of such vastly different magnitudes that scaling the problem for an analog computer would at least be very cumbersome if not next to impossible. It is also an option when there no access to an analog computer at all or to one of appropriate size for a certain problem.

A typical example of such a problem is the classic simulation of bubble formation in the cooling loop of a nuclear reactor. The time constants involved differ by up to eight orders of magnitude while the radius of a bubble changes by five to six orders of magnitude. All of this in conjunction with variables to the fourth power makes the problem unsuitable for a classic analog computer.[2]

A problem like this could be solved quite efficiently on a parallel DDA such as TRICE but machines like this were rare while digital computers became increasingly common in the early 1960s. Consequently, the idea of simulating an analog computer was self-evident. [SELFRIDGE 1955] describes a system allowing "*a digital computer to operate as a differential analyzer*".[3] The digital computer used in this work was the *defense calculator*, later called the *IBM 701*,[4] one of the earliest commercially available digital computers. Although this system was incredibly slow from today's perspective it gave rise in the 1960s to the development of about 30 such simulation systems implemented in various programming languages on a variety of digital computers.[5]

One of the first practical systems was introduced in 1963 and became known as *MIDAS*.[6] Typical for the 1960s, MIDAS was operated in batch mode. A simulation run had to be prepared as a stack of punch cards, which were fed to the digital computer executing the simulation program and returning the results on fan-fold paper – not a very convenient or appealing mode of operation, particularly for people accustomed to the high degree of interactivity offered by analog computers.

---

1 JOHN VON NEUMANN used this wording when writing about generating (pseudo-)random numbers using digital computers.
2 See [FRISCH et al. 1969, pp. 19 f.].
3 A similar system is described by [STEIN et al. 1959].
4 See [BASHE et al. 1986, pp. 135 ff.].
5 See [BRENNAN et al. 1967, p. 243].
6 See [HARNETT et al. 1963].

In 1964 *PACTOLUS*[7] was introduced for the *IBM 1620*, a very small digital computer.[8] Since the 1620 was typically only used by a single user, it allowed for some degree of interactivity, which greatly helped the case of simulating analog computers. Based on experiences gained with this setup other systems allowing for increased degrees of interactivity were developed in due course. Typical examples are *DSL-90*, *MIMIC*, and most importantly *CSMP*, which is described in more detail in section 12.3.[9]

12.1 Basics

The overall structure of a typical simulation system is shown in figure 12.1. The model description is first analysed and then translated into a suitable intermediate representation such as an abstract syntax tree. This step is followed by a sorting run, which is necessary since the analog computing elements are simulated in a sequential fashion, abandoning the full parallelism of an analog computer. The order in which these individual steps are performed has a great impact on the result.[10]

One advantage of systems like this over a DDA is the possibility of using sophisticated numerical integration schemes, such as linear multistep methods, thus increasing the accuracy of the results and often having shorter run times.[11]

12.2 DDA programming system for the IBM 7074

An interesting system was developed by GEORGE JOSEPH FARRIS in the first half of the 1960s at the Institute for Chemical Engineering at Iowa State University of Science and Technology[12] for the *IBM 7070/7074* computer series.[13] The rationale behind this development was described as follows:

---

[7] See [BRENNAN et al. 1964].
[8] See [BASHE et al. 1986, pp. 508 ff.].
[9] [KORN 2005] describes such early developments from the late 1960s onwards.
[10] Cf. [GILOI 1975, pp. 35 f.]. Typical algorithms for this are described in [STEIN et al. 1970], a description of the implementation used for CSMP can be found in [SPECKHART et al. 1976, pp. 12 ff.].
[11] See [PRESS et al. 2001, pp. 123 ff.] for details on such methods in general.
[12] See [FARRIS 1964].
[13] See [BASHE et al. 1986, p. 578] and [BENDER et al.] for details on this early transistorised digital computer system.

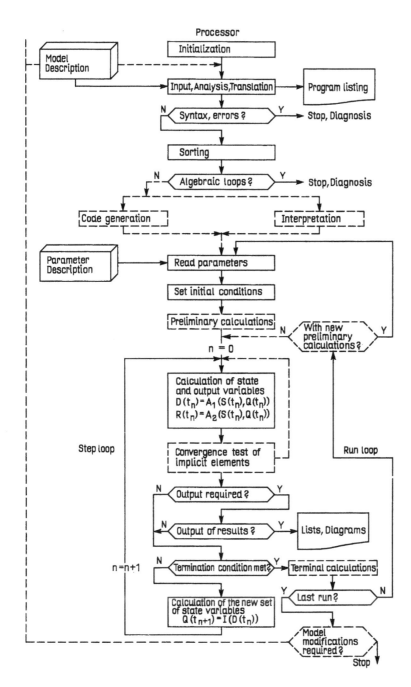

Fig. 12.1. Flow diagram of a typical digital analog computer simulation system (see [GILOI 1975, p. 99])

"In any problems arising in engineering, particularly in the simulation of processes and equipment, it may be faster, simpler, and more meaningful to use an analog method of solution that to reduce the problem to its most compact mathematical form and write a digital computer program for solving the resulting equations."[14]

This simulator, called *DIAN* as it combined some of the typical advantages of stored program *digital* and *analog* computers, eliminated the need to scale the problem equations by employing a floating point value representation. It, too, was run in batch mode with every component of the simulated analog computer with its associated inputs and outputs represented by a single punch card.

The central algorithm for performing the integration steps is interesting as it is not a traditional RUNGE-KUTTA method, which was also considered but turned out to require more computing time while not yielding substantially better results. The integration in DIAN is bootstrapped with a simple EULER iteration step

$$y_1 = y_0 + y_0' \Delta x,$$

which is followed by a single trapezoidal step

$$y_2 = y_1 + \left(\frac{3y_1' - y_0'}{2}\right) \Delta x.$$

From then on explicit GREGORY-NEWTON steps

$$y_{n+1} = y_n + \left(\frac{23 y_n'}{12} - \frac{4}{3 y_{n-1}'} + \frac{5}{12 y_{n-2}'}\right) \Delta x$$

are computed.

The DIAN system is demonstrated in [FARRIS 1964] with an initial value problem and a boundary problem, both simulating a stirred tank reactor described by

$$\dot{C} = a_1 - b_1 C - k_1 C e^{-\frac{E}{RT}} \text{ and} \quad (12.1)$$

$$\dot{T} = a_2 - b_2 T + k_2 C e^{-\frac{E}{RT}}. \quad (12.2)$$

These equations for concentration and temperature would be implemented on an analog computer by the program shown in figure 12.2. This setup can be described by a set of just 21 punch cards as input for DIAN.[15] The simulation was run for integration limits of 0 and 3000 seconds at 10 second intervals with intermediate results for temperature and concentration being printed every 300 seconds of simulation time. This required a total of 30 second of actual run time on the IBM 7074 system.

---

14 See [FARRIS 1964, p. 1].
15 See [FARRIS 1964, pp. 71 f.].

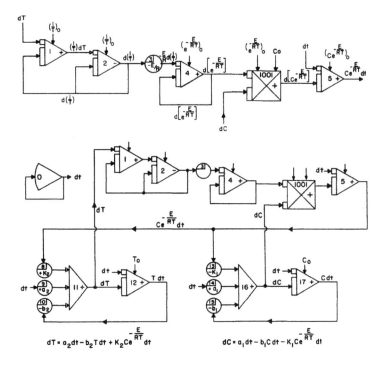

Fig. 12.2. Simulated analog computer setup for equations (12.1) and (12.2) (see [FARRIS 1964, pp. 23 f.])

12.3 CSMP

CSMP, the *Continuous System Modelling Program* is one of the best known and most influential early analog computer simulators.[16] A CSMP program consists of three sections of code:

INITIAL: This (optional) section contains everything necessary prior to an actual simulation run. This includes initial conditions for integrators as well as static scaling options, etc.
DYNAMIC: This section contains the description of the actual simulation circuit, which is then iteratively solved by the digital computer.
TERMINAL: This section is also optional and allows to specify certain actions such as output of values, plotting of functions, etc., after a simulation run has been completed.

---

[16] See [SPECKHART et al. 1976] for a thorough description and introduction to CSMP.

The following simple example problem shows the actual process of using the CSMP system. A mass-spring-damper system described by

$$m\ddot{y} + d\dot{y} + sy = 0$$

is to be simulated. First, the program variables are defined:

$$\begin{aligned} \text{MASS} &= m \\ \text{POS} &= y \\ \text{POS0} &= y_0 \\ \text{D} &= d \\ \text{S} &= s \\ \text{VEL} &= \dot{\text{POS}} \\ \text{ACC} &= \dot{\text{VEL}} \end{aligned}$$

Using the initial conditions $y(0) = 1$, $\dot{y}(0) = 0$ and $m = 1.5$, $d = 4$, and $s = 150$ yields the CSMP program shown in figure 12.3, consisting only of an (implicit) DYNAMIC section.[17] Output is controlled by the TIMER statement in line 8.[18] A typical simulation output listing the values for TIME, POS, VEL, and ACC is shown in figure 12.4. At its heart are the two integrations

$$\text{POS} = \int_0^t \text{VEL}\, dt + \text{POS0 and}$$

$$\text{VEL} = \int_0^t \text{ACC}\, dt$$

A more complex example, the simulation of a motorized cable reel, is described in [BRENNAN et al. 1967]. The system to be simulated is shown in figure 12.5. It consists of a motor-driven shaft connected to a reel. The velocity of the cable taken up on the reel or rolling off is measured by a tachometer feeding a motor control system, which is also connected to a manual unwind/reverse control switch and the objective is to control the speed of the cable. Figure 12.6 shows a classic analog computer setup for this system.

The corresponding CSMP program is shown in figure 12.7. As a simple list of numerical values would not give the required insight into complex systems like this, CSMP could also print graphs, as shown in figure 12.8.

---

17 Cf. [SPECKHART et al. 1976, p. 15].

18 PRDEL controls the time interval between two print operations. If omitted, its default is $\frac{\text{FINTIM}}{100}$, see [SPECKHART et al. 1976, pp. 18/24].

```
            CONSTANT MASS = 1.5, D = 4.0
            CONSTANT S = 150.0, POS0 = 1.0
            ACC = (-S * POS - D * VEL) / MASS
            POS = INTGRL(POS0, VEL)
            VEL = INTGRL(0.0, ACC)
            PRINT POS, VEL, ACC
   TITLE    MASS-SPRING-DAMPER SIMULATION
            TIMER FINTIM = 2.0, PRDEL = 0.05
   END
   STOP
   ENDJOB
```

```
                 SIMULATION OF A MASS-SPRING-DAMPER SYSTEM

   TIME            POS          VEL            ACC
   0.0             1.0000E 00   0.0           -1.0000E 02
   5.0000E-02      8.8283E-01  -4.4884E 00   -7.6313E 01
   1.0000E-01      5.7793E-01  -7.3878E 00   -3.8093E 01
   1.7863E-01     -8.2319E 00   4.0884E 00
   2.0000E-01     -2.1177E-01  -7.0839E 00    4.0068E 01
   2.5000E-01     -5.0491E-01  -4.4556E 00    6.2373E 01
   3.0000E-01     -6.4574E-01  -1.1340E 00    6.7598E 01
   3.5000E-01     -6.2098E-01   2.0328E 00    5.6677E 01
   4.0000E-01 ...              ...           ...
```

Fig. 12.3. Simulating a mass-spring-damper system with CSMP

Fig. 12.4. Typical CSMP simulation result (see [SPECKHART et al. 1976, p. 17])

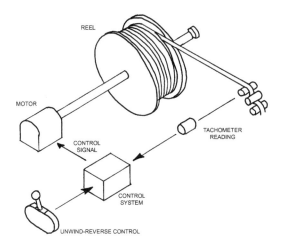

Fig. 12.5. Motorized cable reel system

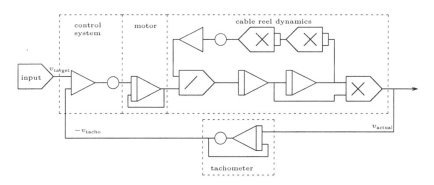

Fig. 12.6. Analog computer setup for the cable reel problem (see [BRENNAN et al. 1967, p. 251])

```
TITLE  CABLE REEL SIMULATION
*
INIT
       K1  = (D ** 2) / (2.0 * PI * W)
PARAM     D = 0.1, W = 2.0
CONST     PI = 3.14159
PARAM  RFULL = 4.0, REMPTY = 2.0
*
DYNAM
       I  = 18.5 * (R ** 4) - 221.0
       TH2DOT = TORQUE / I
       TH1DOT = INTGRL(0.0, TH2DOT)
       R  = INTGRL(RFULL, (-K1 * TH1DOT))
       ERROR = VDESIR - VM
*
PARAM VDESIR = 50.0
       CONTL = GAIN * ERROR
       GAIN = 0.5
       TORQUE = 500.0 * DUMMY
       DUMMY = REALPL(0.0, 1.0, CONTL)
       VACT = R * TH1DOT
       VM = REALPL(0.0, 0.5, VACT)
*
FINISH    R = 2.0
TIMER     DELT = .05, FINTIM = 20.0
          PRDEL = 0.5, OUTDEL = 0.5
*
PRINT  VACT, VM, ERROR, CONTL, TORQUE, R, I
PRTPLT VACT
*
LABEL  PRELIM. TEST OF SYSTEM STAB. (G = 0.5)
METHOD RECT
END
*
PARAM  GAIN = 1.5
RESET  LABEL
LABEL  PRELIM. TEST OF SYSTEM STAB. (G = 1.5)
END
STOP
ENDJOB
```

Fig. 12.7. Cable reel simulation program (see [BRENNAN et al. 1967, p. 260])

```
TIME  VACT
 0.0  0.0                 +
 0.5  3.2774E 00       --+
 1.0  1.1179E 01   ---------+
 1.5  2.2859E 01   ------------------+
 2.0  3.4448E 01   ------------------------+
 2.5  4.4997E 01   --------------------------------+
 3.0  5.3483E 01   ---------------------------------------+
 3.5  5.9391E 01   -------------------------------------------+
 4.0  6.2637E 01   ---------------------------------------------+
 4.5  6.3481E 01   ---------------------------------------------+
 5.0  6.2411E 01   ---------------------------------------------+
 5.5  6.0028E 01   --------------------------------------------+
 6.0  5.6951E 01   -----------------------------------------+
 6.5  5.3734E 01   ---------------------------------------+
 7.0  5.0814E 01   ------------------------------------+
 7.5  4.8486E 01   ----------------------------------+
 8.0  4.6902E 01   ---------------------------------+
 8.5  4.6081E 01   --------------------------------+
 9.0  4.5924E 01   --------------------------------+
 9.5  4.6339E 01   --------------------------------+
10.0  4.7090E 01   ---------------------------------+
10.5  4.8013E 01   ----------------------------------+
11.0  4.8948E 01   -----------------------------------+
11.5  4.9970E 01   ------------------------------------+
12.0  5.0400E 01   ------------------------------------+
12.5  5.0801E 01   -------------------------------------+
13.0  5.0976E 01   -------------------------------------+
13.5  5.0954E 01   -------------------------------------+
14.0  5.0786E 01   -------------------------------------+
14.5  5.0529E 01   -------------------------------------+
15.0  5.0238E 01   ------------------------------------+
15.5  4.9960E 01   ------------------------------------+
16.0  4.9729E 01   ------------------------------------+
16.5  4.9566E 01   ------------------------------------+
17.0  4.9477E 01   ------------------------------------+
17.5  4.9457E 01   ------------------------------------+
18.0  4.9492E 01   ------------------------------------+
18.5  4.9565E 01   ------------------------------------+
19.0  4.9657E 01   ------------------------------------+
19.5  4.9751E 01   ------------------------------------+
20.0  4.9834E 01   ------------------------------------+
```

Fig. 12.8. Cable reel simulation result (see [SPECKHART et al. 1976, p. 17])

12.4 Modern approaches

In the early 1980s simulation systems were developed for the rapidly growing personal computer market. [TITCHENER et al. 1983] describes an early example, which was aimed at applications in computational chemistry. In 1985 BEUKEBOOM et al. described another personal computer based simulation system.[19] This influenced the development of more modern tools such as *MATLAB* featuring (among many others) a plugin called *Simulink*. This contains a graphical block diagramming tool, which allows systems to be described quickly by using what are essentially classic analog computer programming symbols and techniques. Figure 12.9 shows a mass-spring-damper simulation in MATLAB using Simulink.

A more recent development is the open source system PyAnalog[20] It can be used as an analog computer simulator as well as for generating configuration setup

---

19 See [BEUKEBOOM et al. 1985].
20 See https://github.com/anabrid/pyanalog (retrieved 07/03/2022).

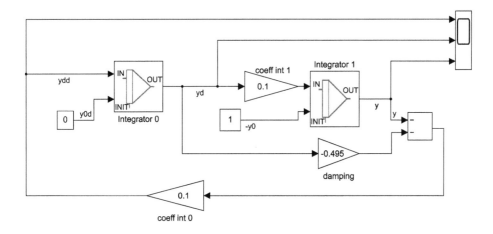

Fig. 12.9. Simulation of a mass-spring-damper system in Matlab

data for reconfigurable analog computers. It can also be coupled with an actual analog computer in a hybrid computer setup.

13 Applications

The following sections give an overview of typical analog computer applications. Due to the constrained space and the multitude and complexity of areas in which analog computers were used, these descriptions will be quite cursory. As ERNST KETTEL puts it, there is next to no scientific field where an analog computer could not be applied successfully.[1]

13.1 Mathematics

Although analog computers do not excel when it comes to accuracy,[2] there are many areas in mathematics where analog computers can be applied with great benefit.

13.1.1 Differential equations

Differential equations are obviously ideally suited for treatment on an analog computer. The simplest case of *linear ordinary differential equations*[3] poses no special problems and can be handled directly with the KELVIN feedback technique or the substitution method described in sections 7.2 and 7.3. The equations of interest are of the general form

$$\sum_{i=0}^{n} a_i \frac{\mathrm{d}^i y}{\mathrm{d} y^i} = f(t).$$

If the coefficients a_i are constant, the treatment of such equations and systems thereof with an analog computer is straightforward. If the a_i are variable, the additional problem of generating these coefficients during a simulation run arises. In most cases the a_i will be generated as solutions of auxiliary differential equations. Nevertheless, the scaling process is often difficult in the case of variable coefficients.

Non-linear differential equations are also challenging with respect to the scaling process. The fact that the structure of the computer setup often depends on the selected scaling factors complicates their solution as rescaling often not only requires the change of coefficients but also a change of the computer setup. A wealth of information about these challenges can be found in [GILOI et al. 1963,

---

[1] Cf. [KETTEL 1960, p. 165].
[2] There are ways to improve accuracy of solutions obtained by analog computers by using a hybrid computer approach.
[3] *LODE* for short.

pp. 221 ff.] and [MACKAY 1962, pp. 212 ff.]. The application of hybrid computers in this area is covered in detail in [VALISALO et al.].

Boundary value problems are problems in which boundary values for a differential equation at $t = 0$ and $t = t_{\max}$ are given and these must be satisfied by a solution. These can often be solved easily with an analog computer[4] running in repetitive mode by applying a trial-and-error procedure.[5] Starting with the boundary value at $t = 0$ the equations are solved repeatedly with a human operator or the digital computer part of a hybrid computer system varying the parameters until the second boundary value condition is satisfied by a certain set of parameters.[6]

The treatment of partial differential equations has already been described cursorily in section 7.4. More information can be found in [HEINHOLD et al., pp. 183–197], [MACKAY 1962, pp. 243 ff., pp. 293 ff.], and [HOWE 1962]. An interesting method for generating a flicker-free display of the solutions of partial differential equations on an oscilloscope is described in [AMELING 1962/1], while [FEILMEIER 1974, pp. 215 ff.] deals with the application of hybrid computers for solving this class of equations. [BEKEY et al. 1971] describes a hybrid computer techniques for solving nonlinear parabolic partial differential equations while [HSU et al.1968] show a kind of time-sharing technique for reusing analog computer components within a single large-scale problem.

Another interesting technique, which was developed by LAWRENCE WAINWRIGHT, uses a magnetic tape for storing analog data and is described in [JACKSON 1960, pp. 586 f.]. [ASHLEY et al. 1968] show the application of the classic seperation of variables approach for solving partial differential equations by means of a hybrid computer setup.

13.1.2 Integral equations

In general, an analog computer is not particularly well suited for treating integral equations. However, due to their importance in fields like electrical and commu-

---

[4] See [KORN et al. 1956, pp. 142 ff.] for example.
[5] See [GILOI et al. 1963, pp. 249 ff.] and [HEINHOLD et al., pp. 108 ff.].
[6] See [HEINHOLD et al., pp. 111 ff.] and [FEILMEIER 1974, pp. 202 ff.]. More detailed information about iterative approaches can be found in [HEINHOLD et al., pp. 117 ff.], [FEILMEIER 1974, pp. 197 ff.], and [GILOI et al. 1963, pp. 253 ff.]. Special approximation techniques for solving boundary value problems for partial differential equations are described in [VICHNEVETSKY 1969].

nications engineering, etc., many approaches were developed to solve this class of equations with an analog computer. Integral equations are of the form[7]

$$f(x) + \int_0^a K(t,x) u(t)\,\mathrm{d}t = 0 \quad \text{or} \tag{13.1}$$

$$f(x) + \int_0^a K(t,x) u(t)\,\mathrm{d}t = u(x). \tag{13.2}$$

These equations are called FREDHOLM[8] equations of type 1 or type 2, respectively. $u(x)$ is the unknown, while $f(x)$ and $K(t,x)$, the *kernel*, are known. The representation of this kernel is the main challenge on an analog computer since functions of more than one variable are typically difficult to implement. A hybrid computer can simplify this considerably. In cases where no hybrid computer is available, x has often been restricted to a constant value so that a simple function generator using only one argument variable can be used to implement $K(t, \bar{x})$ with $\bar{x}$ denoting that x is treated as a constant.

Using proper time-scaling, the upper boundary of the integral in equation (13.1) or (13.2) can be identified with computer time τ. The result of a computer run is then compared with either 0 or $u(x)$, depending on the type of equation to be solved. Based on the deviation from the desired result, the parameters defining $u(x)$ can be varied manually or under program control. So basically solving integral equations with an analog or hybrid computer boils down to a trial-and-error technique.

Restricting x to a constant value is often not sufficient – in these cases x is discretised appropriately, using function generators to implement $K(t, x_i)$ for a number of such constant values x_i. Running the analog computer in repetitive operation and using electronic switches to switch between these function generators depending on the actual index i, it is possible to display a family of curves on an oscilloscope.[9] More information about solving integral equations with analog and hybrid computers can be found in [SYDOW 1964, pp. 112 ff.], [MACKAY 1962, pp. 317 ff., pp. 331 ff., pp. 352 ff.], and [CHAN 1969].

It should be noted that the idea of using analog computers to treat integral equations is quite old. NORBERT WIENER[10] suggested a *product integraph*,[11] which

---

[7] A classic comprehensive introduction to integral equations in general can be found in [MUSKHELISHVILI 1953].
[8] IVAR FREDHOLM, 04/07/1866–08/17/1927
[9] This requires at least three independently controlled integrator groups, which in turn required a complex control circuit.
[10] 11/26/1894–03/18/1964
[11] See [BENNETT 1993, pp. 104 ff.].

was later developed and built by Vannevar Bush,[12] F. G. Kear, H. L. Hazen, H. R. Stewart, and F. D. Gage at the Massachusetts Institute of Technology. This mechanical device could evaluate the integral over the product of two time-varying functions:[13]

$$\Phi(x) = \int_a^x f(t)\Phi(t)\,dt.$$

This required two operators to manually follow given curves for $f(t)$ and $\Phi(t)$ (similarly to the setup shown in figure 2.21 on page 35), a tedious and not very precise process, which typically took about four minutes for a run performed at "normal speed".[14] The result was produced by a watt-hour meter performing the required multiplication based on one of the functions represented a time-varying voltage and one by a current.

The problem of manual curve followers was solved with the subsequent development of the *cinema integraph*. It could evaluate equations such as

$$\int_0^\infty f(t)\sin(nt)\,dt \text{ or } \int_0^\infty f(t)\cos(nt)\,dt$$

as well as the more general form

$$\int_0^T f(t)g(t)\,dt.$$

The two functions under the integral were represented by the optical density of photographic films. These were arranged in such a way that light from a suitable and constant light source had to pass both films after which it was measured by a photocell. This rendered the manual curve following process obsolete and allowed much shorter solution times.[15]

[Bekey et al. 1967] describe a hybrid computer approach for solving integration equations using the von Neumann and Fisher iteration method.

---

[12] See [Bush et al. 1927].
[13] See [Bush et al. 1927, p. 82 ff.].
[14] See [Bush et al. 1927, p. 75].
[15] The cinema integraph is described in much more detail in [Hedeman 1941] and [Hazen et al. 1940]. Further information on integraphs can be found in [Stine 2014] and [Macnee 1953].

13.1.3 Roots of polynomials

Determining the real or complex roots of polynomials

$$P_n(x) = \sum_{i=0}^{n} a_i x^i$$

with an analog computer is a straightforward process:[16] $P_n(x)$ is generated either by explicit multiplication generating the x^i, or by repeated integration yielding x, $-\frac{1}{2}x^2$, $\frac{1}{6}x^3$, etc. Since analog computers typically have more integrators than multipliers available and since multipliers are expensive and multiplication is not as precise an operation as integration, the latter approach is usually preferred.[17] This proved especially useful in the case of LEGENDRE-, CHEBYSHEV-, or HERMITE-polynomials.[18]

Given a computer setup generating the required polynomial, the analog computer can then vary x in the interval of interest. Using a comparator, the zeros of the polynomial can be detected and the machine can be automatically placed into halt mode to read out the corresponding value of x.

13.1.4 Orthogonal functions

An important problem in control engineering applications and other fields is the approximation of a function $f(t)$ as a sum of orthogonal functions $\varphi_i(t)$:[19]

$$f(t) \sim \sum_{i=1}^{n} a_i \varphi_i(t)$$

Normally, the square error term

$$\int_I \left(f(t) - \sum_{i=1}^{n} a_i \varphi_i(t) \right)^2 w(t) dt$$

is to be minimised over an interval I with $w(t)$ denoting a weighting function. This is satisfied if the $\varphi_i(t)$ form an orthogonal system and the coefficients a_i are defined by

$$a_i = \frac{1}{c} \int_I f(t) \varphi_i(t) dt \qquad (13.3)$$

---

16 See [HEINHOLD et al., pp. 173 ff.] and [ATKINSON 1955].
17 If servo or time division multipliers are available, the powers of x can be generated in a straightforward way.
18 See [HEINHOLD et al., pp. 174].
19 See [HERSCHEL 1962].

where

$$\int_I \varphi_i(t)\varphi_k(t)\mathrm{d}t = \begin{cases} c \neq 0 & \text{if } i = k \\ 0 & \text{if } i \neq k \end{cases}$$

holds. Using an analog computer, the coefficients a_i can be determined based on (13.3).

13.1.5 Linear algebra

Solving systems of linear equations is a task not inherently well suited for analog computers.[20] Nevertheless, several techniques suitable for analog computers were developed, which were typically employed when no digital computer was available. For a long time these approaches were of historic interest only but with the advent of artificial intelligence, especially *artificial neural networks*, these analog techniques might become important again as they offer a chance to drastically lower the required energy for training such networks using approaches like *backpropagation*.

A direct approach implementing the system of linear equations using only summers and coefficient potentiometers is unsuited for an analog computer since it results in *algebraic loops*.[21] Other techniques like the JACOBI[22] method, the GAUSS[23]-SEIDEL[24] method or relaxation approaches are better suited for an analog computer.[25]

As an example a system of linear equations $\mathbf{A}\vec{x} = \vec{b}$ is to be solved by the JACOBI method. First $\vec{x} = \mathbf{B}\vec{x} + \vec{k}$ is computed, with

$$\mathbf{B} = -\begin{pmatrix} 0 & \frac{a_{12}}{a_{11}} & \cdots & \frac{a_{1n}}{a_{11}} \\ \frac{a_{21}}{a_{22}} & 0 & \cdots & \frac{a_{2n}}{a_{22}} \\ \vdots & \vdots & \ddots & \vdots \\ \frac{a_{n1}}{a_{nn}} & \frac{a_{n2}}{a_{nn}} & \cdots & 0 \end{pmatrix} \quad \text{and} \quad \vec{k} = \begin{pmatrix} \frac{b_1}{a_{11}} \\ \vdots \\ \frac{b_n}{a_{nn}} \end{pmatrix}.$$

---

[20] A stored-program digital computer is often a better approach.
[21] An algebraic loop is a positive feedback path involving only summers, i.e., a loop involving an odd number of sign-inverting summers. Such a circuit is inherently unstable and must be avoided at any cost.
[22] CARL GUSTAV JACOB JACOBI, 12/10/1803–02/18/1851
[23] CARL FRIEDRICH GAUSS, 04/30/1777–02/23/1855
[24] PHILIPP LUDWIG VON SEIDEL, 10/24/1821–08/13/1896
[25] More detailed information about these methods can be found in [HEINHOLD et al., pp. 135–152], [VOCOLIDES 1960], [GILOI et al. 1963, pp. 152 ff.], [MARQUITZ et al. 1968], [KOVACH et al. 1962], [JACKSON 1960, pp. 332 ff.], and [MACKAY 1962, pp. 192 ff.]. The effects of parameter variations are analysed in [KAHNE 1968].

Typically, **B** and $\vec{k}$ are determined once, aided by a digital computer. Iteration steps of the form

$$\vec{x}_{n+1} = \mathbf{B}\vec{x}_n + \vec{k}$$

can then be performed on the analog computer.[26] This iterative approach requires pairs of integrators for each component of $\vec{x}$ being used as storage elements, requiring a complex control scheme.[27]

[MITRA 1955] describes a technique for solving systems of linear equations on a specialised analog computer, which not only yields the desired solution but also the eigenvalue of the matrix with the largest absolute value.

A different approach transforms the problem of solving a system of linear equations to the problem of solving a corresponding system of differential equations by

$$\mathbf{A}\vec{x} - \vec{b} = \vec{\varepsilon} \text{ and}$$
$$\dot{\vec{x}} = -\vec{\varepsilon},$$

which converge to the desired solution for $t \to \infty$. This technique is treated in detail in [HEINHOLD et al., pp. 148 ff.] and [ULMANN et al. 2019], which not only examines convergence criteria but also contains practical examples.

13.1.6 Eigenvalues and -vectors

Determining eigenvalues and eigenvectors for a given matrix of degree n, $\mathbf{A} = (a_{ij}), i, j = 1, \ldots, n$, is a common task and can be reduced to solving a system of linear equations of the form

$$(\mathbf{A} - \lambda_i \mathbf{I})\vec{x}_i = 0. \tag{13.4}$$

In the case of complex valued eigenvalues, a system of equations of degree $2n$ is required. In either case, the eigenvalues λ_i correspond to the roots of the characteristic polynomial $\det[\mathbf{A} - \lambda \mathbf{I}]$ with $\mathbf{I}$ denoting the identity matrix.[28]

Although this problem can be readily solved by applying the methods described above, the VON MISES iteration is often employed, as described in [POPOVIĆ 1964]. This method works directly on the components of $\mathbf{A}$ and does not require an explicit characteristic polynomial to be derived.

---

26 The classic spectral radius convergence criterion must be satisfied for this to work.
27 Using the same idea, but starting with an identity matrix, matrix inversions can also be performed by an analog computer.
28 Cf. [PRESS et al. 2001, pp. 368 ff.].

13.1.7 Fourier synthesis and analysis

While transformations from frequency domain into time domain and vice versa are performed nowadays on stored-program digital computers employing *Fast Fourier Transforms*[29] or *Wavelet Transforms* neither of these methods, nor adequately fast stored-program computers, were available until the 1970s. Therefore the decomposition of a signal into its components, specified by frequency, phase, and amplitude, was done regularly with analog computers[30] by a Fourier[31] transform. Basically, every periodic function $f(x)$ with a period of l can be represented by

$$f(x) = \frac{a_0}{2} + \sum_{k=1}^{\infty}\left(a_k \cos\left(\frac{2\pi k x}{l}\right) + b_k \sin\left(\frac{2\pi k x}{l}\right)\right) \qquad (13.5)$$

under certain circumstances where the coefficients a_k and b_k are defined by

$$a_k = \frac{2}{l}\int_0^l f(x) \cos\left(\frac{2k\pi x}{l}\right) dx \quad \text{and} \qquad (13.6)$$

$$b_k = \frac{2}{l}\int_0^l f(x) \sin\left(\frac{2k\pi x}{l}\right) dx. \qquad (13.7)$$

Given a_k and b_k the reconstruction of a signal, the Fourier *synthesis*, is what harmonic synthesizers were built for.[32] On an electronic analog computer, the sine-/cosine-terms required can be implemented by an analog computer setup as shown in figure 8.2 (page 165).

The inverse process, determining a_k and b_k for a given function $f(x)$, the Fourier *analysis*, is more involved. The direct approach, evaluating the integrals (13.6) and (13.7), requires a very exact time base as this determines the bounds of the integration operations.[33]

Another approach uses low-pass filters, each consisting of a single integrator with negative feedback in the simplest case, to determine the coefficients for a given $f(t)$.[34] This has the disadvantage that only one pair of a_k and b_k can be determined at a time. In addition to that, the low-pass filters typically have rather long time constants.

---

29 *FFT* for short.
30 See [Dick et al. 1967], [Giloi et al. 1963, pp. 306 ff.], [Kovach 1952], and [Ratz 1967].
31 Fourier, Jean-Baptiste-Joseph, 03/21/1768–05/16/1830
32 See section 2.6.
33 Cf. [Giloi et al. 1963, pp. 307 f.].
34 See [Giloi et al. 1963, pp. 308 ff.].

Another approach is based on a circuit like that shown in figure 8.2 (page 166), which is subjected to a forcing function. The resonant frequency of this circuit allows it to determine a pair a_k and b_k directly.[35]

13.1.8 Random processes and Monte-Carlo simulations

The study of random processes is an important area of application for analog computers since many commercially interesting processes, ranging from chemical technology to the behaviour of flight vehicles, nuclear reactors, the creation of steam bubbles in coolant loops, etc., are affected by external random events.[36]

Consequently, a suitable source of randomness (noise) is required. Typical random generators are either based on physical noise sources, see section 4.12, or on shift-registers with feedback generating pseudorandom bit sequences, which can be converted to analog voltages by DACs. These can be readily implemented using digital elements most medium to large analog computers already offer. The main advantage of a pseudorandom source is the repeatability of simulation runs since it is possible to generate the same pseudorandom sequence over and over again. Its disadvantage is that controlling and guaranteeing the required degree of randomness can be challenging, depending on the sensitivity of the system under consideration.[37]

Using such noise sources *Monte-Carlo methods* can be employed to a wide variety of problems on an analog or hybrid computer.[38] [BEKEY et al. 1966] describes a hybrid computer technique for parameter optimisation based on a random search. A fascinating approach for solving partial differential equations by Monte-Carlo methods on a hybrid computer is described in [LITTLE 1966].

The investigation of random processes often requires the extension of the classic discrete arithmetic mean to a time varying continuous variable. An interesting approach to this problem is the *exponentially mapped past* developed by JOSEPH OTTERMAN and ROBERT MARIO FANO.[39] Details of this technique can be found

---

[35] See [GILOI et al. 1963, pp. 309 ff.].
[36] Cf. primarily [LANING et al. 1956], and [KORN 1966] but also [GILOI et al. 1963, pp. 357 ff.] and [JACKSON 1960, p. 367].
[37] Detailed information on these topics can be found in [GILOI et al. 1963, pp. 358 ff.], [MEYER-BRÖTZ 1962], [OTT 1964] (computation of correlation coefficients), [SYDOW 1964, pp. 250 ff.] (quality control), [BOHLING et al. 1970], [RIDEOUT 1962, pp. 140 ff.], [KORN et al. 1956], and [FEILMEIER 1974, pp. 275 ff.]. [DERN et al. 1957] describes an early distribution analyser.
[38] See [RIDEOUT 1962, pp. 238 ff.] and [FEILMEIER 1974].
[39] 11/11/1917–07/13/2016

in [OTTERMAN 1960], [FANO 1950], with practical implementations described in [EAI 1.3.2 1964], and [ULMANN 2021/3].

13.1.9 Optimisation and operational research

Optimisation problems are of great commercial importance and were often solved by analog computers. In general, a linear optimisation problem can be represented by a real-valued matrix $\mathbf{A} \in \mathbb{R}^{n,m}$ with $m < n$ and two vectors $\vec{b} \in \mathbb{R}^m$ and $\vec{x} \in \mathbb{R}^n$, component wise satisfying the condition

$$A\vec{x} \leq \vec{b}. \tag{13.8}$$

The goal of the optimisation process is to determine a vector $\vec{x}$ that not only satisfies (13.8) but also maximizes the scalar product $\vec{c}^{\mathrm{T}}\vec{x}$ with $\vec{c} \in \mathbb{R}^n$ representing the *value coefficients*.[40]

Very early approaches used direct analogies such as conductive sheets or contour map function generators[41] that were sampled by a stylus driven with an xy-plotter.[42] Another simple analog computer for optimisation tasks is shown in 13.1. Its basic circuit, shown in figure 13.2, is extremely simple: Each parameter of the optimisation problem is represented by two potentiometers of which the first one represents the parameter itself and will be varied manually during the optimisation process, while the second potentiometer is used to set a fixed weighting factor describing the underlying process. The currents flowing through these potentiometer groups are then summed and displayed on a meter.

[DENNIS 1958] describes techniques and specialised electric networks to

> "*minimise a function of several variables while satisfying some inequality constraints on other functions of the variables.*"

Other approaches to optimisation with analog computers are based on Monte-Carlo techniques as mentioned above, or use gradient methods like *continuous steepest ascent/descent*.[43] Gradient methods are especially well suited for hybrid computers where the digital processor controls the ascent/descent-process performed by the attached analog computer.[44]

---

[40] See [PIERRE 1986, pp. 193 ff.] for general information on optimisation problems. Basic techniques for tackling this class of problems with analog computers are described in [KORN et al. 1956, pp. 147 ff.], [KOVACH et al. 1962], and [DEZIEL 1966]. Recent work on analog computers for optimisation are described in [VICHIK 2015].
[41] See section 4.5.7.
[42] Cf. [PIERRE 1986, pp. 251 ff.].
[43] See [PIERRE 1986, pp. 296 ff.], [LEVINE 1964, pp. 217 ff.], and [ALBRECHT et al.].
[44] Cf. [FEILMEIER 1974, pp. 243 ff.].

 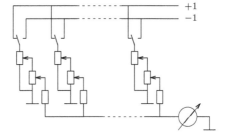

Fig. 13.1. BPRR-2 computer used for the optimisation of industrial processes (see [USHAKOV 1958/2, p. 1961])

Fig. 13.2. Basic circuit of the BPRR-2 (see [USHAKOV 1958/2, p. 1961])

A large area of application was the optimisation of pipeline networks.[45] Figure 13.3 shows an impressive analog computer installation for the simulation and optimisation of such networks.

13.1.10 Display of complex shapes

An intriguing application of analog computers is the display of multidimensional shapes on an oscilloscope. A simple example for such a display has been described in section 8.10. Figure 13.4 shows a more complex example, the projection of a four-dimensional hyper-cube generated by a high-speed analog computer.[46] In addition to this, even stereographic projections were implemented with analog computers.[47]

13.2 Physics

Analog computers are ideally suited for applications in physics due to their underlying principle of operation. The following sections describe some typical uses.

---

[45] See [Montan-Forschung], [LIENHARD 1969], and [NIX 1965].
[46] See [Telefunken/3], [LUKES 1967, pp. 129 ff.], and [MACKAY 1962] for general information on this topic.
[47] [MACKAY 1962, pp. 134 ff.].

254 — 13 Applications

Fig. 13.3. Large analog computer installation for the simulation of pipeline networks

Fig. 13.4. Display of a four-dimensional hyper-cube on an oscilloscope (see [MACKAY 1962, p. 137])

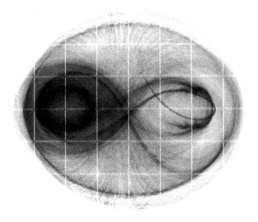

Fig. 13.5. Chaotic orbits in a two-star system (see [ULMANN 2017])

13.2.1 Orbit calculations

Many-body problems arising in astronomy and space flight are well suited to being solved on analog computers. For example, rendezvous simulations were of prime importance during the early days of space flight.[48] Figure 13.5 shows chaotic orbits in a two-star system.

13.2.2 Particle trajectories and plasma physics

Two typical applications from nuclear physics are the calculation of particle trajectories in a magnetic field and beam control in particle accelerators.

[Telefunken/5] describes the simulation of the historic RUTHERFORD[49] experiment – alpha particles being scattered on a thin gold foil – on an analog computer. Basically, an alpha particle approaching an atomic nucleus is subject to a force

$$F = \frac{1}{4\pi\varepsilon_0} \frac{q_\alpha q_k}{r^2},$$

where q_α and q_k denote the charges of the alpha particle and that of the target nucleus. Simplifying this setup by restricting it to two dimensions yields the two components

$$\ddot{x} = a\frac{x}{r^3} \quad \text{and}$$

---

[48] A simple example describing the simulation of a planet's orbit can be found in [Telefunken 1966].
[49] ERNEST RUTHERFORD, 08/30/1871–10/19/1937

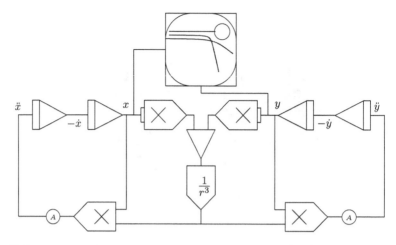

Fig. 13.6. Basic circuit for simulating an alpha particle trajectory (see [Telefunken/5, p. 2])

$$\ddot{y} = a \frac{y}{r^3}$$

with a suitably chosen constant of proportionality a. Integrating twice over $\ddot{x}$ and $\ddot{y}$ respectively yields a coordinate pair (x, y) describing the position of the alpha particle. The resulting circuit is shown in figure 13.6.[50] It should be noted that instead of an explicit division by r^3 a function generator yielding $1/r^3$ is used.

The variable parameter of this simulation is the initial condition input of the second integrator from the right yielding y.[51] This determines the height at which the simulated particle is injected. Using a simple digital circuit controlling an additional integrator, this initial condition can be varied automatically spanning a certain interval to simulate many trajectories starting at various heights, which is adumbrated in the oscilloscope screen depicted figure 13.6.

Figure 13.7 shows the overall computer setup for this simulation. On the right hand side a Telefunken RA 741 tabletop analog computer can be seen with a *DEX 100* digital expansion unit to its left. On top of the DEX 100 is a two channel *OMS 811* oscilloscope. The DEX 100 performs the parameter variation.

The next example shows the simulation of charged particle trajectories in a magnetic field – a problem typically arising during the development and operation of particle accelerators. Large-scale simulations like this were performed

---

50 A more detailed setup can be found in [Telefunken/5, p. 3].
51 This input is not shown explicitly in figure 13.6.

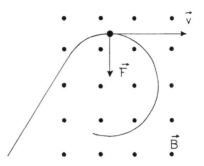

Fig. 13.7. Setup of the particle trajectory simulation (see [Telefunken/5, p. 1])

Fig. 13.8. Trajectory of a charged particle in a magnetic field (see [BORCHARDT, p. 1])

at $DESY$,[52] using at first a large $EAI\ 231RV$ analog computer, which was later replaced by a Telefunken RA 770 based hybrid computer system.[53]

A charged particle moving in a magnetic field is subject to a LORENTZ force[54]

$$\vec{F}_\mathrm{L} = q(\vec{v} \times \vec{B})$$

with q representing the particle's charge, $\vec{v}$ its velocity, and $\vec{B}$ the magnetic flux density. Since $\vec{v}$ is constant, $\vec{F}_\mathrm{L}$ results in a change of direction, so

$$\vec{F}_\mathrm{L} = m\ddot{\vec{r}}$$

holds. This effect is shown schematically in figure 13.8 where a particle is injected diagonally into a field of constant magnetic flux. Figure 13.9 shows the analog computer setup for computing trajectories like this with a typical simulation output shown in figure 13.10. Using a digital control system or the digital computer part of a hybrid computer, various trajectories, corresponding to different particle injection speeds $\dot{x}$, can be generated.

[SHEN 1970] and [SHEN et al. 1970] describe the application of analog computers to problems in plasma physics.[55] Using only analog computer techniques several interesting observations can be made, which can be also observed in real plasma experiments:[56]

---

52 Short for *Deutsches Elektronensynchrotron*.
53 [BORCHARDT et al.], [BORCHARDT 1965], [BORCHARDT 1966], and [BORCHARDT et al. 1965] contain more information about these simulations and the special techniques developed.
54 See [BORCHARDT].
55 The systems used here were an EAI TR-48 and an *Applied Dynamics AD-2-64-PBC*.
56 See [SHEN 1970, pp. 4 f.].

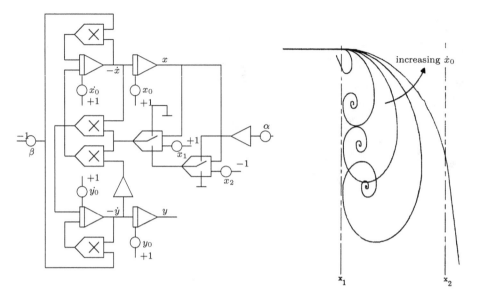

Fig. 13.9. Computer setup for simulating the trajectory of a charged particle in a magnetic field (see [BORCHARDT, p. 6])

Fig. 13.10. Various trajectories of charged particles in a magnetic field (see [BORCHARDT, p. 4])

> "[...] we observe: An 'anomalous drift' motion in nonstatic magnetic field, cyclotron resonance as well as subharmonic cyclotron resonance, [and the] violation of the commonly used definition for the adiabatic invariance of the magnetic moment in a nonstatic but spatially uniform magnetic field."

13.2.3 Optics

A basic yet instructive application of analog computers in the field of optics is described in [RUDNICKI]. This work describes the simulation of a laser interferometer based on a modulated HeNe gas laser. Using two oscillator setups like the one shown in figure 8.2, the laser beam as well as the modulation oscillator are modelled, thus simulating the interferometer on a wave basis.

13.2.4 Heat-transfer

The solution of heat-transfer problems invariably involves partial differential equations typically requiring a large number of computing elements when a discretisation approach as shown in section 7.4.1 is used. A practical application example for a complex heat-transfer simulation is given in [JAMES et al. 1971, pp. 193 ff.] where the design of heat sinks for radioisotope thermoelectric generators to be used in unmanned spacecraft is described. The temperature on one side of the heat sink is determined solely by the decay heat of the radioisotopes. The variable parameter of the optimisation problem is the thickness of the heat sink. This simulation was implemented by associating machine time τ with the position x on a heat sink cross section.

[SWITHENBANK 1960] describes a special purpose analog computer for the simulation of regenerative heat exchangers.

A nice introductory example can be found in [Telefunken 1963/1], which describes the simulation of the radial heat distribution in a loaded power cable using a finite difference approach to model the cable's geometry. [VALISALO et al. 1982] describes a heat distribution analysis of irregularly shaped two-dimensional structures using a Monte-Carlo approach on a hybrid computer.

Two special purpose direct analog computers for the analysis of heat-transfer problems, the *Electronic Analog Frost Computor (EAFCOM)* and the *Heat Exchange Transient Analog Computer (HETAC)*, are shown in figures 13.11 and 13.12. The EAFCOM was used by the US Army Corps of Engineers to simulate soil freezing and associated effects.[57] It is a hardware implementation of a two-dimensional quotient of differences approach to heat conduction.[58] This grid consists of four different types of computing elements: A-elements are simple adders accepting values from their neighbor elements, C-elements are basically coefficient potentiometers, J-elements represent integrators, and Z-elements are used to model inert areas such as heat sources or sinks.

One A-, J-, Z-, and two C-elements form a macrocell representing one layer of soil. These macrocells are connected not only with their direct neighbors but also with more remote elements representing thermal bridges or water flowing between layers.[59]

HETAC, developed at the *University of Virginia*, also is a two-dimensional implementation of a quotient of differences approach for solving the heat-transfer equation.

[KERR 1978] and [KERR 1980] describe analog computer models for the analysis of convection currents developing on inclined heat-exchanger surfaces. Using a small *EAI 380* analog computer solutions for this problem could be generated at a frequency of 500 Hz; this generated a flicker-free display of solution curves

---

[57] See [ALDRICH et al. 1955].
[58] A much smaller and simpler, but nevertheless similar, passive circuit is described in [ULMANN 2020/2]. A recent development, a nanophotonic analog mesh computer, is described in [KAYRAKLIOGLU 2020].
[59] A detailed description of these direct analogies and their application to frost and thaw problems can be found in [PAYNTER].

Fig. 13.11. The *Electronic Analog Frost Computor*, US Army Corps of Engineers, New England Division (see [ALDRICH et al. 1955, p. 259])

Fig. 13.12. The *Heat Exchange Transient Analog Computer, HETAC* for short (see [N. N. 1957/5])

on an oscilloscope. Parameters of the simulation could be changed manually while observing their influence on the underlying system.

An interesting zone melting simulation is described in [CARLSON et al. 1967, pp. 318 ff.]. Applications like this were essential for the rapid development of semiconductor technology, which required high-purity materials like germanium or silicon. These raw materials are usually purified using a zone melting technique, which depends on a variety of parameters, including the diameter of the germanium or silicon monocrystal, the width of the melting zone, the speed at which this zone moves through the crystal, etc. These parameters could be determined using analog computer simulations.

Similar techniques were also applied to the development of building structures. An analog computer setup simulating the "*one-dimensional heat gain through a flat, composite roof section into an airconditioned space below*" is described in [HEADRICK et al. 1969]. The simulation of heat and cooling loads using digital and analog approaches is detailed in [BORDES et al. 1967].

A fascinating one-of-a-kind special purpose analog computer, the *Phytotron*, is shown in figure 13.13. It was used to develop modern and energy efficient greenhouses and is described in detail in [KOPPE et al. 1971] and [EUSER 1966].

Fig. 13.13. The Phytotron analog computer (picture by BERT BROUWER)

[BRYANT et al. 1966] describe a complex large-scale simulation of a co-current laminar double-pipe heat exchanger by means of a digitally controlled trial-and-error technique. The heat exchanger is described by two STURM[60]-LIOUVILLE[61] differential equations, which are solved on the analog computer while the digital system performs the parameter space search. The application of finite FOURIER transforms to analog computer simulations with emphasis on heat transfer problems is described in detail in [LIBAN 1962].

Maybe one of the most arcane applications of analog computers to heat transfer and related problems is described in [VAN ZYL 1964]. This describes the application of analog computer to the solution of non-linear partial equations describing the self-heating of fishmeals.[62]

---

[60] CHARLES-FRANÇOIS STURM, 09/29/1803–12/18/1855
[61] JOSEPH LIOUVILLE, 03/24/1809–09/08/1882
[62] This work was done at the department of Electrical Engineering and the Fishing Industry Research Institute of the University of Cape Town.

13.2.5 Fallout prediction

Surface testing of nuclear weapons required an accurate means of predicting radioactive fallout. A special purpose analog fallout predictor was developed in 1956 and is described in [SKRAMSTAD 1957].[63] The system consisted primarily of a variety of function generators (*winds units*) and an elaborate scanning mechanism driving a two-dimensional oscilloscope display.[64] The underlying forecast problem was simplified considerably by ignoring vertical winds and assuming horizontal wind speeds and directions to be constant during the entire fallout period.[65]

The overall predictor and a typical fallout prediction result are shown in figures 13.14 and 13.15. The brightness of the display corresponds to the expected intensity of fallout in the respective area.

13.2.6 Semiconductor research

Analog computers were also used extensively in early semiconductor research. [APALOVIČOVÁ 1979] describes the simulation of the electric field in the depletion zone of a *Metal Oxide Semiconductor (MOS)* field effect transistor[66] in order to predict the behaviour of different transistor structures. The solution of the underlying elliptic partial differential equations was done by a hybrid computer.[67]

[CARLSON et al. 1967, pp. 332 ff.] describes the analysis and simulation of the dynamic response of a *tunnel diode*.[68] An interesting aspect of this particular analog computer setup is that it contains an algebraic loop, which is stable due to a unique combination of circumstances.[69] Furthermore, it is a good example of a problem that happens too fast to be observed and analysed in a real specimen, so an extremely slowed down machine time on an analog computer is required to gain the necessary insight.

---

[63] Interestingly no direct analogies have been employed.
[64] These functions generators are quite interesting as they generate functions of two variables.
[65] Obviously, such simplifications were common in the past and dated back to earlier manual fallout prediction techniques.
[66] *MOSFET*
[67] Numerical methods for the simulation of semiconductor junctions and field effect transistors can be found in [AKERS 1977].
[68] Also known as ESAKI (LEO ESAKI, *03/12/1925–*) diode.
[69] Normally algebraic loops are to be avoided at any cost due to their inherent instability.

13.2 Physics — 263

Fig. 13.14. Fallout computer (see [SKRAMSTAD 1957, p. 103])

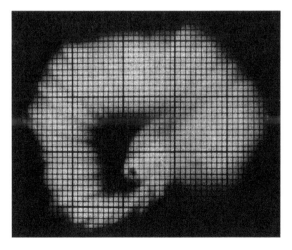

Fig. 13.15. Typcial predicted fallout pattern (see [SKRAMSTAD 1957, p. 102])

13.2.7 Ferromagnetic films

Ferromagnetic films were of great interest in the 1960s as prospective storage elements for digital computers, before static and dynamic semiconductor *Random Access Memory (RAM)* circuits eventually took over. An interesting research problem was the behaviour of such ferromagnetic films under varying environmental conditions. [BORSEI et al.] describes an analog computer approach to simulate the behaviour of a ferromagnetic film subjected to mechanical stress.

13.3 Chemistry

The following sections describe two typical applications of analog computers in chemistry. For applications in chemical engineering refer to section 13.13.

13.3.1 Reaction kinetics

Most chemical processes including enzymatically catalysed processes in biochemistry[70] involve a variety of heavily interdependent reaction steps. Studies regarding such processes were mainly performed on analog computers well until the late 1970s/early 1980s because the underlying differential equations are ideally suited for this approach. A simple example problem is shown in figure 13.16:[71] A substance A is to be converted into a product D involving an intermediate substance B, which in turn is partially transformed into C and vice versa. The rates at which these steps happen are k_1, k_2, k_3, and k_4.

Based on figure 13.16 this reaction can be described by the following four coupled differential equations:

$$\dot{A} = -k_1 A$$
$$\dot{B} = k_1 A - k_2 B - k_4 B + k_3 C$$
$$\dot{C} = k_2 B - k_3 C$$
$$\dot{D} = k_4 B$$

The result of a typical simulation run based on these equations with $k_1 = k_2 = k_3 = k_4 =$ const. is shown in figure 13.17. Starting with 100% of component A, the four parallel reactions set in, finally yielding the desired end product D. The corresponding simulation program is shown in figure 13.18.[72]

---

[70] See [SAURO 2019] for an in-depth treatment of enzyme kinetics.
[71] See [Dornier/2].
[72] Further introductory examples can be found in [ULMANN 2020/1, pp. 109 ff.].

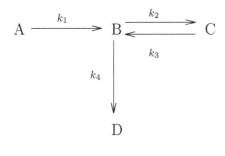 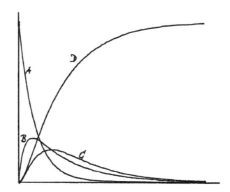

Fig. 13.16. Simple reaction kinetics example involving four substances (see [Dornier/2, p. 3])

Fig. 13.17. Reaction kinetics simulation result (see [Dornier/2, p. 7])

Further interesting examples can be found in [CHENG], where the simulation of polymerization processes using analog computers is described. Techniques to determine process parameters for the chlorination of methane are described in [BASSANO et al. 1976], while [RAMIREZ 1976, pp. 74 ff.] describes the simulation of a large-scale benzene chlorination. [KÖPPEL et al. 2022] is a recent work on the application of analog computers to molecular dynamics.

A large area of application for reaction kinetics studies can be found in biochemistry, medicine, etc., where enzymatically catalysed reactions are of prime importance. A thorough introduction to this topic from the perspective of systems biology can be found in [SAURO 2019].

13.3.2 Quantum chemistry

Direct analog computers proved very useful in the early days of quantum chemistry,[73] where solutions of the SCHRÖDINGER[74] *equation*, which are wave functions, are of prime interest.[75] These solutions determine the *electronic structure* of molecules, which in turn determine their respective properties.

In 1944, GABRIEL KRON[76] suggested a direct analog electronic analog computer to model the behaviour of the SCHRÖDINGER equation under different bound-

---

[73] [ANDERS] gives a good description of the historical development of quantum chemistry.
[74] ERWIN SCHRÖDINGER, 08/12/1887–01/04/1961
[75] See [PREUSS 1962] and [PREUSS 1965] for the basics of quantum chemistry.
[76] 12/01/1901–10/25/1968

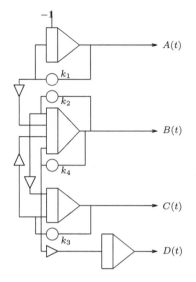
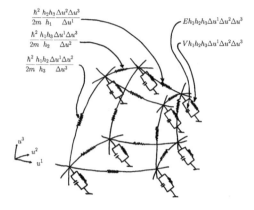

Fig. 13.18. Reaction kinetics simulation program (see [Dornier/2, p. 4])

Fig. 13.19. Analog model of SCHRÖDINGER's equation with three independent variables (see [KRON 1945/1, Fig. 7])

ary conditions. A typical circuit derived for this purpose is shown in figure 13.19.[77] Later, HANS KUHN[78] developed an *acoustical analog computer* containing a multitude of coil springs mounted on a carrier. These springs were interconnected horizontally by leaf springs of equal moduli of resilience. This particular machine is described in detail in [PREUSS 1962, pp. 291 f.] and was used to simulate state transitions of π-electrons in complex shaped molecule chains. An easy treatment of the one-dimensional SCHRÖDINGER equation for educational purposes can be found in [MÜLLER 1986] and [HUND 1982].[79]

13.4 Mechanics and engineering

Problems emerging from mechanics and engineering, ranging from vibrating coupled mass systems to the simulation of surge chambers and many more, are also well suited for treatment by analog computers.[80]

---

[77] See [KRON 1945/2] and [KRON 1945/1].
[78] 12/05/1919–11/25/2012
[79] [ULMANN 2019] shows a more recent implementation.
[80] A wealth of information on these subjects as well as example programs can be found in [MAHRENHOLTZ 1968].

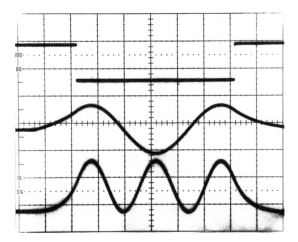

Fig. 13.20. Typical solution for the one-dimensional SCHRÖDINGER equation

13.4.1 Vibrations

The treatment of vibrating systems is a typical application for an analog computer as the examples in sections 8.3 and 8.7 demonstrated. Vibrations are highly undesirable in certain applications, ranging from mere inconveniences to real hazards, while they are a necessity in other areas, such vibratory conveyors. Accordingly, the commercial impact of such studies was, and still is, high. Depending on the complexity of the underlying mechanical system, these problems can be quite challenging as [SANKAR et al. 1979, p. 11.] notes:

> "*The control of vibration in mechanical systems is a serious and challenging design problem.*"

A notorious example for unwanted vibrations is the *Pogo effect* that was experienced in early large rockets, which used liquid fuels.[81] This effect was caused by pressure variations in the feed lines for the rocket engines which caused unsteady combustion. This resulted in varying accelerations of the rocket which in turn tended to amplify these pressure variations and so on. In extreme cases the resulting forces can reach levels capable of destroying the rocket or at least damaging sensitive payloads such as satellites or astronauts. Other examples involve the stability of tall buildings subjected to wind forces, earthquakes, etc. In the case of a vibrating conveyor the isolation of the vibrating machine parts from their mountings is a critical design parameter.

---

81 See [BILSTEIN 2003, pp. 360 ff.] and [WOODS 2008, pp. 83 ff.].

In general, machines can be considered as vibrating multi-mass systems with spring and damper elements coupling the various parts of the overall system. Treating systems like this with analog computers is described in detail in [Dornier/3], [MACDUFF et al. 1958], [TSE et al. 1964], and [TRUBERT 1968, pp. 416 ff.], while [JACKSON 1960] focuses on the analysis of vibrating truss structures.

13.4.2 Shock absorbers

[SANKAR et al. 1979] describes an interesting optimisation problem: A shock absorber is to be designed, which should impose minimal accelerations on the attached mass while simultaneously limiting the distance the damped mass travels under an excitation impulse. The simulations for determining the parameters of this system were performed on a hybrid computer. The task was considerably complicated by the necessity of implementing an accurate model of dry friction requiring many non-linear computing elements.[82]

[MEZENCEV et al. 1978] describes the development of an adaptive hydropneumatic anti-vibration device for marine propulsion systems using an analog computer.

13.4.3 Earthquake simulation

The analysis of the behaviour of buildings excited by LOVE- and RAYLEIGH-waves resulting from an earthquake is of great importance, especially in countries with high seismic activity, such as Japan, where pioneering work in this area was done. [Hitachi 200X, p. 17] describes the simulation of a multi-story building which is modelled as a multi-mass vibrating system where each floor is lumped together as a single mass. The walls act as spring-damper systems connecting adjacent floors.

A simple building with two floors is characterized by the weights of the floors, m_1 and m_2, the moduli of resilience s_1, s_2, and the damping coefficients d_1, d_2. With y_1 and y_2 denoting the horizontal displacements of the respective floors, the building dynamics are described by the following two coupled differential equations

$$m_1\ddot{y}_1 + d_1\dot{y}_1 + d_2(\dot{y}_1 - \dot{y}_2) + s_1 y_1 + s_2(y_1 - y_2) = m_1 a(t) \text{ and}$$
$$m_2\ddot{y}_2 + d_2(\dot{y}_2 - \dot{y}_1) + s_2(y_1 - y_1) = m_2 a(t)$$

with $a(t)$ representing the ground motion induced by the earthquake. In practical applications real seismic data stored on analog magnetic tapes was used to generate $a(t)$.

---

[82] Cf. [Dornier/4].

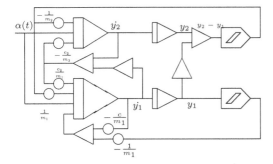
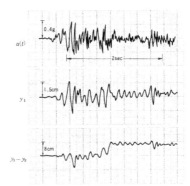

Fig. 13.21. Computer setup to simulate a two-story building excited by horizontal ground movements (see [Hitachi 200X, p. 17])

Fig. 13.22. Results of an earthquake simulation (see [Hitachi 200X, p. 17])

Based on these equations, the program shown in figure 13.21 can be derived.[83] The two function generators on the right hand side are used to implement two hysteresis curves modelling the non-linear stiffness of the building structures. The results of a typical simulation run are shown in figure 13.22.[84]

A special purpose analog computer, $SERAC$,[85] was developed by Hitachi in the late 1960s.[86] This system featured a photoelectric curve follower, which directly generated the excitation signal $\alpha(t)$ based on actual seismograms.

13.4.4 Rotating systems and gears

Rotating systems are of great importance in many areas, ranging from large positioning systems, to centrifuges, to jet engines, etc. These systems are often critical in the sense that human lives depend on their flawless operation.

Bearings are some of the most critical parts in rotating systems. Long shafts, common in turbines and similar machines, are prone to bending vibrations, which

---

[83] The analog computer used in this example, a Hitachi 200X, features summers and integrators each providing normal and inverted outputs, thus simplifying computer setup considerably, which is reflected in the program shown.
[84] A similar simulation of a three-story building is described in [JAMES et al. 1971, pp. 168 ff.].
[85] Short for *Strong Earthquake Response Analog Computer*.
[86] See [Hitachi 1969, pp. 6 ff.].

subject the bearings to additional and sometimes critical loads. Consequently, many studies were performed to simulate the behaviour of such systems.[87]

Gas and steam turbines often use hydrodynamic bearings since traditional bearings would fail rapidly in such applications due to disturbances of the oil film. The simulation of such hydrodynamic bearings on analog computers is described in [MCLEAN et al. 1977] and [RIEGER et al. 1974].

Some rotating systems achieve extremely high speeds. A turbomolecular pump with a rotor weight of 1.7 kg, described in [FREMEREY, p. 33], operates at 51,600 revolutions per minute. Ultra centrifuges for the enrichment of uranium run at even higher speeds. Applications like these require magnetic bearings, of which active and passive variants exist. Passive magnetic bearings were developed in the 1960s and rely on permanent magnets, while active magnetic bearings use electromagnets controlled by a servo loop.[88] Analog computers were not only used to determine the basic parameters of bearing systems like these but were also employed to model the servo circuits for active magnetic bearing systems. Often the analog computer was used in *hardware in the loop* simulation setups where the analog computer is an active part of the system to be analysed. In this case the analog computer was used to control the current energising the electromagnets of a real magnetic bearing in operation. In such a setup the consequences of parameter variations can be directly explored using real hardware.

[OKAH-AVAE 1978] focuses on the effects on bearings caused by transverse cracks in turbine shafts. The goal of this work was to develop methods allowing an early detection of such damages. The simulation results were compared with actual data gathered from a damaged turbo generator and showed a high degree of correlation.

13.4.5 Compressors

A complex centrifugal compressor simulation is desribed in [DAVIS et al. 1974]. Interestingly, this simulation does not just model the macroscopic behaviour of the compressor system itself but also the gas dynamics of medium to be compressed. Another analog centrifugal compressor system simulation can be found in [SCHULTZ et al. 1974].

---

[87] See [RIEGER et al. 1974].
[88] See [FREMEREY] and [FREMEREY 1978].

Fig. 13.23. Curves of motion obtained through the simulation of a four joint linkage (see [CROSSLEY 1963])

13.4.6 Crank mechanisms and linkages

Many systems such as reciprocating pumps or piston compressors use a crank mechanism like a sliding crank to transform a rotary motion into a linear movement.[89] In the simplest case, this type of transformation can be achieved with a scotch yoke mechanism as shown in figure 2.16 in section 2.6. In most cases, more complex mechanisms are required since a scotch yoke develops high forces on the mechanism's bearings. Additionally, many commercial applications require special curve progressions of the linear movement based on the rotary input of the mechanism.[90]

[CROSSLEY 1963] and [CROSSLEY 1965] describe an interesting approach for the simulation of complex path generating linkages using inverse functions on an analog computer. Figure 13.23 shows typical curves of motion obtained by the simulation of a four joint linkage on a Telefunken RAT 700 table top computer. A further example of a linkage simulation can be found in [MAHRENHOLTZ 1968, pp. 179 ff.].

---

[89] Cf. [Hütte 1926, pp. 82 ff.] for basic information on crank mechanisms and [MAHRENHOLTZ 1968, pp. 129 ff.] or [MAHRENHOLTZ 1968, pp. 79 ff.] for examples of analog computer simulations of such systems.
[90] A simple example can be found in [Dornier/5].

13.4.7 Non-destructive testing

Non-destructive testing is frequently cheaper and safer than testing systems to destruction. Typical application areas are listed by [LANDAUER 1975]:

> "*A speeding automobile goes out of control on a test track and crashes violently. A reactor vessel ruptures at a process plant when unstable conditions are reached. Or, a 650-MW generator loses bearing lubricant and goes into uncontrollable vibration. Each of these occurrences represents destructive testing at its best [...] or worst.*"

In the majority of cases, hybrid computers were used in this context since often functions of more than one variable have to be generated. The analog computer was often also closely tied to the system being analysed.[91]

13.4.8 Ductile deformation

In 1956 *General Electric* implemented a comprehensive analog computer based rolling mill simulation for the *Sharon Steel Corporation* to optimise operating and startup procedures for a newly planned rolling mill.[92] As a result of this extensive simulation, a *tune-up time*[93] of only one day, much lower than the seven to ten days normally required, was achieved.

Steel rolling processes can be described by the KARMAN *differential equation*. Its implementation and treatment on analog computers is described in [Hitachi 1968]. Another practical example can be found in [GOLTEN et al. 1967] where a hybrid computer is used. The mathematical principles for the simulation of cold rolling are described in [HEIDEPRIM 1976]. A thorough treatment of the simulation of drag rolling processes can be found in [MAHRENHOLTZ 1968, pp. 122 ff.].

[DARLINGTON 1965] describes a complex analog simulation for a multistand cold rolling mill. An interesting aspect of this work is that the analog simulation is based on coefficients obtained by a digital computation.

The ideal velocity for roller drives can also be determined using analog computers as [ROHDE 1977] and [ROHDE et al. 1981] show. The hybrid computer system used in these simulations is described in detail in [Schloemann-Siemag 1978]. [MAHRENHOLTZ 1968, pp. 117 ff.] focuses on determining optimal parameters for convex dies, while [OVSYANKO] describes the analysis of plastic deformations on an analog computer. [RAMIREZ 1976, pp. 22 f.] covers the simulation of injec-

---

[91] See [LANDAUER 1975] for more details.
[92] See [N. N. 1958/1].
[93] Tune-up time described the timespan elapsing between parameter changes in a complex industrial installation until normal operation is reached again.

tion molding processes. The analysis of the deformation of steel plates subjected to the forces of an explosion by means of an analog computer is described in [JAMES et al. 1971, pp. 175 ff.]. An interesting investigation of the forces exerted by shearing processes is given in [Dornier/6].[94]

13.4.9 Pneumatic and hydraulic systems

Hydraulic elements are of prime importance in many application areas, and consequently much research on these devices was done on analog and hybrid computers. [COHEN 1971] describes the development of a pneumatic relay based on simulation runs on a hybrid computer, while more complex hydraulic switching elements are analysed in [HANNIGAN]. General information about the treatment of hydraulic systems on analog and hybrid computers can be found in [SANKAR et al. 1980]. [NOLAN 1955] and [AMELING 1962/2] focus on the simulation of liquid flows (with an emphasis on complex piping networks).

Large-scale hydraulic problems, such as flood simulations or the behaviour of sewage systems, have been treated with indirect as well as with direct analog computers. Examples for such studies can be found in [Hitachi 1969, pp. 1 ff.] and [PAYNTER ed. 1955, pp. 239 ff.].

An interesting application is the simulation of *surge chambers*, reservoirs placed at the end of pipes to absorb shock waves generated by the closing of valves, etc. Simulating the generation, distribution and elimination of these shock waves is of prime importance as these can destroy even large and sturdy machines and piping systems.[95]

A nice example is given in [MEISSL 1960/1] and [MEISSL 1960/2] where the focus is on the water level in a *differential surge chamber* as described in [JOHNSON 1915] and shown in figure 13.24. Such a differential surge chamber can be described by the following system of differential equations[96]

$$z_2 - z_1 = \varepsilon_1 \left(\frac{F_1}{t}\dot{z}_1\right)^2 \operatorname{sign}(\dot{z}_1),$$

$$F_1\dot{z}_1 + F_2\dot{z}_2 = f(v_a - v), \text{ and}$$

$$\dot{v} = \frac{g}{L}(z_2 - \varepsilon v|v|)$$

---

94 This last example is a rare case in which an explicit differentiation of a variable is necessary.
95 See [VALENTIN 2003, p. 135].
96 See [MEISSL 1960/2, p. 76].

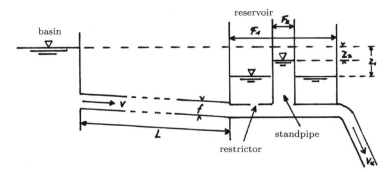

Fig. 13.24. Surge chamber (see [MEISSL 1960/2, p. 76])

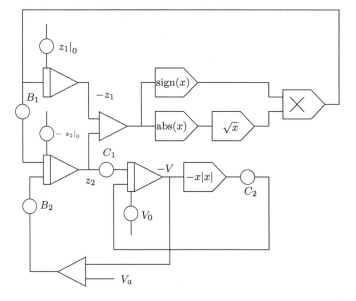

Fig. 13.25. Program for the surge chamber simulation (see [MEISSL 1960/2, p. 76])

with z_1 and z_2 denoting the respective water levels. The corresponding computer setup is shown in figure 13.25. Typical simulation results for z_1 and z_2 of a 100 second run are shown in figures 13.26 and 13.27.[97]

The simulation of the behaviour of the *Appalachian surge tank* operated by the *Tennessee Valley Authority* on an analog computer is described in [JACKSON 1960, pp. 403 ff.]. General information about the use of analog computers for the

---

[97] These simulations were performed on a Telefunken RA 463/2 analog computer, see section 6.1. It should be noted that this particular analog computer only allowed for a maximum of 110 seconds of computing time as its the operational amplifiers were not chopper-stabilised and integration errors due to drift effects became excessive after about 100 seconds.

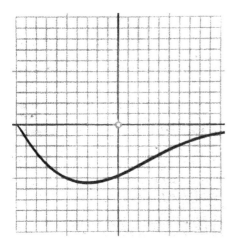
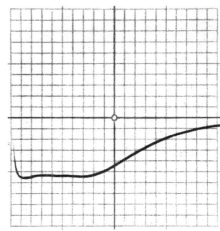

Fig. 13.26. Water-level z_1 for $0 \leq t \leq 100$ s (see [MEISSL 1960/2, p. 77])

Fig. 13.27. Water-level z_2 for $0 \leq t \leq 100$ s (see [MEISSL 1960/2, p. 77])

analysis of water hammer problems can be found in [PAYNTER et al. 1955], [PAYNTER 1955/1], and [JAMES et al. 1971, pp. 184 ff.]. The design of surge tanks for hydro stations is covered in [ANDO 1971]. [SORONDO et al.] describes the simulation of the piping system in a thermal power plant. The focus here is on the effects of the short run-up times of only 3.5 seconds the pumps are capable of. This placed extreme stresses on the piping network which had a total length of about four miles.

The problem of sloshing in a steel converter is treated in [BARTH 1976, pp. 600 f.]. In this simulation, not only the sloshing itself is simulated but also its effect on the tilting control system for the converter.[98] In addition to a direct analogy model consisting of a scaled down version of the converter, an indirect analog model was used for the simulations.

A very interesting one-of-a-kind analog computer was the Dutch *Deltar*, short for *Delta Getij Analogon Rekenmachine*[99] used from 1960 until 1984 at the *Deltawerken*[100] for the planning of dams, storm surge barriers, etc., which is shown in

---

[98] Given the high density of molten steel the resulting forces are quite considerable.
[99] Delta tide analog computer.
[100] The Deltawerken are a complex of large-scale construction projects in the Netherlands aimed at the protection against storm surges, floods, etc.

Fig. 13.28. The Deltar analog computer (picture by ERIK VERHAGEN)

figure 13.28. It was based on the seminal works of JOHAN VAN VEEN,[101] see [VAN VEEN 1947/1], [VAN VEEN 1947/2], and [VAN VEEN 1947/3].[102]

13.4.10 Control of machine tools

Analog computers were also used for the development and implementation of machine tools. Two interesting analog interpolation systems for the control of turbine blade milling machines were developed at the *Lewis Flight Propulsion Laboratory* and are described in [JOHNSON 1962]. One of the systems implements a cubic interpolator using three mechanical integrators to generate a polynomial of third degree thus interpolating a function specified by values given at equidistant sampling points. The other system was based on a direct analogy using a flexible steel ruler to implement a spline interpolation. This ruler was bent into position by a punch card controlled servo system controlling a number of linear actuators.

---

101 12/21/1893–12/09/1959
102 [VAN SANTEN 1966] describes resistor networks for modeling water systems in general.

The bent ruler was then used as a model to control the milling process using a magnetic pickup in a servo loop.

13.4.11 Servo systems

Servo systems are central to nearly every technical system. Consequently the simulation of such systems using analog computers was of high importance and numerous studies deal with related topics. Servo systems are divided into *linear* and *non-linear* systems. Linear servo systems use actuators operating in a continuous way while non-linear servo systems often use actuators supporting only a limited number of states such as valve actuators for reaction control engines, etc. The simulation of such systems is described in [MCLEOD 1962, pp. 92 ff.] and [KORN et al. 1956], while [MCLEOD et al. 1958/2, pp. 297 f.] treats instabilities of servo systems.

[KLITTICH 1974] describes the design and simulation of a positioning system for a large radio telescope antenna. This study was done during the design and development of the Effelsberg radio telescope in Germany. Positioning systems like this are quite complicated as they usually employ multiple motor drive systems for moving the large and heavy (the parabolic dish of this particular installation has a diameter of 100 m with an overall system weight of about 3 200 tons) rotating structure. In systems this large, the elastic properties of the drive train become a problem and can cause underdamped parasitic oscillations. This study not only describes the simulation itself but also compares simulation results with actual measurements taken after completion of the installation in 1971.

A similar study is described in [VON THUN] where the attitude control system of another large radio telescope dish antenna is modelled. It covers effects such as the inertia of the antenna itself, wind load, achievable accuracy, as well as the generation of non-linear antenna movements required for certain tracking tasks.

[BEKEY et al. 1960] describe an interesting technique for generating BODE[103] and NYQUIST diagrams[104] by means of an analog computer.

---

[103] A BODE diagram consists of two function graphs – one showing the amplitude and the other showing the phase shift of a system with the x-axis representing frequency. It was developed by HENDRIK WADE BODE (12/24/1905–06/21/1982).
[104] A NYQUIST diagram shows the *locus curve* of a control system. It was conceived by HARRY NYQUIST (02/07/1889–04/04/1976).

13.5 Colour matching

A very interesting and arcane application of analog computers was the task of *colour matching*. The problem is to find the proportions of a number of pigments so that the resulting colour matches a given colour sample. This is not a simple task as in both cases of additive as well as subtractive mixing an under-determined system of equations must be solved. Colour matching finds application in a variety of applications including blood tests, making false teeth, etc.

This problem can be solved by applying the KUBELKA-MUNK theory, developed in 1931 by PAUL KUBELKA[105] and FRANZ MUNK[106] in 1931.[107] The first simple analog device for solving the resulting equations, which were previously solved by graphical techniques, was developed in 1955.[108]

In 1958, HUGH R. DAVIDSON[109] and HENRY HEMMENDINGER[110] developed *COMIC* (*Colorant Mixture Computer*), a sophisticated specialised analog computer for colour matching. This system was used extensively in various industries well into the 1970s with about 200 machines sold before it was eventually replaced by digital systems.[111] It reduced the time for a typical colour matching task from hours to a maximum of 5 to 20 minutes.

13.6 Nuclear technology

Although many problems related to nuclear reactors and the like are not well suited for an analog computer due to vastly different time-scales, analog and hybrid computers nevertheless played a vital role in the early decades of nuclear technology. These systems were used for basic research as well as training aids, etc. A lot of general information about the application of analog computers in nuclear technology can be found in [E. MORRISON 1962], [FRISCH 1971], and [DAGBJARTSSON et al. 1976].[112]

---

[105] 04/17/1900–1956
[106] 04/29/1900–02/24/1964
[107] See [KUBELKA et al. 1931].
[108] See [ATHERTON 1955] and [HEMMENDINGER 2014].
[109] 05/11/1918–04/02/2010
[110] 04/01/1915–08/16/2003
[111] A hybrid computer for the same task, the Redifon *Redi-colour* was developed in 1965, see [CUTLER 1965].
[112] The necessary basics of reactor physics are covered in [MARKSON 1958].

13.6.1 Research

Analog and hybrid computers were irreplaceable tools in fundamental nuclear technology research – not least because safety constraints often forbid performing experiments on actual reactors. Typical application examples are

- the analysis of neutron generation and distribution in reactor cores,[113]
- the consequences of changes in reactor geometry due to temperature variations and the generation of delayed neutrons,[114]
- heat-transfer problems in pressurized-water reactors,[115]
- xenon-poisoning,[116]
- control system development,[117] etc.

Most of these applications make heavy use of an analog computer's capability to scale machine time in order to speed up or slow down simulation runs, thus allowing the study of phenomena occurring too quickly in real time to be actually observable by changing k_0 of the integrators accordingly.

Figure 13.29 shows a program used to analyse the stability of a research reactor in Hanford (USA). Of special interest was the behaviour of the reactor in the case of a rapid depressurization of the primary coolant loop occurring in conjunction with an increase in reactivity. The resulting power gain in output power spans three decades in a one second time frame, which is about the maximum that can be handled even on a precision analog computer without either having the variables exceeding the range of ± 1 machine units or introducing exceedingly large errors.[118] Simulations like this, covering longer time intervals, often have to be split into smaller steps, typically requiring rescaling between successive steps.

[GALLAGHER et al. 1957] describes basic techniques for nuclear reactor studies, while [LOTZ et al.] treats the simulation of a digital control system for a fast nuclear reactor on a Telefunken hybrid computer based on an RA 770 precision analog computer. [JUST et al. 1962] describes a comprehensive simulation of a multiple-region reactor, while the simulation of steam pressurizing tanks by analog computers is treated in [BOSLEY et al. 1956].

---

[113] See [E. MORRISON 1962]. The simulations described there make extensive use of complex passive feedback networks to model the neutron flux densities, etc.
[114] Cf. [SYDOW 1964, pp. 225 f.].
[115] [SYDOW 1964, pp. 230 ff.].
[116] See [FRISCH 1971, pp. 72 ff.], [SYDOW 1964, pp. 233 f.], and [BRYANT et al. 1962, pp. 49–51].
[117] See [SYDOW 1964, pp. 234 ff.], [FRISCH 1968], [BREY 1958], and [SCOTT 1958].
[118] See [JONES 1961].

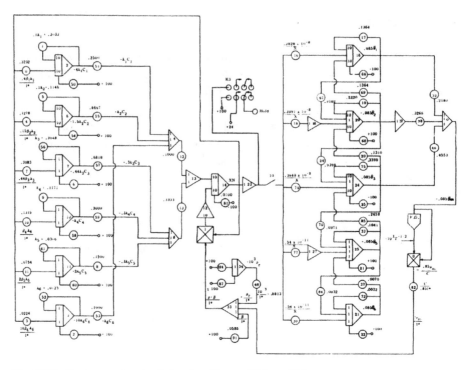

Fig. 13.29. Nuclear reactor simulation (see [JONES 1961, p. 17])

Another interesting simulation is described in [HANSEN et al. 1959], where the dynamic behaviour of a sodium cooled reactor is studied. The simulation is very detailed, even taking hydraulic effects in the piping system, etc., into account.

13.6.2 Reactor/neutron kinetics

Neutron kinetics is a central aspect of reactor design and simulation. The chain reaction in a typical nuclear reactor is based on two groups of neutrons, the overwhelming majority of *prompt* neutrons, which are released basically in the instant when fission takes place, and a tiny fraction of *delayed* neutrons, released with some delay by the decay of unstable fission products.[119] This can be described by[120]

$$\dot{n} = \frac{n}{l^*}(\delta K - \beta) + \sum_i \lambda_i c_i + s \text{ and } \dot{c}_i = \frac{n\beta_i}{l^*} - \lambda_i c_i$$

---

[119] The basics of neutron kinetics are covered in [WEINBERG et al. 1958].
[120] See [FENECH et al. 1973, p. 128].

with n and s representing the neutron *density* and an external neutron source.[121] l^* is the effective neutron *lifetime*, the c_i represent the precursors of the i-th group of delayed neutrons, and β_i is the *decay constant* of this i-th group. Most practical simulations use six of these delayed neutron groups.

Although these equations can be implemented in a straightforward fashion, this requires quite a lot of computing elements. Consequently, specialized computing elements, using a number of RC-combinations in the feedback path of an operational amplifier, were used to model the neutron kinetics of certain nuclear fuels such as ^{235}U or ^{239}Pu. Practical examples of this approach are described in [TYROR et al. 1970] and [ULMANN 2021/2].

[BRYANT et al. 1962, pp. 18–26] describes a reactor kinetics simulation of a reactivity transient caused by the sudden removal of a control rod, which is then quenched by the intrinsic negative temperature coefficient of the reactor under consideration.

13.6.3 Training

Training reactor operators is of prime importance and requires simulators to be as realistic as possible. An early such simulator was built to train the operators of the nuclear reactor aboard the *N. S. Savannah*, the first nuclear powered merchant ship, which was launched on 07/27/1959. This simulator is described as follows in [N. N. 1964/3]:

> "*Through the use of an operator's control console identical to the one aboard the SAVANNAH, the two PACE 231R*[122] *General-Purpose Analog Computers [allow...], trainees [to] acquire operating experience just as if they were actually on board the ship.*
>
> *The analog computers, which are programmed to represent the complete reactor kinetics, as well as primary and secondary heat balances, activate all the recording instruments and dials on the control panel and respond to the student's manipulation of the operating controls just as the real reactor would.*"

In addition to the operator consoles, this simulator featured an additional console where instructors could introduce faults to be handled by the operators in training. A much simpler reactor simulator setup is shown in figure 13.30. This particular system was based on a minor variant of the Telefunken RA 463/2 analog computer.[123]

---

[121] Such external neutron sources are typically used during the startup process of a nuclear reactor.
[122] See section 6.3.
[123] See section 6.1.

Fig. 13.30. AEG reactor simulator based on a Telefunken RA 463/2 (see [GERWIN 1958])

Fig. 13.31. Typical reactor control analog computer modules (General Dynamics)

Figure 13.32 shows the complex *ML-1* reactor simulator developed for the *Army Gas-Cooled Reactor Systems Problem* in the early 1960s, which is described in detail in [CALAMORE et al. 1963]. The analog computer at its heart is an *Aerojet* system with several large consoles.

[FENECH et al. 1973] describes an interesting small-scale analog reactor simulator that is realistic enough to train operators, yet small enough to fit on a medium-scale EAI TR-48 analog computer. A complex large-scale analog computer based atomic power plant reactor is described in [IRWIN et al. 1960]. This system was not only used as a training facility but also as a research tool for further reactor developments.

13.6.4 Control

Analog computer elements were also used as building blocks for actual nuclear reactor control systems – a natural step, given that analog computers had played an important role during the development of early nuclear reactors. Figure 13.31 shows some typical analog control elements. These particular modules were used in the control system of two research reactors installed at the University of Virginia.[124]

A comprehensive description of the simulation of the movement of a control rod in a nuclear reactor can be found in [BRYANT et al. 1962, pp. 6–12]. The rod under consideration is cushioned by a hydraulic system suppressing shocks from rapid stops.

---

[124] Modules like these were, of course, also developed based on analog computer studies, see [CAMERON et al. 1961].

Fig. 13.32. ML-1 reactor simulator – the reactor control panel can be seen on the left next to the Aerojet analog computer (see [CALAMORE et al. 1963, p. viii])

13.6.5 Enrichment

Uranium enrichment plays a central role in nuclear technology. The task of separating different isotopes from elements such as Uranium is extremely difficult due to the incredibly tiny differences in mass on which the separation process is typically based. Classic approaches are separation by ultra centrifuges, by mass spectrometer techniques, and gaseous diffusion.[125] [DELAROUSSE et al. 1962] describes a large complex special purpose analog computer for the simulation of gaseous diffusion plants based on a complex finite difference system.

---

[125] Nowadays, ultra centrifuges are the technology of choice in Uranium enrichment.

13.7 Biology and medicine

There are abundant opportunities for applying analog and hybrid computers in medicine and biology. The earliest attempts to analyse complex biological systems were performed by BRITTON CHANCE[126] using a mechanical differential analyser in 1943 to investigate some aspects of enzyme kinetics.[127] The following sections focus on some later analog computer applications in medicine and biology.[128]

13.7.1 Ecosystems

A simple example of the simulation of a closed ecosystem has already been discussed in section 8.4. A much more complex and interesting simulation is described in [RIGAS et al.] where a food chain spanning from phytoplankton to salmon in an aquatic ecosystem is modelled. This study focuses on external influences such as the effects of fertilizers like phosphates emerging from detergents washed into rivers and eventually the sea, seasonal effects due to incident solar radiation, etc. Interestingly, this study was initially performed on a stored-program digital computer[129] with a graphical display but was then transferred to a mid-range hybrid computer:[130]

> "[...] The initial programming was laborious and the memory requirements were such that the use of the model was restricted to certain hours of the day. Eventually, rental on the CRT^{131} was terminated and the interactive capability was lost. At about the same time, a [...] hybrid computer^{132} became available and the development of an interactive hybrid computer model of the same system became a reasonable alternative."

It turned out that this hybrid approach outperformed its algorithmic predecessor and even the restricted precision offered by the hybrid computer had no perceptible influence on the results obtained in these simulation runs.

---

[126] 07/24/1913–11/16/2010
[127] See [KNORRE 1971, p. 107].
[128] A wealth of information on these topics can be found in [KNORRE 1971] and [RÖPKE et al. 1969].
[129] An IBM 360/67.
[130] See [RIGAS et al., p. 95].
[131] Short for *Cathode Ray Tube* – in this case an IBM 2250 display was used.
[132] An EAI 690 hybrid computer, see [RIGAS et al., p. 95].

13.7.2 Metabolism research

The study of the metabolism of pharmaceuticals is highly important in modern medicine. The development of radioactive tracers made it possible to gather metabolic data from organisms, which formed the basis for extensive analog and hybrid simulations.[133] Many of these studies are *multi-compartment models* where substances are transported between various parts (compartments) of an organism during their metabolism. A number of examples of such studies is described in [HABERMEHL et al. 1969] and [HABERMEHL et al. 1969]. [KNORRE 1971, pp. 201 ff.] describes the simulation of the metabolism of Paracetamol in an organism.

13.7.3 Cardiovascular systems

Multi-compartment models are also used to simulate cardiovascular systems, enabling lesions to be studied in detail on an analog or hybrid computer. A complex system simulation based on seven compartments is described in [BENHAM et al. 1973]. Direct analogies involving intricate hydraulic systems were also used to model cardiovascular systems as shown in [STEWART 1979, p. 43].

Another application area for analog computers was the preprocessing and analysis of ECG data. Operations such as filtering, peak detection, jitter analysis, etc., have been successfully implemented on analog computers.[134]

The following example shows how *respiratory arrhythmia*[135] has been studied on analog computers. The first such study is described in [CLYNES 1960] while a later study can be found in [ALBRECHT 1968]. Picture 13.33 shows the program used in this study. Its purpose is to determine the parameters controlling the heartbeat function $a(t)$.

The integrator in the lower left of this figure is used to generate $a(t)_{real}$ based on real ECG data of a test subject.[136] This function can then be compared to a function $a(t)_{simulated}$ generated by the remaining circuit based on the respiratory activity $r(t)$ of the same test subject.

---

[133] Cf. [HABERMEHL et al. 1969].
[134] These techniques have also been used to process EEG data, see [FEILMEIER 1974, p. 30].
[135] This term describes the effect that the interval Δt between two successive heartbeats is decreased during inhaling and is increased during exhaling.
[136] The integrator's mode of operation is controlled by a signal derived from the ECG signal. Normally the R wave of the signal is used for this purpose. Every occurrence of this R spike resets the integrator, which yields a ramp function at its output linearly increasing with the time elapsing between two successive heartbeats.

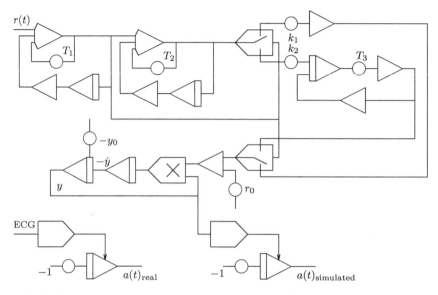

Fig. 13.33. Simulation program for the respiratory arrhythmia (see [ALBRECHT 1968, p. 2])

Based on this, the parameters of the simulator circuit can be tweaked until $a(t)_\text{real}$ and $a(t)_\text{simulated}$ are in good accordance. The resulting parameter set characterizes the behaviour of the sinu-atrial node of the test subject and the effects the subject's vagus nerve has on this.

13.7.4 Closed loop control studies

A nice example of the simulation of closed loop control in organisms is featured in [STEWART 1979, pp. 46 f.], which describes the regulation of CO_2 in the lung. The simulation of the adaptation process of the pupil to changes in environmental brightness is described in [GILOI 1975, pp. 157 ff.], [KÜNKEL 1961], and [LUNDERSTÄDT et al. 1981].

13.7.5 Neurophysiology

A central question in neurophysiology is how action potentials in nerve cells are generated and propagated. In 1952 ALAN LLOYD HODGKIN[137] and ANDREW FIELDING HUXLEY[138] developed a mathematical model describing signal gener-

---

[137] 02/05/1914–12/20/1998
[138] 11/22/1917–05/30/2012

ation and propagation in squid giant axons.[139] This model is described by the following differential equation[140]

$$C\dot{V} = -g_{\text{Na}}m^2 h(V - V_{\text{Na}}) - g_{\text{K}}n^4(V - V_{\text{K}}) - g_{\text{L}}(V - V_{\text{L}}) + I_a,$$

where g_i represents the conductivity caused by the potassium and sodium ions and the unavoidable leakage current in the axon, while $(V - V_i)$ represent differences to the equilibrium potentials V_i. The coefficients $0 \leq m, h, n \leq 1$ are determined by

$$\dot{m} = \alpha_m(V)(1-m) - \beta_m(V)m,$$
$$\dot{h} = \alpha_h(V)(1-h) - \beta_h(V)h, \text{ and}$$
$$\dot{n} = \alpha_n(V)(1-n) - \beta_n(V)n$$

where $\alpha_i(V)$ and $\beta_i(V)$ are determined through experiments. This model explains many experimental observations and formed the basis of later research by RICHARD FITZHUGH[141] and JIN-ICHI NAGUMO,[142] which resulted in the following set of differential equations that was extensively studied using analog computers:[143]

$$\dot{v} = v(a-v)(v-1) + w - I_a$$
$$\dot{w} = bv - \gamma w$$

A more complex model for neural bursting and spiking was devised by JAMES L. HINDMARSH and R. M. ROSE in the first half of the 1980s. This model is described by the following three coupled differential equations:[144]

$$\dot{x} = -ax^3 + bx^2 + y - z + I_{\text{ext}}$$
$$\dot{y} = -dx^2 + c - y$$
$$\dot{z} = r(s(x - x_r) - z)$$

These can be mechanised as shown in figure 13.34.[145] Figure 13.35 shows a typical simulation result for the membrane potential x.[146]

---

[139] These giant axons were often used in early neurophysiology research due to their size reaching up to 10 cm in length and up to 1 mm in diameter.
[140] Cf. [BROUWER 2007, p. 3].
[141] 03/30/1922–11/21/2007
[142] 1929–03/10/1999
[143] More information regarding these analog computer studies can be found in [BEKEY 1960].
[144] See [HINDMARSH et al. 1982] and [HINDMARSH et al. 1984] for details on the actual model.
[145] See [ULMANN 2021/1]. Scaling this model is quite involved and was done in this instance using a numeric solver to explore the value ranges of the various variables.
[146] [OCHS et al. 2021] describes a recent electric circuit implementation of the HINDMARSH-RODE model.

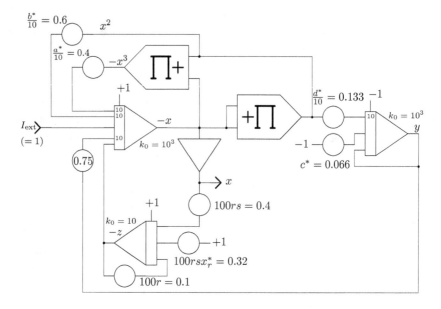

Fig. 13.34. Scaled analog computer setup for the HINDMARSH-ROSE model

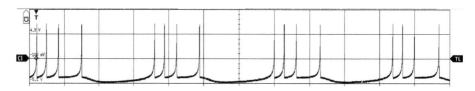

Fig. 13.35. Typical simulation result for a spiking and bursting neuron according to the HINDMARSH-ROSE model

13.7.6 Epidemiology

A simple example for the application of analog computers in epidemiological research is given in [Hitachi 200X, p. 14], which describes a closed system containing a number of individuals, a few of whom are initally infected with a contagious disease. An infected individual can then infect other persons who are not already immune.[147] The output of a typical simulation run is shown in figure 13.37.[148]

---

[147] The analog computer used in this study was a Hitachi 200X featuring non-inverting integrators and summers, which has to be taken into account for the program shown.

[148] A more realistic approach to simulating the spreading of contagious disaeses is the *SEIR*-model, taking susceptible, exposed, infected, and recovered persons into account. This can be easily implemented on an analog computer. The basics of this model are described in [BÄRWOLFF 2021].

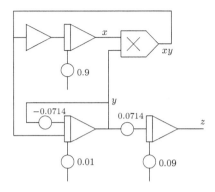
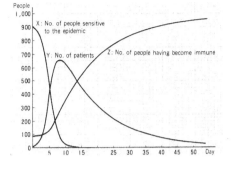

Fig. 13.36. Setup for an infection spreading simulation (see [Hitachi 200X, p. 14])

Fig. 13.37. Results of a simulation run for the infection spreading simulation ([Hitachi 200X, p. 14])

13.7.7 Aerospace medicine

Analog computers were also used extensively in aerospace medicine research. [PAYNTER ed. 1955] describes the analysis of the forces acting on a pilot bailing out with an ejection seat. For this study a special-purpose analog computer was developed to model the effects of accelerations, which typically arise during ejection. The purpose of such *body dynamics studies* is to determine optimal acceleration curves that guarantee a quick and safe separation of the ejection seat from a failing plane, while not harming its passenger.[149]

Analog computers were also used to control large-scale centrifuges used to determine the performance of pilots subjected to high and varying accelerations as would be encountered during emergency situations, rocket launches, or re-entry manoeuvres. A highly complex centrifuge control system based on a large-scale analog computer installation is described in [STONE et al.].

13.7.8 Locomotor systems

Analog computers were also used at the boundary between biology and mechanics in locomotor studies. [TOMOVIC et al. 1961] performed simulations to determine the degree of autarky in typical locomotor systems. Similar studies are described in [BEKEY et al. 1968, pp. 417 ff.].[150]

---

[149] See [PAYNE 1988, p. 272]. The parameter of interest in such studies is the first derivative of the acceleration experienced by a body, the *jerk* or *jolt*.
[150] Based on these a mechanical quadruped was developed.

13.7.9 Dosimetry

Analog Computers were heavily used in nuclear medicine as an example from the *University of Chicago hospital* shows. Here a *Beckman EASE 2132* analog computer was used from the mid 1960s well into the 1970s to perform dosimetry calculations, a task required for cancer treatments. In addition to that, this particular machine was also used for basic research involving nuclear and non-nuclear applications.[151]

13.8 Geology and marine science

Analog computer applications in the fields of geology and marine sciences were often of great commercial value, especially with respect to prospecting tasks.

13.8.1 Oil and gas reservoirs

Oil and gas production relies heavily on hydraulic models of real oil and gas fields. These models are highly complex and require the solution of partial differential equations, typically with four variables. Since good models result in better revenues, several very large and complex direct analog computers have been built to study the effects caused by oil wells to the below-ground distribution of oil and gas.

An impressive example for such a direct analog computer is the *ZI-S* system, which was developed in 1958 at the *All-Union Scientific Research Oil-Gas Institute* in the Soviet Union. This machine consisted of a three-dimensional grid of passive computing elements with about 20,000 vertices. In addition to this, it also contained a large number of operational amplifiers and other active components.[152] Such a large number of components was required to achieve a sufficiently high resolution of the model.

This ZI-S analog computer shown in figure 13.38 was based on

$$\frac{\partial}{\partial x}\left(A_1(x,y)\frac{\partial P}{\partial x}\right) + \frac{\partial}{\partial y}\left(A_2(x,y)\frac{\partial P}{\partial y}\right) = 0$$

and

$$\frac{\partial}{\partial x}\left(A_1(x,y)\frac{\partial P}{\partial x}\right) + \frac{\partial}{\partial y}\left(A_2(x,y)\frac{\partial P}{\partial y}\right) = A(x,y)\frac{\partial P}{\partial t}$$

---

151 GREG PARKHOUSE, personal communication to the author.
152 See [USHAKOV 1958/1].

 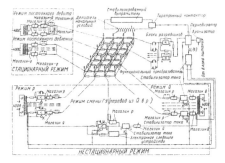

Fig. 13.38. Overview of the ZI-S computer ([USHAKOV 1958/1, p. 1812])

Fig. 13.39. Structure of the ZI-S computer ([USHAKOV 1958/1, p. 1813])

respectively, where x and y represent coordinates in a single layer of the oil or gas reservoir, while P and A_i denote pressures and hydraulic conductivities.

Using this system, reservoirs with a radius of up to 120 km were studied. These simulations considered up to 750 wells depleting the reservoir. The passive analog computer simulating the reservoir itself can be seen in the middle of figure 13.39. The reservoir is modelled with a three-dimensional mesh of resistors and capacitors, which implement lag phenomena, etc. The surrounding modules implement the wells and the instrumentation of the simulation circuits.[153]

13.8.2 Seismology

Prospecting typically relies heavily on the analysis of seismograms to detect belowground reservoirs. This task traditionally requires the fastest stored-program computers available and was thus done, or at least supported by, analog and hybrid computers in the years before the mid 1970s.

[EVANS 1959] describes a special-purpose analog computer developed and used to generate seismic weathering time corrections for refraction and reflection studies.[154] The implementation of this analog computer is quite similar to the BPRR-2 system described in section 13.1.9. Indirect analog computers were used by [SUTTON et al. 1963] for filtering tasks, as well as spectrum analyses on seismic data sets.[155]

---

[153] The rather unique developments of analog computing the Soviet Union are described in detail in [ABRAMOVITCH 2005].
[154] This correction takes the properties of a low-velocity zone, a *weathering layer*, into account.
[155] This study also contains a comparison of spectral data generated by an analog computer approach with data generated on a digital computer.

13.8.3 Ray tracing

Acoustic ray tracing is a common task in marine applications using sonar data and of high commercial and especially military value. This task is quite complex due to varying salt concentrations, temperature differences between water layers, etc.

[LIGHT et al. 1966] describes a ray tracer based on a small EAI TR-10[156] analog computer. The task of tracing a single acoustic wave from a transmitter through a complex medium like open sea to a pickup involves the solution of a second degree differential equation. The analog computer used in this setup contained 17 operational amplifiers and a special function generator to model the structure of the ocean bed reflecting acoustic waves. A typical simulation run took about 15 minutes to complete. Figures 13.40 and 13.41 show the analog computer setup as well as a typical simulation result.

13.9 Economics

ALBAN WILLIAM PHILLIPS,[157] a New Zealand economist, was one of the first people to build economic models using analog computers. Being a trained engineer, the idea of applying engineering principles to economic problems was an obvious approach. Consequently, in 1949 he developed and built *MONIAC*,[158] a hydraulic analog computer, whose structure is shown in figure 13.42.[159]

This system, of which eventually ten were built, was affectionately known as the *financephalograph* and was not only used for early research in mathematical economics but also, due to its structural clearness, for teaching purposes. The flow of money in an economy is represented by coloured water flowing through various hydraulic computing elements. MONIAC features nine control variables representing tax load, import/export subsidies, and the like.[160] Recent detailed descriptions of the overall system can be found in [RYDER 2009] and [RYDER et al. 2021].[161]

In later years, PHILIPPS used an electronic analog computer at the *National Physical Laboratory (NPL)* as well as a stored-program digital computer, for his

---

[156] Cf. section 6.4.
[157] 11/18/1914–03/04/1975
[158] Short for *Monetary National Income Automatic Computer*.
[159] See [PHILLIPS 1950], [Fortune 1952], [SWADE 1995], and [CARE 2006].
[160] MONIAC also inspired some caricatures like that shown in [SWADE 1995, p. 16] as well as the analog computer sitting in the basement of the bank in TERRY PRATCHETT's (04/28/1948–03/12/2015) novel *Making Money*.
[161] [ENGEL et al. 1964] and [ZHANG 2015] give a comprehensive introduction to the description of economic systems by means of differential equations.

13.9 Economics — 293

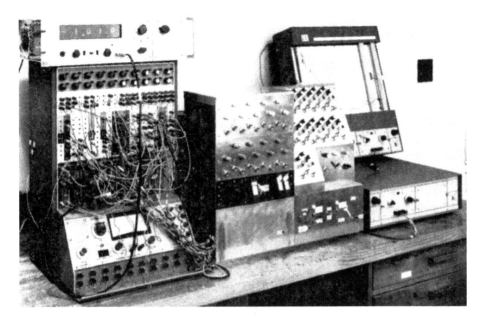

Fig. 13.40. The ray tracer developed by LIGHT, BADGER, and BARNES (see [LIGHT et al. 1966, p. 724])

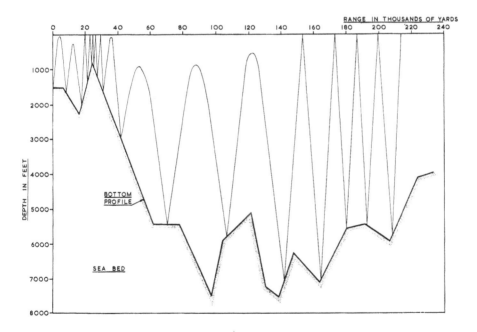

Fig. 13.41. Result of a simulation run of the ray tracer (see [LIGHT et al. 1966, p. 724])

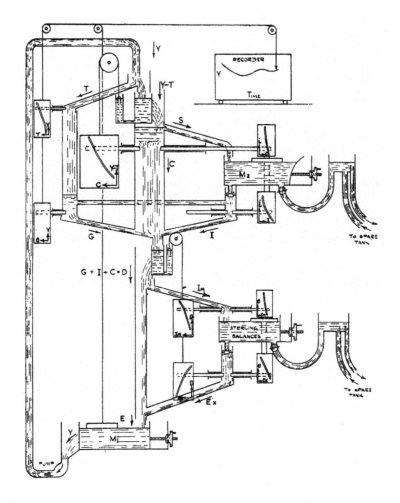

Fig. 13.42. PHILLIPS' hydraulic simulator for economic dynamics (see [PHILLIPS 1950, p. 302])

research.[162] Interestingly, he still preferred the hydraulic MONIAC despite its shortcomings with respect to accuracy, which were more than outweighed by its high clarity. He characterized his models as being "*exposition[s] rather than accurate calculation[s]*".[163]

Analog computer models for the study of economic problems were still used in later years as a number of publications like [HÜLSENBERG et al. 1975], [HERSCHEL 1961], [KREGELOH 1956], and [JACKSON 1960, pp. 363 ff.] show.

---

[162] See [SWADE 1995, p. 17].
[163] See [SWADE 1995, p. 17].

An interesting little preprogrammed "personal analog computer" was the *ISEC 250*.[164] This machine was aimed at stock investment analysis and used an *ISEC index*, a weekly index table distributed by the manufacturer, as well as real-time inputs based on current market behaviour entered through means of some potentiometers and switches, to give an *"instant evaluation of any programmed stock"*.[165]

[STROTZ et al. 1951], [STROTZ et al. 1953], and [ENGEL et al. 1964] contain a wealth of practical examples for the application of analog computers to problems in economics such as the effects of marketing and production inertias, national income models, a nonlinear theory of the business cycle, the effects of policy decisions, and many more. The analog simulation of a colonial socio-economic system is described in detail in [HOWARD 1961].

13.10 Power engineering

In the early 20th century, the field of power engineering developed a demand for computational power that could only be fulfilled by analog and later hybrid computers. As with nuclear reactors, many questions regarding power systems cannot be analysed using real equipment due to the inherent dangers of such experiments.[166] Consequently, simulation techniques were, and still are, the only way to solve a number of vital problems in this field. The following sections describe typical simulation tasks.

13.10.1 Generators

A study of the dynamic behaviour of generators feeding large power grids containing only asynchronous machines as loads, is described in [JUSLIN 1981].[167] The main problem of such power grids is that asynchronous machines exhibit vastly different load characteristics depending on their state of operation. A starting asynchronous machine behaves like a transformer with its secondary shorted, an effect vanishing with increasing motor speed. This study's focus is on severe faults influencing large parts of the simulated power grid, which will result in oscillatory behaviour that might be unstable.

---

[164] See [isec/1] and [isec/2].
[165] See [isec/1, p. 33].
[166] See [NORONHA].
[167] These constraints are not unrealistic in a mostly industrial environment.

[GILBERT 1970] describes the simulation of a generator with a parasitically loaded speed controller to be used in aerospace applications. This model consists of the turbine driven alternator, a voltage regulator and exciter, a parasitic load speed controller, and the load itself.

13.10.2 Transformers

Although a transformer might appear simple at first sight, its behaviour is non-trivial to model and predict. Many studies into the dynamic characteristics of transformers have been conducted. [GILOI 1960] and [Telefunken 1963/2] cover the simulation of a transformer under mixed resistive/inductive or resistive/capacitive loads. In this study, which also takes the dynamic B-H magnetisation curve[168] of the transformer into account, the primary and secondary currents are simulated.

13.10.3 Power inverters and rectifiers

Much more complex than this is the simulation of power inverters and (controlled) rectifiers. Devices like these are essential for high-voltage direct current systems, which were first developed in the early 20th century. Inverters are also commonly used to drive induction motors, etc.

The analysis of a controlled three-phase rectifier by an analog computer model is described by [EYMAN et al. 1976], while similar studies can be found in [BLUM et al.], [TISDALE], and [NAVA-SEGURA et al.]. The latter focuses on fault situations such as shorted thyristors. [SHORE 1977] describes a hybrid computer based simulation as well as an online digital computer control setup for a high voltage direct current link.

The conversion of a direct current into alternating or three-phase current by power inverters has been analysed in analog computer studies such as [KRAUSE 1970], where basic simulation techniques in this field are described, and [FORNEL et al. 1981], where a power inverter loaded by an asynchronous machine is modelled. [BELLINI et al. 1975] describes the analysis of an inverter driving electromechanical actuators, while [WINARNO 1982] focuses on the special requirements posed by large photovoltaic systems. The analog model of a complex high-voltage direct current system is described in [SADEK 1976].

[BORCHARDT et al. 1977] gives a detailed description of the simulation of a high-power inverter system driving the magnets of a large synchrotron at DESY.

---

[168] This magnetization curve is implemented by a diode function generator. It should be noted that this study contains a good example of the implementation of implicit functions.

This study was done with the synchrotron already in operation as the need for significant changes in the power supply system for the magnets arose during its operation. The system itself could not be used to take measurements during parameter changes since this would have shortened its beam-time. Thus, an extensive simulation was performed on a *HRS 860* Telefunken hybrid computer consisting of a RA 770 precision analog computer and an attached TR-86 stored-program digital computer. This model even took care of intrinsic effects of heavily loaded thyristors and other parts of the inverter system and proved to be so accurate that the conversion of the system was performed successfully mainly relying on the simulation results from this study.

Another interesting analog computer study regarding the behaviour of inverters during severe fault scenarios is described by [BORCHARDT et al. 1969]. The simulation focused on the development of suitable protection measures for large inverter systems. [BORCHARDT et al. 1976] describes an analog computer model of a quadrupole magnet power supply used for ejecting the beam from a synchrotron.

13.10.4 Transmission lines

Long power lines are elaborate structures exhibiting a complex behaviour with phenomena such as traveling waves, etc. A summary of typical hybrid computer approaches for modeling power lines can be found in [BAUN 1970]. A study with special emphasis on traveling wave phenomena is described by [THOMAS 1968]. An interesting aspect of simulations like these is that time delay units are required to accurately model the delays encountered in long power line systems. Magnetic tape units and capacitor wheels[169] have been used successfully for this.

13.10.5 Frequency control

While power grids are complicated dynamic systems already, things get even more complicated when multiple power grids are coupled together, as [PAYNTER 1955/2, p. 229] notes:

> "[...] the [...] problem is fundamentally complex due to the multiplicity of generally disparate machines and regulators coupled together by a large number of 'elastic' links."

Early work on the analysis of coupled power grids using direct as well as indirect analog computers was begun in 1947 at MIT. [KRAUSE et al. 1977] describes a study of shaft distortion in generators caused by misaligned grid frequencies when

---

[169] Cf. section 4.11.

disjunct grids are coupled. Special topics regarding the synchronization and control of such coupled power grids are treated in [SYDOW 1964, pp. 214 ff.].

13.10.6 Power grid simulation

Equally complex is the study of whole power grids. Such studies are of great importance due to the dependency of our culture on the ubiquitous availability of electric energy. Since no analytic solutions are possible for complex systems like these, simulations based on analogies are a suitable approach to gain deeper insight into the intricate behaviour of such systems. A good overview of the historical development of such simulation techniques can be found in [KRAUSE 1971] and [KRAUSE 1974].

Driven by the rapidly rising complexity of the United States' power grid, the first special purpose analog computer, a *network analyser*, was conceived in 1925. The purpose of such analysers has been described by [NOLAN 1955, pp. 111 f.] as follows:

> "*Network analyzers have been used to solve quickly the many and various problems concerned with the operation of power systems. They are practical, adjustable miniature power systems. They can be used to analyze results during the progress of a system study and therefore play an active part in system planning as well as checking the performance of completed systems.*"

Early network analysers were direct analog computers, modeling the power grid by means of discrete and mostly passive computing elements. Later studies made extensive use of indirect analog computers such as the *Beckman EASE 1032* system shown in figure 13.43, which was used by the *Bonneville Power Administration* for power grid simulations. A typical problem arising in such simulations on indirect analog computers are algebraic loops.[170] Basic techniques for avoiding such loops in power grid simulations are described in [GILOI et al. 1963, pp. 326 ff.] and [GILOI 1962].

An example of a power grid simulator can be found in [MICHAELS] and [MICHAELS et al.] where a high-voltage distribution system consisting of five power stations and about a dozen transmission lines is simulated. This system was used to train switch room operators. An even more complex system based on a hybrid computer is described in [ENNS et al.]. This hybrid simulation was able to determine the state of a power grid containing 181 power lines from a given

---

[170] Direct analog computers can tolerate such loops due to their lack of amplifying stages coupling the passive computing elements.

Fig. 13.43. EASE 1032 analog computer used at the Bonneville Power Administration for power grid simulations (see [N. N. 1957/2])

set of initial conditions in 20 seconds, while the same computation using only the digital computer part of the hybrid installation required 30 minutes of run time.

The effect of transients caused by breaking conductors, short-circuits, etc., are studied in [THOMAS et al. 1968]. These simulations also required time delay units, which were implemented with a capacitor-wheel containing 24 capacitors. The simulation of transients caused by lightning strikes in electric power transformation substations is described in detail in [HEDIN et al.].

The effects of capacitive and inductive crosstalk between high-voltage power lines have been studied by simulating a power grid consisting of one 500 kV section and two 250 kV sections on a large analog computer installation. These studies are described in [THOMAS et al./1] and [THOMAS et al./2]. This analog computer contained 180 operational amplifiers, 350 coefficient potentiometers, and 24 time delay units – telltale signs of the complexity of the problem under consideration.

Fig. 13.44. Power station simulator (cf. [WHITESELL et al. 1969, p. 85])

A large power grid simulator, the *Hybrid Computer Power Simulator (HCS)*, was proposed in 1976 and is described in [JANAC 1976]. This system would have consisted of a special purpose analog computer containing specialized computing elements representing typical analogies found in such simulations. Instead of a traditional patch panel, an automatic patch panel, an *autopatch*, was proposed. Unfortunately, this machine was never built and autopatch systems did not prove commercially viable due to the extremely large number of switching elements required.[171]

13.10.7 Power station simulation

[WHITESELL et al. 1969] describes an analog computer based training system for power station operators. It modelled the "*two separate, but tightly interconnected, steam and electric power generating stations at the Whiting refinery*". Learning how to control such a complex system is a daunting task and learning-by-doing is not an option in the demanding environment of a refinery. The impressive analog computer setup for this simulator is shown in figure 13.44.

---

[171] An interesting study on autopatch systems was performed by GEORGE HANNAUER on behalf of NASA, cf. [HANNAUER 1968].

13.10.8 Dispatch computers

There are a lot of optimisation problems ranging from scheduling questions to the supply of combustibles, etc., to be solved for running power systems efficiently. An analog computer study aimed at the computation of optimal schedules for a hydrothermal power plant is described by [PERERA 1969].[172]

Special purpose analog computers, *Electronic Dispatch Computers (EDC)*, have been developed to perform such optimisation tasks and are described by [WASHBURN 1962]. The savings made possible by the application of analog computing techniques were remarkable and justified the development of large analog computer installations:[173]

> "*Estimated annual fuel savings obtainable with EDCs may approach US$50 per megawatt installed. For a typical 1,000-megawatt system this saving can amount to US$30,000 to US$50,000 per year and thus may warrant an investment of US$250,000.*"

Figure 13.45 shows an example of an electronic dispatch computer, the *Goodyear Electronic Differential Analyser (GEDA)*. This power dispatch computer saved US$ 200,000 annually (see [N. N. 1957/4]). A not-too-complex optimisation example regarding a thermal plant is described in [HEINHOLD et al., pp. 224 ff.].

13.11 Electronics and telecommunications

The rapid development of electronics and especially telecommunication technology since World War II resulted in an ever-increasing demand for component and circuit simulations. As [KETTEL et al. 1967, p. 3][174] put it, the more complex electronic circuits get, the more important it is to gain insight in their respective behaviour by mathematical models and analogies instead of building breadboard prototypes. The following sections give some examples of such work.

13.11.1 Circuit simulation

An interesting analysis of the behaviour of a transistorised circuit subjected to short transients is described by [RANFFT et al. 1977]. This simulation makes extensive use of the ability of an analog computer to perform temporal expansions

---

[172] This study is based on the *coordination equations* developed by CHANDLER, DANDENO, GLIMN, and KIRCHMAYER, see [CHANDLER et al. 1961].
[173] See [WASHBURN 1962, p. 5-155].
[174] This work contains a good overview of typical applications of analog and hybrid computers in these fields.

Fig. 13.45. GEDA power dispatch computer (see [N. N. 1957/4])

since the effects of interest, which are analysed by the simulation, take only a few nanoseconds (10^{-9} s). This study strikingly shows the remarkable complexity of such circuit simulations. As [RANFFT et al. 1977, p. 76] put it,

> "[s]tandard analog computers are normally not well suited for simulation of transistor circuits."

[BALABAN, p. 771] remarks that

> "a six to eight transistor circuit can be patched on a large analog computer."

ROLAND RANFFT and HANS-MARTIN REIN set out to develop a special-purpose analog computer aimed at the simulation of electronic circuits:[175]

> "*Its advantages, when compared to the usual digital computer simulation, are low costs, simple operation, and short simulation time, allowing a fast man-machine dialog. This simulation method has already been successfully used in the design of several high-speed integrated circuits.*"

The main advantage of this system was not only its high simulation speed but also the high degree of interactivity made possible by the analog computer. The projected cost for an implementation containing 20 transistor models, 12 SCHOTTKY[176] diodes, 24 resistors, and two current/voltage sources was only US$ 3,600. The same amount of money would have had to be spent to perform 100 to 200 transient analyses on a rented stored-program digital computer.

Another simulator, *HYPAC*, short for *Hybrid PACTOLUS*,[177] is described by [BALABAN]. This system was basically a hybrid computer with a special setup on the analog part simulating basic electronic circuit elements. Using the digital computer of the hybrid computer setup, the analog computer's elements were time-shared between the various circuit elements of the circuit to be simulated. This made it possible to simulate circuits of nearly unlimited complexity given enough time.

An extensive simulation study of mixer circuits using a Telefunken RA 463/2 analog computer[178] can be found in [SIERCK 1963]. This study was unique at its time as it was the first all-embracing simulation of such circuits and showed that the results achieved by means of analog computers were meaningful and could be used for actual circuit design.

[BECK et al. 1970] describes a hybrid computer approach to circuit design with a number of practical examples.

13.11.2 Frequency response

Before spectrum analysers were introduced as independent devices into laboratories, determining the frequency response of electronic circuits was a laborious task that could be simplified considerably by an analog computer. Depending on the complexity of the circuit to be analysed and the frequency range of interest, the frequency/phase response of a prototype circuit could be measured and processed

---

[175] See [RANFFT et al. 1977, p. 76].
[176] WALTER HANS SCHOTTKY, 07/23/1886–03/04/1976
[177] See chapter 12 for more information about *PACTOLUS*.
[178] See section 6.1.

directly on an analog computer. It is also possible to simulate the device on the same analog or hybrid computer performing the processing of the data obtained. The advantage of the latter approach is the possibility to change the time-scaling, which is not possible when working with an actual prototype circuit.

Typical analog computer setups for performing a spectral analysis are described in [BARD 1965, pp. 212 f.] and [GILOI et al. 1963]. These analog computer programs are often based on a maximum detector circuit,[179] which is basically a delay circuit. Using a sweep generator as shown in section 8.2 the frequency response of a circuit, be it real or simulated, can then be determined.[180] Another example for such a technique can be found in [SCHÜSSLER 1961] where a sampling process is used to generate a frequency response plot.

13.11.3 Filter design

The design of filter circuits is not too different from the problems described above. An example of such a simulation can be found in [LARROWE 1966], which describes the design of a quadrature band-pass filter.[181] [WADEL 1956] describes the simulation of digital [sic] filters on an analog computer.

A very complex example of filter design is given in [GILOI 1961]. This study describes the development of a WIENER filter on an analog computer, which requires the solution of integral equations.[182] The technique used in this paper is to have the machine time running backwards, which was done by using an analog tape drive on which values were prerecorded in a preliminary step and played back in reverse direction for the actual determination of the filter parameters. A similar technique used to determine solutions of optimum control problems is described in [ANDERSON et al. 1967].

13.11.4 Modulators and demodulators

Modulator and demodulator circuits were also developed by means of analog computer simulations. The design of phase detectors and frequency discriminators is described in [MANSKE 1968]. In this study the circuit under development as well as the processing of measurement data has been implemented on an analog computer. The analog simulation of a frequency modulation based communication

---

179 Cf. [SYDOW 1964, pp. 265 f.].
180 The methods described in section 13.1.7 can also be used in this context.
181 Such a filter basically consists of two separate filters yielding a signal pair with a phase difference of $\pi/2$.
182 See section 13.1.2.

system is described in [HU 1972]. [KETTEL 1960, pp. 170 f.] describes the study of modulators and demodulators for telemetering applications on an analog computer.

13.11.5 Antenna and radar systems

Around 1950 J. A. HAMMER developed an analog computer for the simulation and study of radar[183] antenna systems at the *Nederlands Radar Proefstation*, which is described in detail in [HAMMER 1956], [PRONK 2019], and [GOLDBOHM 1993]. Radar antennas pose some special problems such as *sidelobes*,[184] which must be minimised to obtain a beam as narrow as possible. The antennas under consideration here were slot antennas with up to 41 slots. Calculating the resulting radiation pattern is a formidable task, which justified the construction of a special purpose analog computer. The HAMMER *computer* as it became known used electronic as well as electromechanical computing elements effectively implementing a complex series analyser. Computing the E-field of an antenna took 12 seconds on this machine while contemporary digital computers required up to 700 seconds for the same task.

[MITCHELL 1960] describes an analog computer for the simulation of the directivity characteristics of complex antenna arrays. The electromechanical analog computer built for this purpose was capable of simulating arrays with 50 to 200 aerials.

[BICKEL et al. 1957] describes a complex radar simulation system taking into account a variety of adverse effects such as *target scintillation*,[185] various fading effects, and receiver noise.

13.12 Automation

There are many applications for analog and hybrid computers in the fields of industrial process measurement and control technologies. Obviously, the development of new processes and new control systems can profit from analog computer studies. Analog computers were sometimes even used as integral parts of complex control systems as [KETTEL 1960, p. 165] noted:

---

183 Short for *radio detection and ranging*.
184 Sidelobes are local maxima of emission of an antenna except the desired mainlobe.
185 This term describes the effect that the amplitude of the signal reflected by a target depends heavily on its attitude and can change quickly.

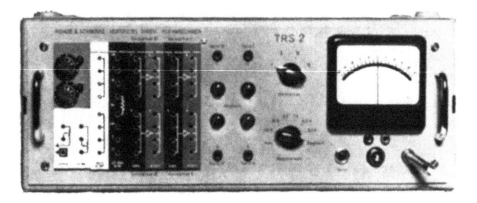

Fig. 13.46. Data capturing and preprocessing system TRS-2 (Rhode & Schwarz, see [N. N. 1961])

> "Not only can an analog computer serve as a computer to simulate control systems, but can also be part of a control system itself. As such it allows the optimisation of the control system in place."[186]

13.12.1 Data processing

An example for such an embedded analog computer is the *TRS-2* data acquisition and processing system shown in figure 13.46. This system was developed and manufactured by *Rohde & Schwarz* and contained standard analog computer components manufactured by EAI for their TR-10 and TR-20 tabletop analog computers. The system could be equipped with standard analog computer modules as well as with user specific modules implementing programs for special applications.[187]

13.12.2 Correlation analysis

[WIERWILLE et al. 1968] describes the use of a hybrid computer for performing autocorrelation analyses on vibration data gathered during rocket tests. A fully digital implementation is also described and both methods are compared.

---

186 "*Ein vorhandener Analogrechner ist nicht nur ein Rechengerät, das z. B. die Berechnung eines Regelsystems erlaubt, sondern es kann ebensogut, vor allem wenn es klein und transportabel ist, als elektrischer Regler in ein Regelsystem eingefügt werden, damit in Verbindung mit der echten Regelstrecke der Regler optimal dimensioniert werden kann.*"

187 Cf. [N. N. 1961]. The system shown in figure 13.46 contains from left to right two coefficient potentiometers, a couple of fixed resistors, and two chopper-stabilised dual operational amplifiers.

13.12.3 Closed loop control and servo systems

The earliest example of an analog computer used to analyse and demonstrate closed loop control systems is GEORGE A. PHILBRICK's Polyphemus, described in section 3.2. Another early analog process simulation system, the *Electro-Analogue*, is described by [JANSSEN et al. 1955]. This process and control simulator consisted of three largely independent subsystems: A model of the control circuit, a model of the process being controlled, and an output system containing an oscilloscope display, some panel meters, and recorders. An interesting aspect is the implementation of time delays, which are often required in control system simulations to model transport delays, etc. These time delay units were implemented as a string of capacitors and inductors with switch-selectable taps representing various delay times.

Further examples can be found in [JAMES et al. 1971, pp. 215 ff.] and [WORLEY 1962], covering bang-bang servos, proportional controllers, etc. The simulation and development of governors is described in detail by [AMMON et al. 1959]. [LOTZ 1969, pp. 211 ff.] describes the simulation of a control system for a hydraulic system. [SYDOW 1964] describes a complex control system study for a paper machine. The emphasis of this study is on the control of 40 steam heated cylinders, which must run at precisely controlled speeds to avoid tearing the paper running through the machine. [KOENIG et al. 1955] describes the treatment of flyweight governors used in a hydroelectric power station by means of an analog computer.

The analysis of damped, non-linear servo systems using an analog computers is described in detail by [CALDWELL et al. 1955]. Non-linear servo systems are of great importance for practical applications but are notoriously complex to model and analyse. Additional information on such systems can be found in [HURST]. The determination of parameters for a non-linear control system using an EAI 231R analog computer[188] is covered by [SURYANARAYANAN et al. 1968].

13.12.4 Sampling systems

Many control systems are, in fact, sampling systems where input signals are sampled at discrete time intervals. These values are then used to control an underlying process. The simulation and analysis of these sampled data systems is difficult to perform on an analog computer since the sampling process has to be implemented by trick circuits such as bucket-brigade circuits consisting of chained integrators under individual control. Typically, hybrid computers were used to model sampling systems since their digital part is well suited to implement the sampling

---

[188] See section 6.3.

process, while the underlying process can be modelled on the analog system. Typical examples for this can be found in [SCHNEIDER 1960] and [SIMONS et al.].

13.12.5 Embedded systems

Analog computers were used early on as embedded components of complex control systems, as described in [KORN et al. 1956, pp. 109 f.]:

> "D-c analog representation of automatic control systems has proved to be a powerful aid in the design of prototype models and pilot plants and in the determination of starting procedures and of optimum controller settings after changes in raw materials or other conditions. But the utility of d-c analog techniques for automatic control applications is not restricted to computations of this type. Special-purpose d-c analog computers, which may often be conveniently assembled from the standard components of commercially available machines, can themselves serve as control-system elements in many applications suited to their characteristics."

An example for such modules has already been shown in figure 13.31 in section 13.6.4. Other examples include the computing elements developed by Telefunken for the RA 800 analog computer,[189] which were also sold as building blocks for industrial applications.[190]

The basics of analog process computers are covered in [LUDWIG et al. 1974], while [WEITNER 1955] describes the actual implementation of a closed loop model. The design and implementation of a special purpose amplifier with a characteristic of $\sqrt[3]{x}$ is described in [KINZEL et al. 1962/1] and [KINZEL et al. 1962/2]. This amplifier was actually used to control injection parameters in large diesel engines.

13.13 Process engineering

Analog computers were invaluable tools for process engineering applications and consequently most chemical companies set up dedicated analog computing centers to aid their development departments. The importance of analog computers in this area can be seen by the following quotation from [HOLST 1982, pp. 316 f.]:

> "Analog computing capacity [at Foxboro] has increased from some 8-10 operational amplifiers in 1938 to more than 150 thirty years later."

---

[189] Cf. section 6.4.
[190] See [N. N. 1960].

Table 13.1 gives an overview of the analog computing capacity installed at some major chemical companies in 1961.[191]

A wealth of information and examples regarding the use of analog computers in process engineering applications can be found in [WAGNER 1972], [WORLEY 1962], [HOLST 1982], and [RAMIREZ 1976]. A small, yet powerful, process simulation system based on an EAI *TR-48* analog computer[192] was introduced at the *Systems Engineering Conference* in 1964 by EAI.

A highly complex simulation of a solvent recovery process is described by [LEWIS 1958]. It requires 120 summers/integrators, 120 limiters, 48 summers, 28 servo multipliers, 16 photoformers, 300 coefficient potentiometers, and 186 external precision capacitors.[193] Another example for a complex process engineering simulation can be found in [GRAEFE et al. 1974] where a copper melting process is modeled, even taking the discontinuous flow of materials into account.

13.13.1 Mixing tanks, heat exchangers, evaporators, and distillation columns

A common and basic element in every chemical plant is the mixing tank, which can be simulated quite straightforwardly on an analog computer as described in [RAMIREZ 1976].

The simulation of heat exchangers and evaporators is more complex since these are described by partial differential equations.[194] The simulation of an evaporator on an EAI 590 hybrid computer system is described by [OLIVER et al. 1974], while a more complex example involving a heat exchanger reactor can be found in [CARLSON et al. 1968].[195] The very complex simulation of a FISCHER-TROPSCH[196] reactor for coal liquefaction is treated in [GOVINDARAO 1975]. The main challenge in this type of simulation is the fact that gaseous, liquid, and solid phases of the reactants coexist simultaneously in the reactor system.[197] This particular simulation required 33 integrators, twelve sample and hold elements, 13 summers, ten inverters, 18 multipliers, 80 coefficient potentiometers, and a digital control system.

---

[191] Equally large installations were used in other countries including the Soviet Union (see [USHAKOV 1958/1] and [USHAKOV 1958/2]).
[192] This medium-scale analog computer contains 48 operational amplifiers.
[193] Such complex simulations were the exception.
[194] Basic information about this problem can be found in [BILLET 1965].
[195] Cf. [SCHÖNE 1976/2] for a general presentation of the mathematical modeling of heat exchangers.
[196] FRANZ FISCHER, 03/19/1877–12/01/1947, and HANS TROPSCH, 10/07/1889–10/08/1935
[197] The basics of the simulation of chemical reactor systems are described in [STARNICK 1976].

Company	Year	Number of amplifiers	Manufacturer
Dow Chemical Co.			
Midland Division	1954	20	Beckman (Berkeley)
	1961	140	EAI
Texas Division	1956	30	Daystrom (Heath)
	1961	80	Philbrick
E. I. du Pont de Nemours & Co.			
Newark, Del.	1950	30	Beckman (Berkeley)
	1955	50	Beckman (Berkeley)
	1958	120	EAI
	1960	300	EAI
Experimental Station, Wilmington, Del.	1960	70	Computer Systems
Monsanto Chemical Co.			
St. Louis, Mo.	1957	116	EAI
	1958	24	EAI
	1959	88	EAI
Ohio Oil Co.			
Denver, Colorado	1957	56	EAI
Humble Oil & Refining Co.			
Baytown, Texas	1960	80	EAI
	1961	80	EAI
Baton Rouge, La.	1959	80	EAI
	1960	40	EAI
Esso Research & Engineering Co.			
Florham Park, N.J.	1959	40	EAI
	1959	40	EAI
	1960	80	EAI
	1960	80	EAI
American Oil Co.			
Whiting, Ind.	1955	–	EAI
	1957	168	EAI
Standard Oil Co.			
Cleveland, Ohio	1955	90	Beckman (Berkeley)
	1957	10	Beckman (Berkeley)
	1961	170	Beckman (Berkeley)
Union Carbide Olefins Co.			
South Charleston	1956	30	EAI
	1958	60	EAI
	1959	60	EAI
Thiokol Chemical Corp.			
Brigham City, Utah	1959	168	EAI
Phillips Petroleum Co.			
Bartlesville, Okla.	1959	80	EAI
	1960	80	EAI
Chemstrand Corp.			
Decatur, Ala.	1960	80	EAI
Shell Oil Co.			
Shell Chemical Corp.	1960	120	EAI
Development Corp.	1956	24	Goodyear
	1957	24	Goodyear
	1960	10	Donner Scientific
	1960	10	Donner Scientific
Hercules Powder Co.			
Wilmington, Del.	1960	44	Beckman (Berkeley)
Daystrom, Inc.			
La Jolla, Calif.	1960	100	Computer Systems

Table 13.1. Analog computer installations for research in petrochemistry in 1961 (see [CARLSON et al. 1967, p. 356])

Even more complex are simulations of fractionating columns containing many column trays. A simulation involving three trays is described in [N. N. 1958/2], while [WILLIAMS et al. 1958] describes a more complex system with five trays.[198] The complexity of such simulations is caused by the complex mixing processes and phase transitions taking place at the column trays. A six tray fractionating column simulation is described in [RAMIREZ 1976, pp.28 ff.].[199]

13.13.2 Adaptive control

The implementation and analysis of an adaptive control system[200] is described in [POWELL]. The system covered in this study operates in two timescales: A short timescale subsystem implementing a predictor and a slow running subsystem implementing the actual control loop using the values generated by the predictor as its inputs.

13.13.3 Parameter determination and optimisation

Another important application area for analog computers was the determination and optimisation of process control parameters. The complexity of such control problems often exceeds that of other optimisation tasks and typically uses hybrid computers to implement gradient methods. Examples of these techniques and applications can be found in a variety of sources such as [KOPACEK], [TROCH 1977], [WOŹNIAKOWSKI 1977], [GRÖBNER 1961], [KRAMER 1968], [KRAMER], and finally [JAMSHIDI 1976].

[PICENI et al. 1975] describes the evaluation of operating parameters for industrial processes in general. An interesting aspect of this study is the comparison of a priori and a posteriori values to determine a measure for the quality of the underlying model. Optimal process control by analog computer studies is the focus of [MIURA et al. 1967].[201] An interesting optimisation technique restricting its parameters to binary values (0 and 1) is described in [O'GRADY 1967].

---

[198] A similar simulation is described in [WORLEY 1962, pp. 5-80 ff.].
[199] More realistic simulations involving many more column trays are usually out of reach of classic analog computer installations. Using hybrid computers, a resource-sharing approach can be implemented, thus allowing the treatment of problems with large numbers of trays.
[200] The development of adaptive control systems began in the 1960s.
[201] Annotations to this can be found in [ROSKO 1968], while similar studies are described by [FEILMEIER 1974, pp. 259 ff.] and [MICHAELS et al. 1971].

Fig. 13.47. Process plant simulator ([LIEBER et al. 1969, p. 78])

A very interesting optimisation scheme based on random generators[202] is developed in [KORN et al. 1970]. The basic idea is to perform random parameter variations and test their respective performance. If one of these variations results in a better overall performance, this parameter set is used as the base for the next variation cycle, otherwise the old parameters are restored.[203] Interestingly, it can be shown[204] that parameters determined by such a stochastic process are often superior to parameters that are the result of a deterministic process.

13.13.4 Plant startup simulation

The startup of complex chemical plants is a very complex and error prone process. Consequently, analog computers were used to simulate startup processes and to train operators. Figure 13.47 shows an analog computer coupled with typical plant monitoring and control devices. This setup, described in detail in

---

[202] See section 4.12.
[203] This method can be seen as a precursor to genetic programming.
[204] See [TACKER et al.].

[LIEBER et al. 1969], allowed the study of complex startup processes in oil refineries and the training of operator personnel.

13.14 Transport systems

Transport systems of all sorts can benefit from the application of analog and hybrid computers as the following sections demonstrate.

13.14.1 Automotive engineering

Until the 1980s, analog computers were a crucial part of the engineering departments of all major vehicle manufacturers. A simplified basic application was described in section 8.7, where the dynamic behaviour of a vehicle, modelled as a two-mass system, was simulated. A more realistic, but still simplified, three-mass simulation is described in [SYDOW 1964, pp. 245 ff.] while [MCLEOD et al. 1958/6, p. 1992] covers the specific problems of the simulation of shock absorbers for heavy lorries and excavators.[205] [GUMPERT 1972] describes an analog computer simulation of self-excited vibrations in cars.[206]

[Dornier/7] describes the application of an analog computer for data processing in a runout inspection. The ability to perform real-time spectral analysis using the analog computer is of particular importance in this type of application.

13.14.1.1 Steering systems
Steering systems are a central part of every car and analog computers were often used to simulate steering systems giving the freedom to vary and optimise parameters prior to building actual prototype systems. A special purpose analog steering simulator developed by General Motors is shown in figure 13.48. It allowed to analyse the dynamic behaviour of a steering system with respect to yaw, roll, and slip moments. The importance of this system is described by [MCLEOD et al. 1958/6, p. 1995] as follows:

> "As a result of this work, an understanding has been gained of the effects of the various car parameters on the car's lateral response. In addition, a number of general observations have been made: Yaw and sideslip are strongly coupled, but there is only a weak coupling linking the roll to the yaw and sideslip. However, this weak coupling is often the reason that a particular automobile is stable or unstable, particularly in yaw."

---

[205] Similar studies were performed for railway vehicles, too.
[206] In this study a soviet made MN-7 analog computer was used.

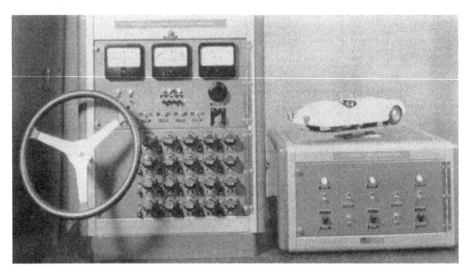

Fig. 13.48. Steering simulator (see [MCLEOD et al. 1958/6, p. 1995])

[ZIMDAHL 1965] describes an automatic control system enabling cars to follow a buried cable by means of an inductive pickup.

13.14.1.2 Transmissions

An interesting example of the application of analog computers in automotive engineering, namely the simulation of a four-speed automatic transmission, is described in [Dornier/1]. This example is especially noteworthy due to the high number of comparators required to model the discrete gearshift points.[207]

The simplified motor driven car underlying this simulation is described by

$$m\ddot{x} = F_{\text{drive}} - F_{\text{drag}} - F_{\text{rolling drag}} - F_{\text{road slope}}.$$

To simplify matters further, $F_{\text{rolling drag}}$ and $F_{\text{road slope}}$ are assumed to be zero, yielding

$$m\ddot{x} - F_{\text{drive}} + F_{\text{drag}} = 0$$

or, in more detail

$$m\ddot{x} - \frac{\eta i J_k}{r} M(n) + C c_w A v^2 = 0, \qquad (13.9)$$

---

[207] A similar, although more complex, simulation of an automatic transmission was implemented at *Daimler Benz* using a Telefunken RA 800 (see section 6.4). This analog computer system was also used to develop one of the first anti-lock braking systems (BERND ACKER, personal communication, 08/23/2007).

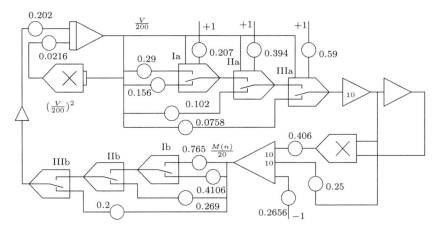

Fig. 13.49. Program for the simulation of a four-speed automatic transmission (see [Dornier/1, p. 5])

with $m = 1120$ representing the car's mass, $\eta = 0.9$ the efficiency factor, $i = 4.11$ the axle ratio, r the wheel diameter, J_k the four gear transmission ratios, $M(n) = 5.306 + 50n - 81.25n^2$ the rotation speed dependent moment of force, and $Cc_w Av^2$ the drag, approximated by $0.003421v^2$. The gear transmission ratios are defined as $J_1 = 3.8346$, $J_2 = 2.0526$, $J_3 = 1.345$, and $J_4 = 1$. The transmission performs an upshift every time the rotational speed of the motor reaches $n = 6000 \text{ min}^{-1}$ corresponding to the following velocities: $v_1 = 41.4$ km/h, $v_1 = 78.8$ km/h, and $v_1 = 118$ km/h.

Based on equation (13.9) the program shown in figure 13.49 can be derived. A typical simulation output is shown in figure 13.50.

[KLITTICH 1966] contains a comprehensive treatment of the simulation of transmissions in general by means of analog computers.

13.14.1.3 Ride simulation systems

A large-scale automobile ride simulation system is described by [KOHR 1960]. This system, built and used by the engineering mechanics department of General Motors, was capable of carrying two passengers in a car frame mounted on top of a hydraulic actuator system and could simulate the overall behaviour of a car with respect to actual road "waves", which were read from a magnetic tape unit and fed to an analog computer. This computer then determined the actual car motions by implementing a seven degree of freedom suspension system. It also controlled the hydraulic motion simulator. The overall system required only 82 operational amplifiers.

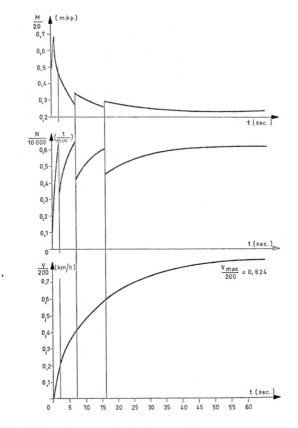

Fig. 13.50. Results of a simulation run for the four-speed automatic transmission (see [Dornier/1, p. 4])

13.14.1.4 Traffic flow simulation

Interestingly, even dynamic systems with strong discrete properties such as traffic flows, were analysed using analog computers. A complex example can be found in [LANDAUER 1974] where the simulation of a transportation system consisting of 40 stations, 600 vehicles, each of which able to transport up to twelve persons, ten crossroads, and 23 junctions is described. The goal of this simulation is the development of an optimal schedule ideally avoiding traffic jams or at least minimising such effects. A much simpler simulation is described in [JACKSON 1960, pp. 371 ff.], which deals with a light signaling system and its parametrisation.

13.14.2 Railway vehicles

Railway vehicle simulations are quite similar to those encountered in automotive applications. Notable exceptions include the simulation of *marshaling humps*[208] and the treatment of motor coach trains. [SYDOW 1964, pp. 238 ff.] describes the simulation of a direct-current railroad engine with emphasis on braking procedures. Two different braking techniques are analysed: Braking by placing a resistive load in parallel to the motor acting as a generator and braking with recuperation of the energy dissipated by the deceleration.

[LOTZ 1970/2] describes an interesting application of an analog computer for fabrication quality control of motor vehicle wheels. One of the most important parameters here is the eccentricity of the wheel. An automated wheel test rig yields acceleration signals, which are then analysed by performing a FOURIER analysis. The results obtained by this are then used as the base for the actual eccentricity computation. All of these steps are done on an analog computer.

13.14.3 Hovercrafts and Maglevs

Analog and hybrid computer systems were also used in the development of hovercraft and maglevs. [LEATHERWOOD 1972] describes the simulation and analysis of a tracked hovercraft[209] with the goal of developing an active control system for the vehicle. [THALER et al. 1980] describes a complex hovercraft simulation system in which the actual simulation was implemented on three stored-program digital computers with a COMCOR CI-5000 analog computer being mainly used as an interface to the operators of the simulator.

The simulation of maglevs is more complex than that of hovercrafts although their basic principle is similar. Most of the additional complexity results from the possibility of coupling several cars together to form a train. Coupled systems like these cannot be treated analytically so that large analog computer installations were used in early maglev developments. An example can be found in [MEISINGER 1978] where a simplified one-mass maglev is modelled. This maglev hovers over a long track, which bends under the weight of the maglev.

---

[208] Such a simulation is described in [GILOI et al. 1963, pp. 45 ff.].
[209] Although this sounds like a contradiction, such vehicles were explored since they promised very low frictional losses and offered better lateral stability than many traditional railway systems.

13.14.4 Nautics

The development of ships and vessels also benefitted from analog and hybrid computers as the following examples show.

13.14.4.1 Dynamic behaviour

A complex simulation of the dynamic response of a ship subjected to waves is described in [MASLO 1974]. Such studies are of high relevance with regard to loading and unloading operations as well as military applications. Parameters of interest are the ship's geometry and dimensions with respect to the wavelengths encountered in the sea, etc. The partial differential equations involved in this type of problems were typically solved on hybrid computers where the discretisation process was left to the stored-program digital computer. This particular simulation required 22 integrators, 42 summers, 28 inverters, 75 coefficient potentiometers, 30 comparators, 15 ADCs, and nine DACs clearly showing the complexity of this type of problem.

[GLUMINEAU et al. 1982] describes another interesting study in which the dynamic response of an oil tanker is analysed. The tanker is to be loaded and is anchored and subjected to wind and wave forces. Since the forces of interest in this system have most of their respective energies concentrated at frequencies of about $1/200$ Hz,[210] the time-scaling ability of the analog computer is used to shorten the simulation time considerably.

13.14.4.2 Propulsion systems

The analysis of a complex turbine fuel system is described in [HORLING et al.]. This simulation was implemented on a hybrid computer, which was capable of yielding results in real-time. A comparative study implemented on a *Control Data CDC 6800* digital computer[211] only achieved 10 % of real-time while a third implementation on a *UNIVAC 1110* was timed at 1 % of real-time, clearly demonstrating the superiority of this hybrid simulation approach.

The simulation of a military vessel powered by twin diesel engines and a gas turbine is described in [THOMPSON]. Of special interest in this study is the behaviour of the vessel while changing from one propulsion system to the other since the diesel engines are used for low and medium speed with the gas turbine only being used at high speeds. For a military vessel it must be guaranteed that

---

[210] Higher frequencies can be mostly neglected due to the long time-constant of the oil tanker itself.

[211] This system was later known as the *CDC 7600*.

there is no loss of thrust in all modes of operations and under all conditions and circumstances.

13.14.4.3 Ship simulation

While the simulations described above focus on specific aspects of ships' behaviours, some applications, especially the training of skippers and the like, require complex and comprehensive full ship motion simulation systems.[212] A simple submarine simulation with only one degree of freedom is described in [JACKSON 1960, pp. 384 ff.].

The simulation of an assault boat LCM-6[213] is described in [KAPLAN]. The focus of this study is on the dynamic behaviour of the boat due to forces exerted by waves. The simulation of the waves is done on the digital portion of a hybrid computer while the analog computer is used to implement a six degree of freedom simulation of the ship.

[GRISWOLD et al. 1957] describes an early mechanical ship motion simulator

> "*intended to duplicate the essential features of the motions of a ship, so that these motions may be studied with reference to their influence on the handling, servicing, and launching characteristics of a missle or of any other device where these ship motions will affect performance.*"

Only limited degrees of freedom are taken into account in this simulator, namely, roll, pitch, and heave motions. The actual system was built by *Chrysler Corporation Missile Operations* and featured a large platform capable of supporting a missile under test with a weight of up to 150 000 pounds, that could by hydraulically moved according to simulated (or actual measured) ship motions.

[KASTNER 1968] describes an electromechanical analog computer for the analysis of the motion of the sea. This motion is represented by an input voltage obtained by measurements on a research vessel. The analog computer then computed the FOURIER coefficients of this signal and its power density spectrum.

In his Ph.D. thesis,[214] KASTNER describes an analog computer setup for the simulation and analysis of a capsizing ship.

[TAKAISHI 1965] describes the analog simulation of a rolling vessel with large, partially filled tanks for fluids. Special emphasis is put on the effects of non-linear damping effects on the rolling of the vessel.

---

[212] The basic requirements for such systems are described in [MCCALLUM].
[213] Short for *Landing Craft Mechanized*, colloquially also known as *Mike Boat*.
[214] See [KASTNER 1968/2].

13.14.4.4 Torpedo simulation

The development of torpedoes relied heavily on analog simulation techniques since tests involving real devices are extremely costly. Analog computers were used at the *Naval Undersea Center* for the simulation of torpedoes from the early 1950s onwards.[215]

The complexity of this area of application is illustrated by the amount of analog computing equipment in use in the 1970s. All in all, two *EAI 8800* hybrid computers[216] each with 350 operational amplifiers, three EAI 231R analog computers[217], each with 250 operational amplifiers, and two stored-program digital computers, a UNIVAC 1110 and a UNIVAC 1230, were used. The first of these digital computers was coupled to 64 ADCs and 120 DACs while the second had 32 ADCs and 24 DACs.

This impressive installation allowed two torpedo simulations to be run simultaneously in real-time. These simulations also included models of the various seeker heads as well as the noise emission by the targets and the torpedoes.

13.15 Aeronautical engineering

Applications of analog and hybrid computers in the field of aeronautical engineering are manifold. Many of the driving forces for analog computer development actually came from this application area as the problems to be solved were extremely complex and money typically was not a limiting factor during the Cold War. Consequently, some of the first commercially available analog computers were developed and built by companies such as *Boeing, Goodyear, Short Brothers*, and *Reeves*, all strong players in aviation and space flight.

Simulations in aeronautical engineering and space flight often pose special requirements, which must be borne in mind when analog and hybrid computers are to be used. First of all many problems involve variables with large domains, often too large to be handled by an analog computer directly.

To complicate things further, many simulations involve various coordinate transformations and coordinate system rotations, requiring either special analog computing elements such as resolvers,[218] or many sine/cosine function generators,

---

215 Cf. [LOWE].
216 These fully transistorised machines used a machine unit of ±100 V, very unusual for transistorised analog computers, and were mainly used in applications where precision and compatibility with older external equipment was of prime importance.
217 See section 6.3.
218 See section 4.10.

multipliers, etc.[219] In addition to this, many aerospace simulations also involve functions of more than one variable, which either requires a vast amount of analog computer hardware or a digital computer as part of a hybrid computer.[220]

The importance of analog computers in these areas of application has been emphasized by [LEVINE 1964, p. 2]:

> "EHRICKE[221] believes that the accelerated development of American Missiles would not have been possible without [analog] computers."

Figure 13.51[222] shows an early *REAC* 100[223] used by *NACA*[224] as early as 1949. For its time this machine was remarkably large and complex.

Two interesting sources are [BARNETT 1963] and [EAI 1964], the first giving an account of the NASA-AMES hybrid computer facilities and their application to problems in aeronautics, the latter detailing on the application of hybrid computers in aerospace engineering in general. The following sections show some typical application examples of analog computers in this field. Basic information on these topics can be found in [BAUER 1962/1].

13.15.1 Landing gears

The development of landing gear for high-performance aircraft is an astonishingly complex and dangerous task if it is based on experiments with real flight hardware.[225] As early as 1953, analog computers were used to simulate the dynamic behaviour of this aircraft subsystem at the *Wright Air Development Center*.

---

219 [HOWE 1957] details on coordinate systems and transformations suitable for flight simulator systems.
220 This is one of the reasons why the idea of hybrid computers initially emerged at companies such as *Ramo-Wooldridge Corporation* and *Convair Astronautics*, cf. section 9.
221 KRAFFT ARNOLD EHRICKE, 03/24/1917–12/11/1984, see [EHRICKE 1960].
222 See https://www.nasa.gov/centers/ames/multimedia/images/2010/iotw/reeves.html, retrieved 08/07/2022
223 This first *Reeves Electronic Analog Computer* system was developed in 1947 by *Reeves Instrument Corporation*.
224 NACA; the *National Advisory Committee for Aeronautics* was established on March 3rd, 1915, to direct research and development in the field of aerospace technology. In 1958, the *National Aeronautics and Space Act* was passed by congress as a reaction to the fear of the United States falling behind the Soviet Union from a technology perspective. This became the foundation of *NASA* into which NACA was incorporated subsequently.
225 A good example for this is the landing gear of the *Me 262* jet, which was plagued by a variety of problems, all of which could, and often did, result in fatal accidents.

Fig. 13.51. Very early REAC analog computer (most probably a REAC 100) used by NACA as early as 1949

The first simulations of landing gear were performed on analog computers under the auspices of Professor W. J. MORELAND.[226] These were based on systems of differential equations with four degrees of freedom and 14 parameters. An interesting detail of these studies is that the shock absorbers used, and modelled, often had different dynamic behaviour depending on the direction of movement of the piston, further complicating the simulation task.[227]

Since no hybrid computers were available in these early days, the necessary parameter variations were done manually, which nevertheless quickly yielded feasible landing gear configurations as the following quote shows:[228]

---

[226] See [MCLEOD et al. 1958/6, p. 1995].
[227] Cf. [JOHNSON 1963, p. 223].
[228] See [MCLEOD et al. 1958/6, p. 1996].

"For the past three years all new landing gear designs have been evaluated according to Professor MORELAND*'s method, thus eliminating the dangers that accompany violent shimmy."*

13.15.2 Aircraft arresting gear systems

Aircraft arresting gear systems aboard aircraft carriers are highly complex systems. They must withstand the immense forces exerted by aircraft being quickly decelerated. Further, they must also exhibit a force characteristic that prevents damage to aircraft structure during its deceleration. Consequently, control of the braking system used to tension the arresting cable is a difficult task that has been successfully modelled and analysed using analog computers. Such a simulation, focusing on two types of aircraft, a jet fighter and a light bomber, is described in [CARLSON et al. 1967, pp. 296 ff.].

The opposite system, a take-off catapult for aircraft carriers, is described and analysed in [ADDICOTT et al. 1968].

13.15.3 Jet engines

The simulation of a turbo jet engine with afterburner on an analog computer is described by [JACKSON 1960, pp. 426 ff.]. This study focuses on the design of a control system for the variable-geometry exhaust of this particular engine. The simulation requires function generators for functions of more than one variable, substantially complicating the computer setup.

A simplified jet engine simulation is covered in [SCHWEIZER 1976/1, pp. 422 ff.]. A study to test the compatibility of a particular jet engine and a special inlet is described in [COSTAKIS 1974]. A comprehensive treatment of the simulation of a turbojet engine can be found in [ROCKCASTLE et al. 1956]. Another jet engine simulator is described in [SKRAMSTAD 1957, pp. 92 ff.].

13.15.4 Helicopters

The analysis of bending and torsion effects in helicopter rotor blades using an analog computer is described in [MCLEOD et al. 1958/4, pp. 1222 f.]. Mechanical systems such as rotor blades are described by partial differential equations and are thus not well suited for direct treatment with an indirect analog computer.

Accordingly, the system used in this study was a direct analog computer similar in its structure to the HETAC and EAFCOM.[229] It was described as follows:[230]

> "The direct analog computer consists of an assemblage of passive electrical circuit elements (resistors, capacitors, inductors and transformers), amplifiers, signal generators and control equipment. [...] The electrical analogy for the bending of beams has great importance in the direct analog method of dynamic analysis for lifting surfaces, since many lifting surfaces can be replaced by lifting lines with bending and torsional flexibility. This is certainly true of the helicopter rotor blade which characteristically has a very large span-to-chord ratio."

The system consisted of about 100 inductors and 50 resistors and capacitors representing seven discrete sections of a rotor blade.

13.15.5 Flutter simulations

A complex direct analog computer for the "*study of supersonic flutter of elastic delta wings*" is described in [BASIN 1954]. Here, the pressures acting on various positions on the wing are derived as power series, which can be handled quite easily. From this the resulting lift is computed, forming the base for dividing the wing into small cells which are then represented by electrical analogs consisting of passive elements.[231]

[SMITH et al. 1959] describes a large and complex flutter simulator capable of up to six degrees of freedom, which is shown in figure 13.52. A typical simulation result obtained for a problem with four degrees of freedom is shown in figure 13.53. It shows the response amplitude "*as a function of excitation frequency at three different air-speed settings.*[232]

Flutter simulations are complicated as this effect causes parts of an aircraft's structure to oscillate in a variety of modes while being coupled mechanically as well as by aerodynamic effects. This system was basically a large indirect analog computer – probably the first of its kind for this type of problems.

An even more complex successor of this system is described in [HICKS 1968]. Vacuum tubes as well as transistors were used to implement this system, which

---

[229] See section 13.2.4.
[230] Cf. [MCLEOD et al. 1958/4, p. 1222].
[231] Flutter effects in aircraft became a major problem early in World War II. In 1940 KONRAD ZUSE (06/22/1910–12/18/1995) was approached by ALFRED TEICHMANN (11/24/1902–09/21/1971) one of the leading German airframe specialists who realized that a relay based digital computer could be successfully applied to flutter problems. This led to the development of ZUSE's Z3 computer.
[232] See [SMITH et al. 1959, p. 12].

Fig. 13.52. Front view of the R.A.E. electronic simulator for flutter investigations in six degrees of freedom (see [SMITH et al. 1959, p. 44])

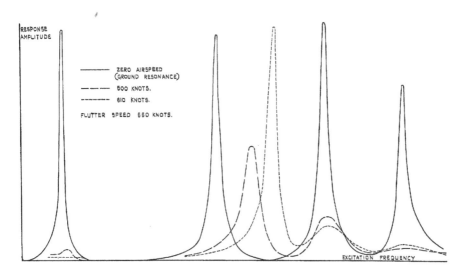

Fig. 13.53. Typical simulation obtained by the R.A.E. flutter simulator [SMITH et al. 1959, p. 42])

was capable of simulating the behaviour of up to twelve coupled linear differential equations of second order.

13.15.6 Flight simulation

One of the most fascinating applications for analog computer are flight simulators, in which aircraft are modelled with typically five or six degrees of freedom. This allows the behaviour of various aircraft configurations to be studied, without the necessity of building complex, costly, and sometimes outright dangerous prototypes risking the lives of test pilots.[233] Analog computer based flight simulation systems are characterized as follows by [KORN et al. 1956, p. 119]:

> "*The operation of a flight simulator on the ground is vastly cheaper as well as safer than operation of actual aircraft.*"

Consequently, the *US-Air Force* acquired a GEDA analog computer[234] in the early 1950s, which would eventually become the first in a long series of increasingly complex air force flight simulation systems.[235]

A very simple flight simulation in the vertical plane, taking only the movements of the elevator into account, is described in [WASS 1955, p. 39]. Figure 13.54 shows the basic setup of this simulation. Here, u denotes the speed along the x-axis, Θ is the inclination, while q represents the angular rate with respect to the y-axis. Based on this, a simple simulation with the elevator angle η as its input can be derived from the following simplified set of equations:

$$m(\dot{u} + w(0)q) = -mg\cos(\Theta(0))\Theta + uX_u + wX_x$$
$$m(\dot{w} + u(0)q) = -mg\sin(\Theta(0))\Theta + uZ_u + wZ_w$$
$$B\dot{q} = wM_w + qM_q + \eta M_\eta$$

$\Theta(0)$ is the initial angle of inclination, Θ the change of inclination, $u(0)$ and $w(0)$ represent the initial speed components along the x- and z-axis. B is the moment of inertia, Z_u, Z_w and X_u, X_w represent the aerodynamic forces along the z- and x-axis per unit of u and w. Finally, M_w, M_q, and M_η are the moments about the y-axis per unit of w, q, and η.

Based on these equations, the program shown in figure 13.55 can be derived. Using η as its input value the movements of the aircraft in the vertical plane can be determined. This computer setup can now be extended to include a simple autopilot as shown in figure 13.56. The autopilot shown will hold the line of flight constant.[236] The term for the required elevator setting to be determined by this simple autopilot is[237]

$$\eta = k \iint \dot{w} - u(0)q \, dt \, dt.$$

---

233 The basics of such flight simulation systems are described in [BAUER 1962/1], [WASS 1955, pp. 39 ff.], and [KORN et al. 1956, pp. 115 ff.].
234 See section 13.10.8 and [HANSEN 2005, p. 147].
235 The development of analog flight simulation at NASA's flight research center is described in [WALTMAN 2000]. Typical analog computer application examples at *Autonetics* (this was a division of *North American Aviation* that primarily developed avoinics systems and inertial navigation systems) are described in [MCLEOD et al. 1958/1, p. 122].
236 A more detailed description of the simulation of autopilots on analog computers can be found in [KORN et al. 1956, pp. 128 ff.].
237 See [WASS 1955, p. 43].

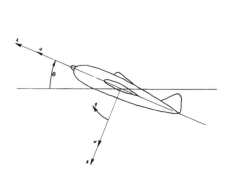 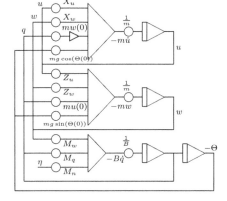

Fig. 13.54. Longitudinal motion of an aircraft (see [WASS 1955, p. 40])

Fig. 13.55. Simulation circuit for the longitudinal motion component of an aircraft (see [WASS 1955, p. 40])

The variables $\dot{w}$ and $u(0)q$ in this equation are readily available from the previous computer setup.

A much more complex simulator system is described in [NOSKER 1957] – the *Dynamic Systems Synthesizer*. This system is especially noteworthy as it abandoned the classic patch panel, which was replaced by a punch card configuration system shown in figure 13.57, thereby allowing rapid reconfiguration of the system. It also featured a *trigonometric resolver*, which generates $\sin(\varphi)$ and $\cos(\varphi)$ with φ given as the quotient of two values x and y.

Of course most of the analog and hybrid computer based flight simulators[238] were much more complex than these examples might suggest. One machine, *TRIDAC*,[239] is particularly interesting due to its sheer size and complexity. It was inaugurated on 10/08/1954 at the Royal Aircraft Establishment, Farnborough, and was described as follows in *The Times*:[240]

> "He was introducing Tridac [...] which is ten times larger than anything else of its kind in this country and one of the biggest computers in the world. It has been installed by the Ministry of Supply at a cost of about £ 750 000 and within its massive form – the equipment would fill six ordinary three-bedroomed houses – there are mechanical computing elements which need 400-horse power to drive them. The total electricity con-

---

[238] A good introduction to the mathematical modeling of aircraft and missiles can be found in [SCHWEIZER 1976/5], while [SCHWEIZER 1976/4] describes the implementation of a flight simulation system on a hybrid computer.
[239] Short for *Three-Dimensional Analogue Computer*. See [WASS 1955, p. 213] for additional information.
[240] See [The Times 1954].

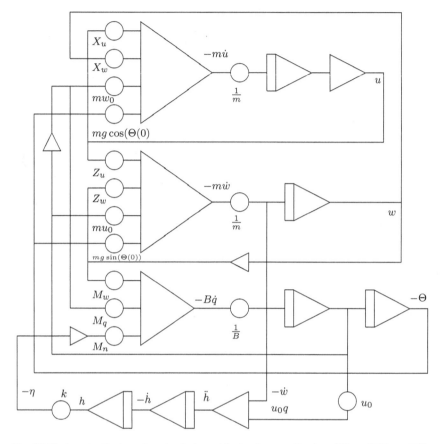

Fig. 13.56. Longitudinal motion simulation with simple autopilot (see [WASS 1955, p. 43])

sumption would light a small town. As an example of its capacity, in 20 seconds Tridac could achieve as much as 100 girls using calculating machines in eight hours."

Figure 13.58 shows the basic structure of TRIDAC. In contrast to general purpose analog computers, most of TRIDAC's computing elements were grouped into building blocks performing basic operations required for flight simulation. These building blocks could then be configured and coupled together arbitrarily.

The central element was the *aerodynamical unit* surrounded by blocks of integrators, which integrated rotational and translational movements of the simulated flight vehicle. There were also resolver blocks, which derived the necessary signals for an *aircraft position computer*. Units to simulate autopilots, the dynamic behaviour of control surface motors, etc., were also available. All in all, TRIDAC

Fig. 13.57. Punch card programming system of the Dynamic Systems Synthesizer (see [NOSKER 1957, 156])

contained about 650 chopper-stabilised operational amplifiers and a plethora of hydraulic function generators and resolvers.[241]

LACE, the *Luton Analogue Computing Engine*, is described in detail in [GOMPERTS et al. 1957] and [JONES et al. 1957]. It was a general purpose analog computer, which found many applications in the field of flight simulation and was built by the *Guided Weapons Division* of the *English Electric Company*, Luton.

In cases where an analog or hybrid simulation required hardware-in-the-loop, *flight tables* were employed. The earliest such flight table was developed in Peenemünde (Germany) during the development of the A4 rocket.[242] The idea was to mount the gyro system of a rocket or aircraft system under investigation on a platform that could be tilted under control of an analog computer to simulate the movements of the rocket and thus evaluate the performance of the gyro and other guidance system components. While this first flight table, being used until the mid 1950s at the Redstone Arsenal during the development of the Redstone

---

241 Just the hydraulic subsystem for TRIDAC required 400 hp (about 300 kW) while the overall TRIDAC system was rated at 650 kW.
242 See [LANGE 2006, pp. 249 ff.], [TOMAYKO 1985, p. 233], and [HOELZER 1992, p. 18].

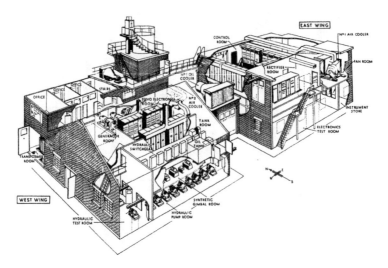

Fig. 13.58. Structure of the TRIDAC simulator (see [Wass 1955, Fig. 141])

missile, only offered one degree of freedom, later systems could be moved in three axes.[243]

Not all problems required such large analog computers for their solution. Figure 13.59 shows an inertia coupling simulation performed at the NACA *High-Speed Flight Station* in 1955 during the development of the *Bell X-2* plane. RICHARD E. DAY,[244] one of the pioneers of analog flight simulation at NACA and later NASA, can be seen on the control stick, with two GEDA analog computers in the background. The high speed of analog computers made it possible to implement man-in-the-loop simulations where the dynamic response of a simulated aircraft could be experienced directly.

TASS,[245] a large hybrid computer based simulation system, is shown in figure 13.60. This system consisted of several EAI 231R analog computers, a *DOS-350* digital subsystem,[246] and several digital computers such as an *EAI 8400* and a *DDP-24*, etc.[247]

---

[243] A plethora of information about flight tables can be found in [BAUER 1962/2].
[244] His obituary, which is very worth reading, can be found at https://www.nasa.gov/centers/dryden/news/X-Press/stories/2004/073004/new_day.html, retrieved 02/23/2022.
[245] Short for *Tactical Avionics System Simulator*.
[246] An EAI 231R coupled with a DOS-350 forms a HYDAC 2000 computer system, see section 6.3.
[247] A detailed description of the system can be found in [KENNEALLY et al. 1966].

Fig. 13.59. Inertia coupling simulation (RICHARD E. DAY at the control stick, see [WALTMAN 2000, p. 6])

Fig. 13.60. The *Tactical Avionics System Simulator* (photo from BRUCE BAKER, reprinted with permission)

Another impressive analog computer system, a *Beckman EASE 2133*, was operated at the *MBB*[248] Aircraft Division and is shown in figure 13.61. This system featured 520 operational amplifiers, 72 multipliers, 64 function generators, ten sine-/cosine-function generators, 24 limiters, 240 servo coefficient potentiometers, 112 manual coefficient potentiometers, and a variety of digital control elements.

The simulation of a complex aircraft such as the *VJ 101C-X2* vertical takeoff plane required even larger analog computers. During the development of this particular plane, an analog computer installation containing more than 1 000 operational amplifiers was used at MBB.[249]

The accuracy of early analog computer simulations was demonstrated in a striking, yet tragic, way in 1956 when test pilot MEL APT[250] died as the result of a stability problem of the X-2 plane that had been predicted previously by a simulation:[251]

> "*We showed him if he increased AOA[252] to about 5 degrees, he would start losing directional stability. He'd start this, and due to adverse aileron, he'd put in stick one way*

---

[248] MBB, *Messerschmitt-Bölkow-Blohm*, was one of the largest German aerospace companies.
[249] A list of analog and hybrid computer installations in use for aeronautical research in 1971 can be found in [PIRRELLO et al. 1971].
[250] 04/09/1924–09/27/1956
[251] See [WALTMAN 2000, p. 138] and [HANSEN 2005, p. 148].
[252] Short for *angle of attack*.

Fig. 13.61. Beckman EASE 2133 analog computer installed at MBB Aircraft Division (see [MBB])

and the plane would yaw the other way [...] We showed APT *this, and he did it many times."*

At that time test pilots did not show much trust in the results of such simulations, which clearly were not "the right stuff" as RICHARD E. DAY remembers:[253]

"*Well, the simulator was a new device that has never been used previously for training or flight planning. Most pilots had, in fact, expressed a certain amount of distrust in the device.*"

During MEL APT's test flight, he encountered the critical region at Mach 3 where roll-coupling set in, as predicted by the analog model, and the X-2 entered a fatal spin. This fatal accident was then analysed by the very same analog computer model.[254] As a result, flight simulation played an increasingly central role in the following years in aircraft development, as well as in pilot training. Eventually, NEIL A. ARMSTRONG[255] spent about 50 to 60 hours of simulator training for each of his flights with the *X-15*, while every real flight only lasted about ten minutes:[256]

---

[253] See [WALTMAN 2000, p. 138].
[254] See [WALTMAN 2000, p. 140].
[255] 08/05/1930–08/25/2012
[256] See [HANSEN 2005, p. 148]. The analog X-15 flight simulation system is described in [COOPER 1961].

"[...] our simulators in the space program were so much more sophisticated and accurate, and our preparation was so much more intense, that we convinced ourselves that the pilots could handle whatever situation we might encounter in flight."

A description of the simulation techniques used during the X-15 project can be found in [DAY 1959]. [MITCHELL et al. 1966] covers additional details of the simulation implementations. Analog computer studies regarding the X-15's reaction control system are described in [STILLWELL 1956], while [STILLWELL et al. 1958] deals with basic aspects of such simulations.

A central problem in the early days of analog flight simulation was how to display information in the cockpit. Simple, yet readily available, analog instruments proved to be insufficient, since pilots were, and still are, used to the special instruments found in real cockpits:[257]

"Presenting the pilot with the necessary flight instruments was a problem of major concern. Originally galvanometer-type meter movements were used. This type of display seemed satisfactory to the analysts and engineers with no actual flight experience. However, serious objections were raised when experienced test pilots were asked to evaluate the simulator's performance. Even though the necessary information was being presented to the pilot, the unrealistic instruments distracted the pilot so much that they impeded his efforts to truly evaluate the flight condition."

Consequently, adapters were required to interface the analog and hybrid computers at the heart of these simulations to traditional cockpit instruments. Later systems used universal graphic display units as described in [SCHWEIZER 1976/6, pp. 531 ff.].[258]

To create a realistic impression for the pilot, a flight simulator not only requires a cockpit, which is ideally mounted on a hexapod platform providing realistic movement, but also needs a realistic representation of the outside world. Figure 13.62 shows the overall structure of the hybrid flight simulation system installed at *Martin Marietta's Simulation & Test Laboratory (STL)*. The picture generation system is based on a large terrain model, which is traversed by a video camera system capable of moving in three axes.[259] This particular model measures 80 by 40 feet. A scale of 1200:1 was used for aircraft simulations while a scale of 225:1 was used for helicopter simulations.[260] Figure 13.63 shows the terrain model in detail

---

257 See [MCLEOD et al. 1958/5, pp. 1387 f.].
258 An interesting study of the impact of instrumentation on a pilot's performance can be found in [SCHWEIZER 1976/3, pp. 561 ff.].
259 The basics of such picture generation systems are described in [SCHWEIZER 1976/7, pp. 394 ff.].
260 See [BAKER 1978].

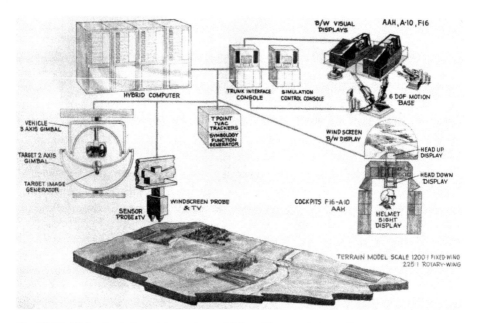

Fig. 13.62. Flight simulation with terrain model (see [BAKER 1978, p. 9])

with a person standing on the far right giving an impression of its size.[261] The hybrid computer system used at Martin Marietta consisted of a three-CPU *Sigma-5* digital computer and six EAI 231RV analog computers with 1 496 operational amplifiers in all.[262]

[SHERMAN et al. 1958] describes the simulation and analysis of the exit phase of flight of a hypersonic aircraft. A very simple analog simulation of the flight of a paper glider is described in [ULMANN 2020/3], which is based on [LANCHESTER 1908] and [SIMANCA et al. 2002].

13.15.7 Airborne simulators

Even more fascinating than these ground-based flight simulators are *airborne simulators*, also known as *in-flight simulations*. Here, an analog or hybrid computer aboard an aircraft was used to actively change the behaviour of the plane in a

---

[261] BRUCE BAKER remarks on this model (see [BAKER 1978, p. 2]): "*Topography is rolling hills modelled after West Germany.*"
[262] See [BAKER 1978, pp. 5 f.].

Fig. 13.63. Terrain model (see [BAKER 1978, p. 10])

way that made the simulation of a wide variety of other aircraft possible.[263] Such simulators are also known as *variable-stability aircraft*.

A basic requirement for such a simulator is an aircraft with a fly-by-wire system. The signals generated by the pilot controls in the cockpit are then fed to the analog computer, which in turn generates the necessary output signals to control the aircraft's actuator systems. Normally two cockpits and two pilots were used with one set of controls wired directly to the plane's control motors, etc., while the second set of controls delivered the input signals to the analog computer.

One of the earliest such systems is the *General Purpose Airborne Simulator* (*GPAS*), which was based on a *Lockheed JetStar* aircraft. Development of this system began in 1960[264] and many studies were performed with this particular airborne simulator.[265]

Airborne simulators were not without risks as one of the GPAS programmers, BOB KEMPEL, remembers:[266]

---

[263] Cf. [SCHWEIZER 1976/7, pp. 396 f.].
[264] See [WALTMAN 2000, pp. 59 ff.] and [BERRY et al. 1966] for a detailed description of the overall system. More information about this type of simulator can be found in [ARMSTRONG et al. 1962], [MCFADDEN et al. 1958], and [KIDD et al. 1961].
[265] [SZALAI 1971] describes such a study in which the JetStar was validated for the simulation of the handling qualities of large transport aircraft.
[266] See [WALTMAN 2000, p. 64].

"*I remember the incident when we were airborne and we [LARRY CAW and BOB KEMPEL] were looking at different feedback schemes. [...] Well, as you know, signs were sometimes confusing.* FITZHUGH FULTON[267] *was the pilot. The sign on beta was wrong, and we ended up with a dynamically unstable airplane because of it. We turned on the system for* FITZ *to evaluate, and the airplane immediately began an oscillatory divergence!* LARRY *and I were in the back hollering to* FITZ *to turn it off, but* FITZ *was intrigued with the thing so he wanted to watch it as it diverged or maybe just teach us a lesson. He finally punched the thing off and* LARRY *and I sighed in relief.* LARRY *changed the beta-input sign, and we proceeded with the test. [...] The JetStar was a fun airplane to fly in, but I always had a feeling of impending doom or something else going wrong.*"

The vibrations were so violent, that an observer in a chase plane worried about an impending crash: "STAN *told me [...] he just looked out of the window to see where they would crash as he believed the wings would be torn off.*"[268]

13.15.8 Guidance and control

Analog computers and computing elements were frequently used as integral elements of guidance and control systems aboard aircraft and missiles, with the A4 rocket being the first rocket to be controlled by a fully electronic analog onboard computer.

An example of a mechanical analog subsystem is shown in figure 13.64. It is a mechanical resolver module used in the *PHI-4* dead reckoning computer of the *Starfighter*. The *resolver assembly* basically contains a rotating ball, which splits a flight velocity signal into x- and y-components, both which are then fed to further subsystems of the onboard computer.

One of the most influential analog computer based missile guidance systems was developed for *Nike* missiles. Due to space constraints, a *command guidance system* was developed. At its heart was a large ground based analog computer fed with target data from various radar sources. The system then generated steering commands for the missile.[269] This approach not only made the missiles lighter and cheaper but also allowed for improvements of the guidance system without requiring modifications to already deployed missiles.

The analog computer computed a course, which was significantly better than a plain pursuit curve, thus rendering evasive manoeuvres of the target ineffective.[270]

---

[267] 06/06/1925–02/04/2015
[268] See [WALTMAN 2000, p. 64].
[269] Cf. [Department of the Army 1956].
[270] See [N. N. 1956, p. 4].

Fig. 13.64. Resolver module of the PHI-4 navigation system that was used in the Starfighter

Such a control system requires some kind of memory to store past movements of the target and the missile. Remarkably, the Nike analog computer used differentiators with large time-scale factors to implement these memory functions.[271]

In contrast, the *Polaris* missile, whose development started in 1956, employed an onboard guidance system based on a DDA.[272] These missiles were launched from ballistic missile submarines, thus a self-contained onboard guidance system was required as no command guidance system could be implemented for a submerged submarine. The required high accuracy of the missile system further forced the development of a digital guidance system. Since the researchers at MIT's *Instrumentation Laboratory* mostly had an analog computer background, the decision to develop a DDA was obvious.[273]

The resulting bit-serial DDA featured a word length of 17 bit corresponding to a resolution of about 10^{-5} and twelve words of memory implemented as shift registers. The first incarnation of this DDA had a volume of 11,000 cm^3, contained

---

[271] See [N. N. 1956, p. 3]: *"[T]he computer needs a memory. [...] It must receive and remember position data for 4 seconds until it knows exactly the direction and speed of the motion involved. The memory of the computer lies in its differentiating circuits."*
[272] See section 10.
[273] See [HALL 1996, pp. 38 ff.].

about 500 NOR-gates[274] implemented with germanium transistors, and required about 80 W of electrical power. A later implementation based on silicon transistors and an improved module packaging technology only required a quarter of this volume and only about half of the electrical power.

The trigonometric operations performed by this onboard computer for steering the Polaris missiles were based on *CORDIC*,[275] an algorithm developed by JACK E. VOLDER.[276] This algorithm was subsequently extensively used in early scientific pocket calculators such as the *HP-35*, introduced by Hewlett-Packard in 1972. It is still widely used to compute trigonometric functions and logarithms on digital computers.

Another *High Speed Differential Analyzer* (HSDDA) for use in airborne systems was developed in the early 1960s by the Guidance and Control Systems Division of *Litton Systems, Inc.*, and is described in detail in [Litton 1963].

13.15.9 Miscellaneous

The applications described in the preceding sections are not exhaustive – there were many more – sometimes quite arcane – areas of application of analog and hybrid computers in aerospace engineering. An example of this is a study on parachute stability using analog as well as digital computers, which is described in [LUDWIG 1966]. This paper focuses on the strongly non-linear oscillations induced by parachute systems.

13.16 Rocketry

The development of rockets offered an equally fertile ground for the application of large analog and hybrid computer installations as the following sections show.

13.16.1 Rocket motor simulation

Liquid-propellant rocket motors exhibit an extremely complex dynamic behaviour that is impossible to study using purely analytic methods. The particular value of (analog) computer simulations for the development of rocket motors has been described by [SZUCH et al. 1965, p. 2] as follows:

---

[274] A NOR-gate is an OR-gate with an inverter connected to its output.
[275] Short for *Coordinate Rotation Digital Computer*.
[276] See [VOLDER 1959].

"A computer simulation, when properly used, can be a powerful tool in guiding an engine development program. It provides an easy and economical means for evaluating various design approaches and forewarns the designer of possible problem areas before costly hardware is developed and subsequently scrapped. Once a qualitative design is established, the system may be 'tuned' for optimum performance by varying the system parameters about their design values."

This study describes the simulation of the *M-1* LH_2/LOX-rocket motor,[277] which was intended to be used in the *NOVA* rocket and was eventually superseded by the highly successful family of *Saturn* rockets. This study is quite remarkable since apart from the analog computer, a digital computer was used to study details of the dynamic behaviour of the rocket motor that could not be treated directly on the analog computer due to its limited precision.

The analog part of this simulation required about 250 operational amplifiers, 50 multipliers, 20 variable, and five fixed diode function generators, demonstrating the complexity of the problem. It turned out that one of the most useful features of the analog computer was its ability to study short-lived effects during engine startup through time-scaling.

Another interesting source is [FOX et al. 1969], which describes an analog computer study of low-frequency fluid dynamics encountered in nuclear rockets, while [HART et al. 1967] focuses on frequency responses and transfer functions of such rockets. [NORUM et al. 1962] describe the simulation of particle trajectories in fluid flow governed by LAPLACE's equation with a detailed error analysis.

13.16.2 Rocket simulation

The requirements for the simulation and analysis of the dynamic behaviour of rockets are quite similar to those encountered in aircraft flight simulations, although normally there is no pilot in the loop in a rocket simulation. The basics of such simulations are described in [JACKSON 1960, pp. 390 ff.]. The number of computing elements necessary for such simulations is as equally impressive as for aircraft flight simulators. An early rocket simulation performed on a *REAC* analog computer required 304 operational amplifiers, 369 coefficient potentiometers, 25 servo multipliers, eight resolvers, eight diode function generators, and two random signal generators.[278] The following quotation from [BILSTEIN 2003, p. 72] emphasizes these vast hardware requirements:

---

277 LH_2 denotes liquid hydrogen, *LOX* is liquid oxygen.
278 See [JACKSON 1960, p. 393].

Fig. 13.65. Laboratory for three-dimensional guided missile simulation (see [BAUER 1953, p. 195])

> "For modifications and installation of new equipment, MSFC spent over $ 2 000 000 after acquiring the site in the summer of 1962. The array of digital and analog computers for test, checkout, simulation and engineering studies made it one of the largest computer installations in the country."

The advantage of high speed operation of analog computers, compared with their contemporary program digital computer counterparts, has been noted by [BIGGS, p. 6]:

> "Because of the random nature of some of the missile system inputs the prediction or extrapolation work may require a large number of runs. AGWAC should compute these runs in real time or something like real time, whereas fast digital machines at present available would be at least one hundred times slower."[279]

The setup of an early laboratory for three-dimensional guided missile simulation is described by [BAUER 1953]. This development was the objective of *Project Cyclone* at Reeves Instrument Corporation. The simulation system was put into operation in 1952 and consisted of 13 REAC systems, each containing seven integrators, seven summers, six inverters, and 23 coefficient potentiometers. In addition to this, 14 servo units containing a mix of servo multipliers and servo function generators, and three cabinets for additional devices like uncommitted operational amplifiers,

---

[279] AGWAC contained more than 400 operational amplifiers, 20 fully electronic multipliers, six servo multipliers (each having about 20 ganged potentiometers), and some resolvers as well as other special computing elements (see [BENYON 1961]).

limiters, etc., were used. Most of this installation is shown in figure 13.65. It was described in the original press release as follows:[280]

> " The Navy revealed a guided missile 'launching site' right in the heart of New York city today – but the launchings are all on paper and nobody gets hurt. The hitherto top secret research program – 'Project Cyclone' – is centered in what has become known as 'the house of 91st street'. There is a sort of brainlike machine [which] is fed flight problems and comes up with mathematical computations. From these figures, scientists can tell what a certain type of guided missile can do in flight – even though the missile doesn't exist except on paper. With the paper work out of the way, exact specifications and plans are read and the missiles can be build. It's a big money-saver, the Navy said. It would cost millions to build the missiles, try them out and then correct their shortcomings. It would take a lot of time, too. The Reeves Instrument Corp., a subsidiary of Claude Neon, Inc., created the computing center in 1946 at the direction of the Navy. Capt. J. R. RUHSENBERGER, a naval research expert, said project Cyclone has given the United States a big head start in the development of guided missiles. 'We are not tipping out hand to our enemies by unveiling project Cyclone', he told newsmen. 'We are showing the American taxpayer through you what has been going on these past 'peacetime' years."

A comprehensive description of typical large-scale problems simulated at project Cyclone can be found in [BAUER et al. 1957]. Some of the conclusions of this report were:[281]

> "The feasibility of solving full, three-dimensional problems has been established. Accuracy has not been impaired in problems requiring as many as 300 amplifiers. The time spent in maintenance, trouble-shooting and checking is not disproportionate to the size of the problem."

A comprehensive treatment of mathematical vs. physical simulations can be found in [HINTZE 1957]. [ELFERS 1957] gives details on the simulation of spacecraft, eventually leading to the first US satellites in orbit. Putting the first successful US satellite, *Explorer I*, into orbit required the attitude control system of the *Jupiter C* rocket to tilt the rocket gradually into the required injection attitude. [HOSENTHIEN et al. 1962, pp. 454 ff.] describes the attitude control system simulations performed for this project.

Part of a large-scale analog computer system for the simulation of the dynamic behaviour of the *Saturn V* rocket is shown in figure 13.66. This *General Purpose Simulator (GPS)* was used to implement a twelve degree of freedom simulation of the first stage of a Saturn V rocket, taking effects such as dynamic winds, bending of the rocket structure, and fuel sloshing into account.

---

[280] Associated Press, July 26th, 1950.
[281] See [BAUER et al. 1957, p. 239].

Fig. 13.66. The GPS-system, used to simulate the Saturn V rocket's dynamic behaviour ([TEUBER 1964, p. 26])

Using real wind data stored on analog tape drives, this analog computer could simulate the performance of a S-IC first stage up to 3 000 times faster than real-time. The system contained 50 integrators, 50 summers, 350 coefficient potentiometers, 20 quarter square multipliers, and 15 function generators, each containing an additional 70 operational amplifiers. To make simulations with stochastic inputs possible, the system also featured random generators, high- and low-pass filters, etc. It allowed 50 complete flight simulations per second, making it possible to generate a flicker-free display of the solutions on oscilloscopes.[282]

The simulation of a two-stage satellite launch vehicle on an analog computer is described in [JACKSON 1960, pp. 261 ff.]. The focus of this study is primarily on possible abort scenarios. The possibility of a piloted rocket flight was analysed in a study described in [WALTMAN 2000, pp. 34 f.]. A centrifuge was used in these simulations to exert gravitational forces of up to 14 G[283] on the pilot while studying his performance in controlling a simulated rocket in various adverse flight regimes.

[KRAFT et al. 2002, p. 209] remembers that these simulations of the dynamic behaviour of rockets and associated systems also resulted in frequent changes to the systems based on results of these simulations:

---

[282] Cf. [TEUBER 1964].
[283] 14 times of earth's gravitation.

> "When we first used the word dynamic to describe the simulators, it meant that they acted pretty much like the real thing. But in practice dynamic also applied to design because we discovered things in simulations that needed to be changed."

13.16.3 Real-time data analysis

Another interesting application of analog computers in the development of guided missiles is the real-time analysis of flight data transmitted in-flight by telemetry systems from a missile to a ground station. The need for such a system was recognised in the late 1940s which led to a sophisticated system consisting of two *UNIVAC 1103A* digital computers, a three-axis flight table, and several analog computers. This system and its purpose is described in [STEINHOFF et al. 1957]:[284]

> "The function of the loop [consisting of the aforementioned digital and analog computers] is to perform the analysis of the missile behavior by comparison of the actual behavior with the simulated behavior. If one has been completely accurate in setting up his missile simulation then the performance of the actual missile and the simulated missle would be identical. This perfect match is highly unlikely, so the analysis task is that of making adjustments in the simulated missile so as to effect simulated performance matching with the actual performance [...] "

13.16.4 Spacecraft manoeuvres

Although the limited precision of electronic analog computers places severe constraints on orbital simulations and similar calculations, analog simulations were nevertheless used to obtain at least rough estimates for orbital elements, etc. If the orbit radii are restricted to a small range and if the number of circuits is small (less then 10 e. g.), solutions obtained by analog computers are often sufficiently exact to gain valuable insight into spacecraft manoeuvres.[285]

Basic information on such simulations can be found in [SCHWEIZER 1976/2], while [Telefunken/7] describes the computation of the orbital elements for the early passive satellite *Echo-1*. A trajectory optimisation for a Mars mission is described in [GILOI 1975, pp. 161 ff.]. The much more complicated fine-positioning of a geostationary satellite is simulated on an analog computer and described in [Telefunken/6]. Figure 13.67 shows the simulation results of the final positioning phase of the satellite.

[REINEL 1976] describes the analog simulation of the onboard attitude control system of the European communication satellite *Symphony*. This simulation re-

---

[284] See [STEINHOFF et al. 1957, p. 243].
[285] See [Telefunken/6, p. 1].

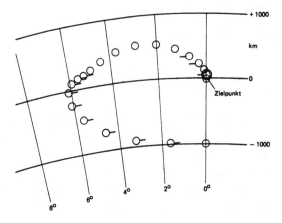

Fig. 13.67. Positioning of a satellite into a geostationary orbit (see [Telefunken/6])

quired a flight table as well as a sun and earth simulator to generate proper input signals to the satellite's sensor elements, which were part of the simulation setup.

A comparative study regarding the application of a *reaction wheel* for attitude control of a satellite was performed on a hybrid computer as well as on several digital computers and is described in [GRIERSON et al.]. While the hybrid computer was able to solve the equations in real-time, a *Cray-1* vector processor, the fastest digital computer of its day, only achieved a tenth of this speed.[286]

[KRAUSE 1964] describes the simulation of the dynamic behaviour of a rotating space station with astronauts on board. To compensate for oscillations in roll direction induced by movements of the astronauts, control systems using flywheels or even "*corrective motion of the astronauts*"[287] were suggested and extensively simulated.

A comprehensive study of large angle attitude manoeuvres of bodies in space can be found in [KRAUSE et al. 1963]. Attitude manoeuvres like this are commonplace in rendezvous operations, preparation of re-entry of a spacecraft, soft landing on the moon, midcourse corrections, etc. This study also takes stability issues of an asymmetric spacecraft into account.

---

[286] The only other digital computer capable of competing with the Cray-1 solving problems like this was an *AD-10*, developed by *Applied Dynamics*. This system was optimised for the solution of problems typically solved on analog computers.
[287] See [KRAUSE 1964, p. 39].

Fig. 13.68. Analog computer setup used during the development of the Mercury control stick (author's archive)

13.16.5 Mercury, Gemini, and Apollo

Analog and hybrid computers played a central role in the *Mercury*, *Gemini*, and *Apollo* space flight programs of the United States. Apart from applications like those described above, sophisticated training simulators were built to gain an understanding of the special requirements and problems posed by manned space vehicles. Figure 13.68 shows such a simulation system, which was used in the early phases of the Mercury program to develop the control stick used in the Mercury capsules. A complex mission simulator for the Mercury project, based on an EAI 231R analog computer,[288] was developed at Bell Laboratories.

Following Project Mercury, one of the main goals of Project Gemini was to work out procedures for docking spacecraft in orbit, a central requirement for the subsequent Apollo moon landing project. Simple rendezvous simulations on

---

[288] See [222, p. 569].

analog computers were performed as early as in the late 1950s.[289] The importance of such simulations cannot be overestimated as [HANSEN 2005, p. 248] explains:

> "Without extensive simulator time, it is doubtful that any astronaut could ever have been truly ready to perform a space rendezvous. 'Rendezvous simulation in Gemini was really quite good', ARMSTRONG notes. 'We achieved fifty to sixty rendezvous simulations on the ground, about two-thirds of which were with some sort of emergency.'"

The simulation of a pilot controlled rendezvous manoeuvre is described in detail in [BRISSENDEN et al. 1961]. This study not only focused on the performance of the pilot during the manoeuvre but also helped developing suitable display instruments for use during the piloted approach. In addition to this, boundary conditions of trajectory errors, which could make a successful rendezvous manoeuvre impossible were determined. A highly complex rendezvous simulation system was put into operation at the *TRW Systems Group*[290] in the 1960s. This system featured a TV camera motion unit similar to that used in the terrain generator described in section 13.15.6.[291] [FOX et al. 1963] describe a six degree-of-freedom feasibility study of a manned orbital docking system. This study, too, has its focus on pilot performance.

A comparable hybrid simulation system was developed by *McDonnel Aircraft Corporation* and *IBM*, which was used for re-entry simulations during Project Gemini.[292]

An interesting in-flight simulator system was developed to train the pilots of the *Lunar Excursion Module (LEM)*. This system, affectionately known as the *flying bedstead* due to its peculiar appearance, was in fact a flight vehicle mimicking the dynamic behaviour of the LEM under the constraints of earth's gravitation. This was done by an onboard special purpose analog computer. The resulting system proved very valuable as a research and training device:[293]

> "The notation of attacking the unique stability and control problems of a machine flying in the absence of an atmosphere, through an entirely different gravity field, 'That was a natural thing for us, because in-flight simulation was our thing at Edwards', ARMSTRONG relates. 'We did lots and lots of in-flight simulations, trying to duplicate other vehicles, or duplicate trajectories, making something fly like something else.'"

---

289 See [WALTMAN 2000, p. 107].
290 *Thompson-Ramo-Wooldridge*
291 See [BEKEY et al. 1968, p. 392].
292 See [BEKEY et al. 1968, p. 394].
293 See [HANSEN 2005, p. 314].

13.17 Military applications

Although analog and hybrid computer based flight and rocket simulations were used extensively in a military context, too, there were also problems unique to military applications that were tackled using analog simulation techniques. An interesting example can be found in [EAI Primer 1966, pp. 13 ff.], which describes the simulation of a setback leaf system used as a safety mechanism in a projectile fuse. This mechanism arms the fuse only after a sufficient acceleration profile has been experienced by the projectile.

Studies of projectile trajectories are described in [KORN et al. 1956, pp. 110 ff.] and [JOHNSON 1963, pp. 175 ff.]. Of prime importance is the accurate representation of aerodynamic properties and forces acting on the projectile, often involving functions of more than one variable, which required a large number of computing elements.

The implementation of a simplified naval gunnery problem on an analog computer is covered by [WASS 1955, pp. 57 ff.], while [DEMOYER 1980] describes a highly complex interactive anti-aircraft gun fire control simulation developed in the late 1970s.

The Cold War gave rise to the development of one of the earliest military computer games: *HUTSPIEL*. Developed in 1955, it is a

> "theater-level war game [...] directed to the study of the effects on a defense of stabilized positions in Western Europe of various employments of tactical atomic weapons and conventional air support,"[294]

implemented on several GEDA systems. Two combatants, *blue* representing NATO troops and *red* representing Soviet forces, face each other in HUTSPIEL. One minute of computer time corresponds to one day of battle action with a pause after each simulated day to give the opponents the chance to plan their next moves. The current states of blue and read troops is represented by about 40 individual variables each.

This remarkable system is described in detail in [CLARK et al. 1958].[295] Figure 13.69 shows the overall setup during an actual run of the game.[296]

---

[294] See [CLARK et al. 1958, p. 1].
[295] More background information can be found in https://if50.substack.com/p/hutspiel-and-dr-dorothy-k-clark, retrieved 08/28/2022. More information on Dr. DOROTHY KNEELAND CLARK can be found in [N. N. 2020].
[296] See [CLARK et al. 1958, p. 16].

Fig. 13.69. The HUTSPIEL system

13.18 Education

One of the main advantages of analog computers in education is their intuitive way of operation. Being mathematical machines at their heart, they avoid the additional abstraction layer of the algorithmic approach imposed by digital computers. This was first noted by WARREN WEAVER after the dismantling of the ROCKEFELLER differential analyser:[297]

> "[I]t seems rather a pity not to have around such a place as MIT a really impressive Analogue computer; for there is vividness and directness of meaning of the electrical and mechanical processes involved [...] which can hardly fail, I would think, to have a very considerable educational value. A Digital Electronic computer is bound to be a somewhat abstract affair, in which the actual computational processes are fairly deeply submerged."

---

[297] See [OWENS 1986, p. 66].

A further advantage of analog computers is their unmatched degree of interactivity which allows students to gain insight into the behaviour of complex dynamic systems in a very hands-on way.[298]

More general information about the application of analog computers in education can be found in [MARTIN 1969] and [MARTIN 1972]. Educational examples in mathematics are described in [Dornier/8], [Dornier/9], [RASFELD 1983], [BLUM 1982], and [Dornier/10]. The simulation of mechanical systems is covered in [PARK et al. 1972] and [SPIESS 2005]. Educational applications in process control are the focus of [NISE] and [MEDKEFF et al. 1955], while [TABBUTT 1967] and [HAMORI 1972] describe some examples from chemistry. [MÜLLER 1986] contains a collection of physical problems and their solution by means of an analog computer.

[TABBUTT 1969] describes the plan for a time-shared analog computer to be used in science education. This system would have a number of graphical display terminals connected to it, each of which would serve an individual student who could experiment with parameter settings for a given science problem. This would make use of the high speed of an analog computer – the proposed system would be able to support up to 20 student terminals to be used at once without noticeable delay in the system's response. [MARTIN 1970] describes the development of analog/hybrid terminals for use in a teaching environment.

A modern educational analog computer, *THE ANALOG THING*,[299] was introduced to the market in 2022 by anabrid GmbH, a German research company and manufacturer of analog computers. It comes with a booklet containing several introductory examples.[300]

13.19 Arts, entertainment, and music

Analog computers also found extensive applications in arts, entertainment, and music as shown in the following sections.

13.19.1 Arts

In the 1940s, BEN LAPOSKY[301] developed the first analog computer solely intended for artistic purposes. In the late 1950s, the computer graphics pioneer

---

[298] An early analog computer aimed at engineering education is the *Comdyna GP-6*, developed in the late 1960s by RAY SPIESS, see [SPIESS 1992] and [SPIESS 2005].
[299] See section 6.6.
[300] See [FISCHER 2022].
[301] 09/30/1914–2000

JOHN WHITNEY built an electromechanical analog computer from parts of an *M-5* antiaircraft director that had been used during World War II and was sold for scrap afterwards.[302]

HEINRICH HEIDERSBERGER[303] was charged with the creation of a large wall painting for the School of Engineering in Wolfenbüttel (Germany).[304] His first ideas for this work were based on LISSAJOUS figures.[305] These figures are described by

$$x = a\sin(\alpha t + \varphi) \text{ and}$$
$$y = b\cos(\beta t).$$

Based on this idea, HEIDERSBERGER built a mechanical analog computer, the *Rhythmograph*, which could generate much more complex figures than plain LISSAJOUS figures. Figure 13.70 shows him working on the Rhythmograph. A typical picture generated by this setup can be seen in figure 13.71. The device exposed photographic paper to a fine spot of light that was deflected by its swinging booms and mirrors. The Rhythmograph was itself a piece of art. After setting its parameters, which influenced the periods of the various mechanical oscillators, the Rhythmograph was set into motion by electromagnets and then started to "paint" a picture by light.[306]

An electronic analog computer, built by FRANZ RAIMANN, was used in the early 1960s by the scientist, author, and artist HERBERT W. FRANKE[307] to generate pictures such as the one shown in figure 13.72 on an oscilloscope screen. In contrast to HEIDERSBERGER's Rhythmograph, this analog computer offered the high degree of interactivity that is characteristic of electronic analog computers, giving the artist more freedom to experiment with the mathematical parameters of an artwork. Figure 13.73 shows HERBERT W. FRANKE working with this analog computer.[308]

In 1979, BENJAMIN HEIDERSBERGER[309] started the development of an electronic analog computer resembling the Rhythmograph. This machine contains three oscillators, which can be controlled separately with respect to frequency, phase, and damping. Additional modules implement the necessary control cir-

---

[302] See [SOMERS 1980] and http://en.wikipedia.org/wiki/John_Whitney_(animator), retrieved 03/03/2013.
[303] 06/10/1906–07/14/2006
[304] See [HEIDERSBERGER, p. 7] and [HOFFMANN 2006, p. 35].
[305] These are named after JULES ANTOINE LISSAJOUS (03/04/1822–06/24/1880).
[306] More information about this can be found in [HOFFMANN 2006].
[307] 05/14/1927–07/16/2022
[308] See [Hobby 1969, p. 43] and [DEKEN 1984] for more examples and background information.
[309] Son of HEINRICH HEIDERSBERGER.

Fig. 13.70. HEINRICH HEIDERSBERGER working on the *Rhythmograph* (Archiv Nr. 9179/1, self-portrait, Wolfsburg 1962, reprinted with permission from BENJAMIN HEIDERSBERGER)

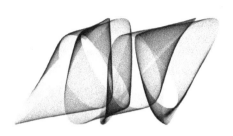

Fig. 13.71. Rhythmogramm #229, *Arabeske*, 1950s/60s (reprinted with permission from BENJAMIN HEIDERSBERGER)

Fig. 13.72. "Dance of the electrons" – an artwork of HERBERT W. FRANKE, 1961/1962 (reprinted with permission of the artist)

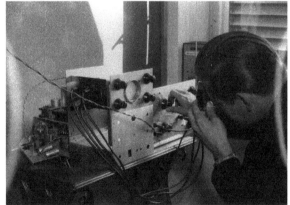

Fig. 13.73. HERBERT W. FRANKE with his special purpose analog computer (reprinted with permission of the artist)

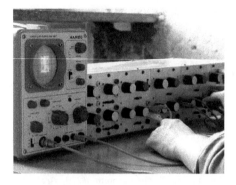 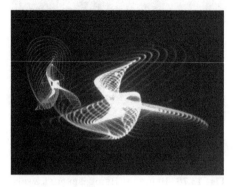

Fig. 13.74. BENJAMIN HEIDERSBERGER's analog computer (reprinted with permission)

Fig. 13.75. Graphic, generated with BENJAMIN HEIDERSBERGER's analog computer (reprinted with permission)

cuitry for repetitive operation, function generators, multipliers, etc. The machine is shown in figure 13.74 while figure 13.75 shows an example of a typical artwork generated with it.

13.19.2 Entertainment

Probably the first special purpose analog computer exclusively built for entertainment purposes was the *Cathode-Ray Tube Amusement Device* patented in 1948 and described as follows in [GOLDSMITH et al. 1948]:

> "*This invention relates to a device with which a game can be played. The game is of such a character that it requires care and skill in playing it or operating the device with which the game is played. Skill can be increased with practice and the exercise of care contributes to success.*"

The goal of this game is to hit targets displayed on an oscilloscope screen by firing a beam, which can be controlled in its length and angle by the player. To generate a flicker-free picture, the analog computer operates in a high-speed repetitive mode. Clearly, this device was motivated by analog training devices for gunners developed during World War II. The patent specifications suggest the use of small, adhesive aircraft-shaped stickers on the oscilloscope tube to denote the targets.

Ten years later, WILLIAM HIGINBOTHAM[310] independently developed a two-player game aptly called *Tennis for Two*, which was demonstrated during an

---

[310] 10/25/1920–11/10/1994

Fig. 13.76. Cartoon still showing a rabbit driving a car ([MIURA et al. 1967/2, p. 143])

open house presentation of the *Brookhaven National Laboratories*; previous events had shown that the public was not too interested in purely static displays as HIGINBOTHAM remembers:[311]

> "*I knew from past visitors days that people were not much interested in static exhibits, so for that year I came up with an idea for a hands-on display - a video tennis game. [...] It was wildly successful, and* HIGINBOTHAM *could tell from the crowd reaction that he had designed something very special. 'But if I had realized just how significant it was, I would have taken out a patent and the U.S. government would own it!' he said.*"

An analog electronic replica of HIGINBOTHAM's game, which was originally implemented on a small *Systron Donner* tabletop analog computer, has been developed by the *Museum of Electronic Games & Art*.[312]

A much more complex game based on an electronic analog computer is the "Golf Game Computing System" described by [RUSSEL et al. 1971]. This golf simulator is characterized by the unusual high simulation fidelity. Even effects like the change from laminar to turbulent air flow around the ball, spin, and many more were taken into account for the trajectory computation.

[RUSSELL 1978] describes a simple homebrew analog computer implementing a moon landing simulation.[313] A very complicated billiard simulation is described in [LOTZ 1970/1].

A fascinating application of analog computer was the animation of cartoons described in [MIURA et al. 1967/2]. Figure 13.76 shows a rabbit driving a car, animated on a Hitachi *ALS-2000* analog computer running in repetitive mode with a repetition rate of 3 kHz.

---

[311] See [N. N. 2006].
[312] See http://www.m-e-g-a.org/de/research-education/research/t42-tennis-for-two/, retrieved 03/03/2013.
[313] A similar example can be found in [FISCHER 2022, p.12].

13.19.3 Music

Although analog synthesizers can be justifiably regarded as being special purpose analog computers, not much use has been made of general purpose analog computers in this area. One notable exception of this is the work of the Dutch composer HANS KULK, who has used *Hitachi-240* analog computers to create music since the late 1980s.[314] The high degree of interactivity offered by analog computers is exploited extensively in his work by using special input devices such as a three-dimensional manual controller, which enables the composer to become part of the composition and sound generation process by changing various parameters of the analog computer setup with simple hand movements. The advantages and the possibilities of using analog computers in musical applications has been summarized by KULK as follows:[315]

> "One main conclusion is that the use of an analog [...] computer in analog sound synthesis is very effective, [yet] has been highly ignored by the electronic music community for long, but is slowly gaining interest as the concepts of analog technique become valuable tools in current developments. The 1970s dual Hitachi 240 set-up in this sound synthesis lab will continue to be an inspiring and open system with still many hours of fun to come."[316]

13.20 Analog computer centers

A typical phenomenon in the heyday of electronic analog and hybrid computers were computer centers set up by the major manufacturers of these systems such as Electronic Associates Inc., Beckman, etc. These computer centers served a dual purpose: First, potential customers were able to get hands-on experience with the latest machines and could solve practical problems in order to decide whether a particular analog computer met their needs. Second, these computer centers offered analog computer time on a rental basis for customers who either had no analog computer installation at all or required additional computing power for the solution of certain problems.

The first analog computer center in Europe was set up in 1957 in Brussels by EAI.[317] Figure 13.77 shows part of the EAI *Pace 96* analog computer which

---

[314] This early work was stimulated by [APPLETON et al. 1975].

[315] A more recent discussion on the application of analog computers in music, involving HAINBACH, HANS KULK, and BERND ULMANN, has been recorded on 08/30/2020 and can be viewed at https://www.youtube.com/watch?v=bgyzeyatS-0 (retrieved 08/07/2022).

[316] Personal communication to the author.

[317] This was, in fact, the world's third analog computer center – two earlier computer centers were setup in Princeton (New Jersey) and Los Angeles (California).

Fig. 13.77. EAI Pace 96 analog computer, installed in Europe's first analog computer center in Brussels (see [N. N. 1957/3])

was installed in Brussels and contained 96 operational amplifiers, 120 coefficient potentiometers, 20 servo multipliers, four quarter square multipliers, five function generators, two digital voltmeters, a two-channel xy-plotter, two single channel xy-plotters, two six-channel recorders, and two electronic noise sources.[318] EAI's computer center in Brussels was always equipped with their latest systems. In 1964 a HYDAC 2000,[319] shown in figure 6.8, was installed.

One year later, Beckman opened their first analog computer center in Los Angeles, which mainly offered computer time for paying customers. Figure 13.78 shows the EASE 1132, the first computer installed there. One year before its inauguration, this computer center was announced in [N. N. 1957/6]:

> "*It will be available to business, industry and educational institutions for solution of complex problems relating to aircraft, jet engines, guided missiles and industrial processes.*"

Figure 13.79 shows one of the large US domestic analog computer centers, equipped with several fully equipped EAI 231RV analog computers and many additional components.

---

[318] This model was a precursor of the highly successful model EAI 231R.
[319] See section 6.3.

Fig. 13.78. EASE 1132 analog computer, installed in the *Beckman/Berkeley Computation Center*, which opened on February 28, 1958 in Los Angeles (see [N. N. 1957/1])

Fig. 13.79. EAI computation center (see [EAI 1964, p. iv])

14 Future and opportunities

In the 1970s, analog and hybrid computers were largely displaced by the upcoming mini- and microcomputers. Although these initially couldn't compete with analog computers in terms of computational power, they quickly became cheaper to buy, operate, and maintain. In addition to this, stored-program digital computers could be run in a time-sharing mode, which gave multiple users concurrent access to a single machine. Changing from executing one program to another can be done in a matter of milliseconds on a digital computer, while it can take a substantial amount of time on a classic analog computer, with its central patch panel and many, often manually set, coefficient potentiometers and function generators.[1]

Another advantage of digital computers is their ability to trade solution time against problem complexity. As long as memory size is not a limiting factor, problems of varying complexity can be solved with time typically increasing with problem complexity. An analog computer on the other hand has to grow with the complexity of the problem to be solved. If there are not enough computing elements for a given problem, the problem cannot be solved on a purely analog computer.[2]

Eventually, analog computers were relegated to museum pieces and the very idea of analog computing vanished from curricula in schools and universities.[3] The idea of setting up an (electronic) model to solve a problem by a measurement process applied to an electronic circuit became obsolete, despite its advantages in terms of computational speed and energy efficiency over today's dominant algorithmic approach.

Digital computers have become many orders of magnitude faster and smaller since the 1970s, but this progress with respect to computational power, integration density, and energy efficiency is slowing down considerably as physical limits are being reached. Integration densities are eventually limited by the size of atoms, maximum clock frequencies are limited by energy consumption, which grows superlinearly with clock frequency, and the speedup obtained with massively parallel digital computers is described by AMDAHL's[4] *law* as an asymptotic process. Additionally, the overall energy consumption of today's information and communication sector (*ICT*) continues to soar, with computationally-intensive applications,

---

[1] The tedious reconfiguration process on classic analog computers was a major contributor to their eventual demise, see [TISDALE 1981, p. 1].

[2] If there are enough elements, the solution time on an analog computer is basically independent from problem size!

[3] This change in technology and computational paradigm is a near-perfect example for a *paradigm shift* as described by THOMAS SAMUEL KUHN (07/18/1922–06/17/1996) in his seminal work [KUHN 1996].

[4] GENE MYRON AMDAHL, 11/16/1922–11/10/2015

such as artificial intelligence, being the main contributors.[5] Modern digital processors also require a multi-level cache structure to alleviate the exceedingly long access times to main memory. This, together with the requirement for ever increasingly complex control circuitry, results in only a comparatively small portion of the overall transistor functions on a digital processor being involved in actual computations.

Consequently, analog computing will experience a renaissance and will return to stay, as it perfectly complements digital computers in the form of analog coprocessing, effectively creating modern hybrid computers. The main advantages offered by developing and employing modern, highly integrated analog computers are

- their unmatched degree of parallelism and thus high computational power,[6]
- their high energy efficiency,[7]
- no algorithmic (mainly sequential) control whatsoever,[8]
- no main memory, eliminating slow instruction and data fetches.

Today's digital (namely algorithmic) monoculture must be extended by different computational paradigms to deliver the new levels of computational power required to solve current important problems in all areas of science and technology. Analog computing will be one of these paradigms.

In 1975 WOLFGANG GILOI predicted that "*certainly [...] the analog method of hardware parallel processing*" will be preserved. LEE ALBERT RUBEL[9] was even more optimistic, as the following quote from a personal note to JONATHAN W. MILLS of the Indiana University shows:

---

[5] See [LANGE et al. 2020].
[6] In some cases, even direct analogies may be used to great advantage. Some simple yet impressive examples of such problems can be found in [DEWDNEY 1988/1], [DEWDNEY 1988/2], and [HOFFMAN 1979].
[7] See [GUO et al. 2016]. Cf. [KÖPPEL et al. 2021] for this and the preceding topic.
[8] There are additional advantages such as the fact that analog computers do not suffer from many of the effects caused by a binary floating point number representation and numerical integration schemes (see [TISDALE 1981, p. 2]). Additionally, programming an analog computer is a much more natural process than developing a numerical algorithm for the solution of a problem. [TISDALE 1981, p. 3] notes aptly: "*Digital languages [...] obstruct the user's contact with the physical analogy. In other words, the digital programmer becomes preoccupied with the programming task and unfortunately loses sight of the analogy he hopes to create. Hybrid techniques go hand-in-hand with Laplace and Fourier expressions. The analogy is not only established, but the engineer's knowledge and skill grows quickly, giving rise to innovations and breakthroughs.*"
[9] 12/01/1928–03/25/1995

"The future of analog computing is unlimited. As a visionary, I see it eventually displacing digital computing, especially, in the beginning, in partial differential equations and as a model in neurobiology. It will take some decades for this to be done. In the meantime, it is a very rich and challenging field of investigation, although (or maybe because) it is not in the current fashion."

[GUZDIAL et al. 2013] are also quite enthusiastic (p. 12):

"For exascale computing, reliability, resilience, numercial stability and confidence can be problematic when input uncertainties can propagate, and single and multiple bit upsets can disturb numerical representations. [...] Could analog computing play a role? Please note that I am not advocating a return to slide rules or pneumatic computing systems. Rather, I am suggesting we step back and remember that the evolution of technologies brings new opportunities to revisit old assumptions. Hybrid computing may be one possible way to address the challenges we face on the intersecting frontiers of device physics, computer architecture, and software."

It should be noted that parallel DDAs based on *FPGAs (Field Programmable Gate Arrays)* will probably also find widespread application in the future as they, too, exhibit the perfect parallelism of analog computers, thereby outperforming classic digital computers.[10]

Hybrid value representation is another interesting development for modern analog computers, which could solve the problem of problem scaling, thereby simplifying programming considerably.[11]

14.1 Challenges

There are some challenges which must be overcome for analog computing to become a ubiquitous computational paradigm for the 21st century. The first to mention is the need for automatic reconfigurability. The classic patch panel is a thing of the past and must be replaced by electronic switches under the control of an attached digital computer.

---

[10] See [MENZEL et al. 2021]. [NALLEY 1969] suggested a small DDA as a co-processor to speed up Z-transformations in 1969. [HYATT et al. 1968] described an *electrically alterable DDA*, and [ELSHOFF et al. 1970] developed a DDA working on binary floating point arithmetic. [MCGHEE et al. 1970] describes a DDA working on very long integers, an attempt to minimize the computational problems caused by floating point numbers. Practical implementations of DDAs on FPGAs are described in [LAND].

[11] See [O'GRADY 1966], [SKRAMSTAD 1959], and [WAIT 1963].

This idea is by no means new as the Rockefeller differential analyser shows.[12] Another early attempt to replace the cumbersome patch panel was the punch card programming technique used in the Dynamic System Synthesizer.[13]

In the late 1960s, NASA initiated a joint project with EAI to develop a reconfigurable analog computer because the time required for reconfiguration of the large analog/hybrid simulators used in aerospace engineering became increasingly prohibitive for many applications.[14] The naïve approach of using a rectangular matrix consisting of $n \times m$ $(n, m \in \mathbb{N})$ switches[15] is of course not feasible due to the sheer number of switches required even for a medium-scale analog computer. Consequently, a more sophisticated configuration topology will be required for this scheme to be feasible.

[HANNAUER 1968] suggested a system consisting of three layers of switching matrices (called *blocks* in this work): A layer of *input blocks*, followed by *middle blocks*, which in turn feed a layer of *output blocks*. Using such an arrangement, the number of required switched can be reduced significantly albeit at the cost of increased setup complexity.[16]

Based on this idea, a switching matrix extension for an EAI 680 analog computer was developed to demonstrate the feasibility of this proposed topology. This eventually led to the development of the EAI *SIMSTAR* system,[17] which was announced in 1983[18] and is described in detail in [LANDAUER 1983] and [EMBLAY].

Another interesting automatically reconfigurable (*"electronically patched"*) analog computer was the *US Army Material Command Advanced Hybrid Computer System (AHCS)*, which could be accessed through distributed terminals connected to two digital computers, which were in turn tightly coupled with a central analog computer. The switching matrices required more than 10^5 switches and were connected to an Applied Dynamics AD/FOUR analog computer.[19]

Modern reconfigurable analog computers will certainly employ multi-stage switching matrices, as full $n \times m$ switching matrices are still prohibitively complex, even for integrated circuits.

Another challenge is the creation of a suitable, flexible, and versatile software ecosystem for modern hybrid computers. Programming analog computers is so fundamentally different from today's prevailing algorithmic approach that a

---

[12] See section 2.8.
[13] See section 13.15.6.
[14] See [HANNAUER 1968].
[15] n and m denote the number of inputs and outputs of computing elements of the underlying analog computer.
[16] This setup resembles a *Clos network*, see [CLOS 1953] and [BENES 1965].
[17] See [EAI 1986, pp. 9 ff.].
[18] See [EAI 1986, p. 51 ff].
[19] See [HOWE et al. 1970], [HOWE et al. 1975], and [GRACON et al. 1970].

software layer will be required to abstract from the actual model-based programming approach. Ideally, a compiler will accept the problem equations specified in some *domain specific language* (DSL) and automatically generate the required configuration bit stream for the switching matrices of the analog co-processor.

Real hybrid computer operation will require additional software support to facilitate the tight coupling between the digital and analog computer. An interesting problem here are the long interrupt latencies exhibited of modern operating systems, which can easily reach $> 10\mu s$, sometimes exceeding typical solution times on the analog computer.[20] The problem equations could either be automatically scaled by the compiler mentioned above or in an iterative way on a hybrid computer. The analog co-processor could issue an interrupt when it detects an overload condition, which would then cause the digital computer to (partially) rescale the problem and so forth.

14.2 Applications

Applications for modern reconfigurable analog computers implemented as integrated circuits are abundant. As a co-processor, such a system could either directly speed up computations, which would otherwise require excessive amounts of CPU time, or it could generate approximate solutions, which could then be increased in precision by numerical algorithms executed on the digital computer.[21]

Modern highly integrated analog computers will also considerably reduce the energy consumed in solving certain problems. An example for this can be found in [COWAN et al. 2009, pp. 13 f.] where different approaches to solving a particular stochastic differential equation are compared with respect to the amount of energy consumed. The traditional algorithmic approach required about 1.2–40 J of energy when run on a typical microprocessor. The same computation running on a *DSP* (*Digital Signal Processor*) required only 0.04–0.4 J. Using a VLSI analog processor developed by COWAN, the same task could be performed with only 0.008 J – quite a substantial saving.[22]

Ther high energy efficiency will make analog co-processors suitable for medical applications where energy is at a premium. The idea of medical implants

---

[20] See [HERZOG et al. 2018].
[21] See [COWAN 2005, pp. 182 ff.]. A similar approach is described by [KARPLUS et al. 1972]. [HUANG et al. 2016] and [HUANG et al. 2017] focus on the application of analog computers as linear algebra accelerators for *high performance computing* (*HPC*) in general. An interesting memristor based analog co-coprocessor for HPC is described in [ATHREYAS et al. 2018]. See also [KOLMS et al. 2020].
[22] Although these figures are now outdated they serve as a good example of possible energy savings. A more recent treatment can be found in [KÖPPEL et al. 2021].

Fig. 14.1. ZRNA FPAA board (see zrna.org, retrieved 09/04/2022)

using energy harvesting instead of bulky accumulators is intriguing and in the future might affect many lives for the better. The same reasoning makes analog computers interesting for mobile devices and edge computing by performing highly energy efficient signal pre- and post-processing.[23] There are already *FPAAs (Field Programmable Analog Arrays)* on the market which are aimed at such applications.[24] Figure 14.1 shows a board featuring an Anadigm® FPAA, which can be programmed by a comprehensive *API (Application Programming Interface)*.[25] The ability of analog computers to adapt quite seamlessly to the physical world is an additional benefit in many applications.[26]

Another major application area for analog computer will be *machine learning* and *artificial intelligence (AI)* in general. There are already several companies working on *in memory computing*,[27] analog voice recognition, training of artificial neural networks, etc.[28]

14.3 Recent work

Beginning in the early 2000s, MILLS and co-workers developed analog coprocessors, to be used with traditional digital computers, based on direct analo-

---

[23] See [BAI et al. 2015].
[24] See [BASU et al. 2010] for the description of a floating-gate based FPAA.
[25] See https://zrna.org, retrieved 09/04/2022. These FPAAs can also be used as analog analog arithmetic circuits as shown in [MORENO et al. 2020] and [BEIL 2021].
[26] An example for a reconfigurable analog system for low-power signal processing is described in [SCHLOTTMANN et al. 2012].
[27] See [DEMLER 2018].
[28] [CHANNAMADHAVUNI et al. 2021] describes analog AI accelerators. A long short-term memory network design for analog computing is described in detail in [ZHAO et al. 2019]. [SHRESTHA et al. 2022] contains a comprehensive survey on *neuromorphic computing* including analog computing approaches.

Fig. 14.2. VLSI analog computer developed by GLENN EDWARD RUSSEL COWAN (reprinted with permission)

gies, as described by LEE ALBERT RUBEL.[29] These *Extended Analog Computers* (*EAC*s) are similar in their structure to the electrolytic tanks mentioned in section 1.3 but use much more modern materials, such as polymer foils.[30] Co-processors like these can speed up the solution of problems described by partial differential equations significantly.

In 2005 GLENN EDWARD RUSSEL COWAN described a reconfigurable VLSI analog computer targeted as a mathematical co-processor for an attached digital computer,[31] a die-photograph of which is shown in figure 14.2.[32] An application for this particular chip, featuring 416 analog functional blocks, is described in [FREEDMAN 2011]. COWAN writes:[33]

> "This chip is controlled and programmed by a PC via a data acquisition card. This arrangement has been used to solve differential equations with acceptable accuracy, as much as 400× fast than a modern workstation. The utility of a VLSI analog computer has been demonstrated by solving stochastic differential equations, partial differential equations, and ordinary differential equations."

COWAN's VLSI analog computer was an important step towards building modern analog computers for the 21st century, but it was limited in its capabilities as it only contained summers, integrators, multipliers, and coefficient, but had no provisions for non-linear functions such as $\sin(\dots)$, etc. The on-chip computing

---

[29] 12/01/1928–03/25/1995
[30] See [MILLS/1], [MILLS et al. 2006], [MILLS 1995], and [MILLS/2] for more information.
[31] See [COWAN 2005] and [COWAN et al. 2006].
[32] See   http://users.encs.concordia.ca/$\sim$gcowan/phdresearch.html,   retrieved 03/03/2013.
[33] See [COWAN 2005].

elements were also uncalibrated, limiting the precision of solutions obtained using this chip. In 2016 NING GUO unveiled an improved VLSI analog computer.[34] This not only surpasses its predecessor with respect to computational power, but it also includes function generators, can be programmed using a *Python* library, and can be automatically calibrated prior to a computation run, increasing the achievable computational precision.

Maybe no technology of the past has such a great potential for future applications as the paradigm of analog computing. Quoting the inscription of JAMES EARL FRASER's[35] statue "*Heritage*" at the Federal Triangle, Constitution Ave. & 9th, Washington, DC:

> "*The Heritage of the Past is the Seed that Brings Forth the Harvest of the Future.*"

---

[34] See [GUO 2017].

[35] 11/04/1876–10/11/1953

Bibliography

[ABRAMOVITCH 2005] DANIEL ABRAMOVITCH, "Analog Computing in the Soviet Union", in *IEEE Control Systems Magazine*, June 2005, pp. 52–62

[ADDICOTT et al. 1968] E. W. ADDICOTT, R. W. JONES, *Solution of the Catapult Take-off Performance Equations by an Analogue Method*, Ministry of Technology, Aeronautical Research Council, London, Her Majesty's Stationary Office, 1968

[ADLER 1968] HELMUT ADLER, *Elektronische Analogrechner*, VEB Deutscher Verlag der Wissenschaften, Berlin, 1968

[Admiralty 1943] N. N., *Handbook of the Admiralty Fire Control Clock Marks I and I**, Admiralty, S.W.1., Gunnery Branch, 1943

[AKERS 1977] LEX A. AKERS, "Simulation of semiconductor devices", in *SIMULATION*, August 1977, pp. 33–41

[ALAGHI et al. 2013] ARMIN ALAGHI, JOHN P. HAYES, "Survey of Stochastic Computing", in *ACM Transactions on Embedded Computing Systems*, Vol. 12, No. 2s, Article 92, May 2013

[ALBRECHT 1968] PETER ALBRECHT, "Über die Simulation der respiratorischen Arrhythmie auf einem Analogrechner", Technische Mitteilungen AEG-Telefunken, *2. Beiheft Datenverarbeitung*, 1968, pp. 13–15

[ALBRECHT et al.] P. ALBRECHT, H. LOTZ, "Einsatz der hybriden Präzisionsrechenanlage RA 770 zur automatischen Parameteroptimierung nach dem Gradientenverfahren", AEG-Telefunken, AFA 005 0670

[ALDRICH et al. 1955] H. P. ALDRICH, H. M. PAYNTER, "First Interim Report – Analytic Studies of Freezing and Thawing of Soils (for the Arctic Construction and Frost Effects Laboratory New England Division, Corps of Engineers)", in [PAYNTER ed. 1955, pp. 247–260]

[AMELING 1962/1] W. AMELING, "Die Lösung partieller Differentialgleichungen und ihre Darstellungsmöglichkeiten auf dem elektronischen Analogrechner", in *elektronische datenverarbeitung*, No. 5/1962, pp. 197–215

[AMELING 1962/2] W. AMELING, "Der Einsatz des elektronischen Analogrechners zur Rohrnetzberechnung kompressibler und inkompressibler Stoffströme", in *Elektronische Rechenanlagen*, 4 (1962), No. 3, pp. 109–116

[AMELING 1963] WALTER AMELING, *Aufbau und Wirkungsweise elektronischer Analogrechner*, Vieweg Verlag, 1963

[AMELING 1963/2] W. AMELING, "Aufbau und Arbeitsweise des Hybrid-Rechners TRICE", in *Elektronische Rechenanlagen*, 5 (1963), No. 1, pp. 28–41

[AMELING 1964] W. AMELING, "Die Entwicklung verbesserter Ersatzschaltungen mit Hilfe der Differenzmethode", in *Elektronische Rechenanlagen*, 6 (1964), No. 1, pp. 35–41

[AMMON] W. AMMON, *Der elektronische Analogrechner und seine Verwendung in der Industrie*, AEG

[AMMON et al. 1959] W. AMMON, G. SCHNEIDER, "Beispiele zur Lösung technischer Probleme mit dem Analogrechner", in *Elektronische Rechenanlagen*, 1 (1959), No. 1, pp. 29–34

[Analog Devices] Analog Devices Inc., *Universal Trigonometric Function Converter AD639*, https://www.analog.com/media/en/technical-documentation/obsolete-data-sheets/ad639.pdf, retrieved 03/13/2022

[Analog Devices 2006] Analog Devices Inc., *Op Amp Applications Handbook*, Elsevier Newnes, 2006

[Analog Devices 2008] Analog Devices Inc., *Analog Multipliers*, MT-079 Tutorial, 2008

[ANDERS] UDO ANDERS, "Early Ideas in the History of Quantum Chemistry", in http://www.quantum-chemistry-history.com, retrieved 03/24/2008

[ANDERSON et al. 1967] MAX. D. ANDERSON, SOMESHWAR C. GUPTA, "Backward time analog computer solutions of optimum control problems", in *AFIPS '67 (Spring): Proceedings of the April 18–20, 1967, spring joint computer conference*, April 1967, pp. 133–139

[ANDO 1971] JOJI ANDO, "Simulation Study of Surge Tank System for Hydro Power Plant by Using Hybrid Electronic Computer", in *The Second International JSME Symposium Fluid Machinery and Fluidics*, Tokyo, September 1971, pp. 259–147

[ANGELO 1952] E. J. ANGELO, *An Electron-Beam Tube for Analog Multiplication*, Technical Report No. 249, October 27, 1952, Research Laboratory of Electronics, Massachusetts Institute of Technology

[APALOVIČOVÁ 1979] R. APALOVIČOVÁ, "Simulation of MOS Transistor Structure on Hybrid Computer Systems", in Tagungsband *SIMULATION OF SYSTEMS '79*, pp. 1013–1019

[APPLETON et al. 1975] JON H. APPLETON, RONALD C. PERERA, *The Development o and Practice of Electronic Music*, Prentice-Hall, New Jersey, 1975

[ARGYRIS et al. 1995] JOHN ARGYRIS, GUNTER FAUST, MARIA HAASE, *Die Erforschung des Chaos – Studienbuch für Naturwissenschaftler und Ingenieure*, Vieweg, 1995

[ARMSTRONG et al. 1962] N. A. ARMSTRONG, E. C. HOLLEMAN, "A Review of In-Flight Simulation Pertinent to Piloted Space Vehicles", AGARD Report 403, July 1962

[ARSCOTT 1964] F. M. ARSCOTT, *Periodic Differential Equations – An Introduction to Mathieu, Lamé, and Allied Functions*, The MacMillan Company, New York, 1964

[ASCHOFF 1938] V. ASCHOFF, "Der Sternmodulator als Doppelgegentaktmodulator", in *Telegraphen-Fernsprech-Funk- und Fernsehtechnik*, Bd. 27, No. 10, 1938, p. 379–383

[ASCOLI 1947] GUIDO ASCOLI, "Vedute sintetiche sugli strumenti integratori", in *Rend. Sem. Mat. Fis. Milano*, 18:36, 1947

[ASHLEY et al. 1968] J. ROBERT ASHLEY, THOMAS E. BULLOCK, "Hybrid computer integration of partial differential equations by use of an assumed sum separation of variables", in *AFIPS '68 (Fall, part 1): Proceedings of the December 9–11, 1968, fall joint computer conference, part 1*, December 1968, pp. 585–591

[ASMAR et al. 2018] NAKHLÉ ASMAR, LOUKAS GRAFAKOS, *Complex Analysis with Applications*, Springer, 2018

[ASHLEY et al. 1985] HOLT ASHLEY, MARTEN LANDAHL, *Aerodynamics of Wings and Bodies*, Dover Publications, Inc., New York, 1985 (unabridged reprint from 1965)

[ATHERTON 1955] E. ATHERTON, "The relation of the reflectance of dyed fabrics to dye concentration and the instrumental approach to colour matching", in *Journal of the Society of Dyers and Colorists*, 71, 1955, pp. 389-398

[ATHREYAS et al. 2018] NIHAR ATHREYAS, WENHAO SONG, BLAIR PEROT, QIANGFEI XIA, ABBIE MATHEW, JAI GUPTA, DEV GUPTA, J. JOSHUA YANG, "Memristor-CMOS Analog Coprocessor for Acceleration of High-Performance Computing Applications", in ACM JOURNAL ON EMERGING TECHNOLOGIES IN COMPUTING SYSTEMS, Volume 14, Issue 3, July 2018, Article No.: 38, pp. 1–30

[ATKINSON 1955] CYRIL ATKINSON, "Polynomial Root Solving on the Electronic Differential Analyser (A Technique for Finding the Real and Complex Roots of a Polynomial Using an Electronic Differential Analyser)", in *Mathematical Tables and Other Aids to Computation*, Vol. 9, No. 52, Oct. 1955, pp. 139–143

[AUDE et al. 1936] AUDE & REIPERT, *Gezeitenrechenmaschine*, Patentschrift Nr. 682836, Klasse 42m, Gruppe 36, A 78729 IX b/42 m, patentiert vom 6. März 1936 ab

[BADER 1985] HEINZ BADER, *Operationsverstärker – Grundlagen und Anwendungen*, Karamanolis Verlag, 1985

[BÄRWOLFF 2021] GÜNTER BÄRWOLFF, "Modelling of COVID-19 propagation with compartment models", in *Mathematische Semesterberichte*, Springer, Band 68, Heft 2, Oktober 2021, pp. 181–219

[BAI et al. 2015] YU BAI, MINGJIE LIN, "Energy-Efficient Discrete Signal Processing with Field Programmable Analog Arrays (FPAAs)", in *FPGA '15: Proceedings of teh 2015 ACM/SIGDA International Symposium on Field-Programmable Gate Arrays*, February 2015, pp. 84–93

[BAKER 1978] BRUCE BAKER, "Martin-Marietta Aerospace Simulation & Test Laboratory", handout for a talk delivered at a Simulation Councils Conference, San Francisco, 1978, author's archive

[BALABAN] PHILIP BALABAN, "HYPAC – A hybrid-computer circuit simulation program", Bell Telephone Laboratories, Holmdel, New Jersey

[BARD 1965] M. BARD, "Schaltung zur automatischen Aufzeichnung von Frequenzgängen mit Analogrechner und Koordinatenschreiber", in *Elektronische Rechenanlagen*, 7 (1965), No. 1, pp. 29–33

[BARNETT 1963] ROBERT M. BARNETT, *NASA-AMES Hybrid Computer Facilities and Their Application to Problems in Aeronautics*, Ames Research Center, Moffett Field, California, 1963

[BARTH 1976] H.-J. BARTH, "Simulation im Maschinenbau zur Festigkeitsermittlung und zur Untersuchung physikalischer Zusammenhänge", in [SCHÖNE 1976/1, pp. 581–602]

[BASHE et al. 1986] CHARLES J. BASHE, LYLE R. JOHNSON, JOHN H. PALMER, EMERSON W. PUGH, *IBM's Early Computers*, The MIT Press, 1986

[BASIN 1954] MICHAEL ABRAM BASIN, *Electric Analog Computer Study of Supersonic Flutter of Elastic Delta Wings*, Thesis, California Institute of Technology, Pasadena, 1954

[BASSANO et al. 1976] J. C. BASSANO, Y. LENNON, J. VIGNES, "An Identification of Parameters on a Hybrid-Computer", in *Trans. IMACS*, Vol. XVIII, No. 1, Jan. 1976, pp. 3–7

[BATE et al. 1971] ROGER R. BATE, DONALD D. MUELLER, JERRY E. WHITE, *Fundamentals of Astrodynamics*, Dover Publications, Inc., 1971

[BASU et al. 2010] ARINDAM BASU, STEPHEN BRINK, CRAIG SCHLOTTMANN, SHUBHA RAMAKRISHNAN, CSABA PETRE, SCOTT KOZIOL, FAIK BASKAYA, CHRISTOPHER M. TWIGG, PAUL HASLER, "A Floating-Gate-Based Field-Programmable Analog Array", in *IEEE Journal of Solid-State Circuits*, Vol. 45, No. 9, September 2010, pp. 1781–1794

[BAUER 1953] LOUIS BAUER, "New Laboratory for Three-Dimensional Guided Missile Simulation", in AIEE-IRE '53 (Western) Proceedings of the February 4–6, 1953, western computer conference, pp. 187–195

[BAUER et al. 1956] WALTER F. BAUER, GEORGE P. WEST, "A System for General-Purpose Analog-Digital Computation", in *Proceedings of the 1956 11th ACM national meeting*, January 1956, pp. 79–82

[BAUER et al. 1957] LOUIS BAUER, A. KAREN, B. LOVEMAN, "Solution of Large Problems at Project Cyclone", in [White Sands 1957, pp. 211-239]

[BAUER 1962/1] LOUIS BAUER, "Aircraft, Autopilot, and Missile Problems", in [HUSKEY et al. 1962, pp. 5-49–5-64]

[BAUER 1962/2] LOUIS BAUER, "Partial System Tests and Flight Tables", in [HUSKEY et al. 1962, pp. 5-64–5-71]

[BAUN 1970] P. J. BAUN Jr., "Hybrid Computers: Valuable Aids in Transmission Studies", in *Bell Laboratories Record*, June/July 1970, pp. 181–185

[BECK et al. 1958] ROBERT M. BECK, MAX PALEVSKY, "The DDA", in *Instruments and Automation*, November 1958, pp. 1836–1837

[BECK et al. 1970] CHARLES H. BECK, MING H. KUO, *Designer's Manual for Circuit Design by Analog/Digital Techniques*, Systems Laboratory, School of Engineering, Tulane University, 1970

[BEDFORD et al. 1952] LESLIE HERBERT BEDFORD, JOHN BELL, ERIC MILES LANGHAM, *Electrical Fire Control Calculating Apparatus*, United States Patent 2623692, Dec. 30, 1952

[BEECHAM et al. 1965] L. J. BEECHAM, W. L. WALTERS, D. W. PARTRIDGE, *Proposal for an Integrated Wind Tunnel-Flight Dynamics Simulator System*, Ministry of Aviation, Aeronautical Research Council, 1965

[BEHRENDT 1965] E. BEHRENDT, "Wunschautos auf Knopfdruck", in *hobby – Das Magazin der Technik*, Nr. 6/65, pp. 36–41

[BEIL 2021] MARTIN BEIL, *Entwicklung und Anwendung eines Analogcomputers mit FPAA Schaltkreisen*, Bachelorarbeit im Studiengang Elektrotechnik, B-TU Cottbus-Senftenberg, 2021

[BEKEY 1960] GEORGE A. BEKEY, "Analog Simulation of Nerve Excitation", in [JACKSON 1960, pp. 436–444]

[BEKEY et al. 1960] GEORGE A. BEKEY, L. W. NEUSTADT, "Analog Computer Techniques for Plotting Bode and Nyquist Diagrams", in *IRE-AIEE-ACM '60 (Western): Papers presented at the May 3-5, 1960, western joint IRE-AIEE-ACM computer conference*, May 1960, pp. 165–-172

[BEKEY et al. 1966] GEORGE A. BEKEY, M. H. GRAN, A. E. SABROFF, A. WONG, "Parameter optimization by random search using hybrid computer techniques", in *AFIPS '66 (Fall): Proceedings of the November 7-10, 1966, fall joint computer conference*, November 1966, pp. 191–200

[BEKEY et al. 1967] GEORGE A. BEKEY, J. C. MALONEY, R. TOMOVIC, "Solution of integral equations by hybrid computation", in *AFIPS '67 (Fall): Proceedings of the November 14-16, 1967, fall joint computer conference*, November 1967, pp. 143–148

[BEKEY et al. 1968] GEORGE A. BEKEY, WALTER J. KARPLUS, *Hybrid Computation*, John Wiley & Sons, Inc., 1968

[BEKEY et al. 1971] GEORGE A. BEKEY, MAN T. UNG, "On the hybrid computer solution of partial differential equations with two spatial dimensions", in *AFIPS '71 (Fall): Proceedings of the November 16-18, 1971, fall joint computer conference*, November 1971, pp. 401–410

[BELLINI et al. 1975] ARMANDO BELLINI, CLAUDIO CERRI, ALESSANDRO DE CARLI, "A New Approach to the Simulation of Static Converter Drives", in *Annales de l'Association internationale pour le Calcul analogique*, No. 1, Janvier 1975, pp. 3–7

[BENDER et al.] R. R. BENDER, D. T. DOODY, P. N. STOUGHTON, "A Description of the IBM 7074 System", in *IRE-AIEE-ACM '60 (Eastern): Papers presented at the December 13-15, 1960, eastern joint IRE-AIEE-ACM computer conference*, December 1960, pp. 161–-171

[Bendix] N. N., *Bendix Computer – Digital Differential Analyzer D-12*, Bendix Computer, 5630 Arbor Vitae Street, Los Angeles 45, California

[Bendix 1954] N. N., *Operation Manual – Digital Differential Analyzer – Model D12*, Bendix Computer, 5630 Arbor Vitae Stress, Los Angeles 45, California, Copy No. 2, April 1954

[BENES 1965] V. E. BENES, *Mathematical Theory of Connecting Networks and Telephone Traffic*, Academic Press, New York, 1965

[BENHAM 1970] R. D. BENHAM (ed.), "Evaluation of Hybrid Computer Performance on a Cross Section of Scientific Problems", in *AEC Research & Development Report*, BNWL-1278, UC-32, January 1970

[BENHAM et al. 1973] R. D. BENHAM, G. R. TAYLOR, "A PDP 11 Study of the Physiological Simulation Benchmark Experiment", in *DECUS PROCEEDINGS*, Fall 1973, pp. 83–88

[BENNETT 1993] S. BENNETT, *A history of control engineering 1930–1955*, IEE Control Engineering Series 47, Peter Peregrinus Ltd., 1993

[BENYON 1961] P. R. BENYON, "The Australian Guided Weapons Analogue Computer AGWAC", Third International Conference on Analog Computation, Opatija, 4–9 September 1961

[BERKELEY et al. 1956] EDMUND CALLIS BERKELEY, LAWRENCE WAINWRIGHT, *Computers – Their Operation And Applications*, Reinhold Publishing Corporation, New York, Chapman & Hall, Limited, London, 1956

[BERRY et al. 1966] DONALD T. BERRY, DWAIN A. DEETS, *Design, development, and utilization of a general purpose airborne simulator*, presented at the 28th Meeting of the AGARD Flight Mechanics Panel, Paris 10–11 May 1966

[BEUKEBOOM et al. 1985] J. J. A. J. BEUKEBOOM, J. J. VAN DIXHOORN, J. W. MEERMAN, "Simulation of mixed bond graphs and block diagrams on personal computers using TUTSIM", in *Journal of the Franklin Institute*, Volume 319, Issues 1–2, January–February 1985, pp. 257–267

[BICKEL et al. 1957] HENRY J. BICKEL, ROBERT I. BERNSTEIN, "A Realistic Radar Simulator", in [White Sands 1957, pp. 199–210]

[BIGGS] A. G. BIGGS, *Red Duster Acceptance Trials, Scientific Evaluation – A Mathematical Model of the Missile System Suitable for Analogue Computation*, Department of Supply, Australian Defence Scientific Service, Weapons Research Establishment, Report SAD 20, No. 8 J.S.T.U. D3

[Bild der Wissenschaft 1970] N. N., "Das Mathematische Kabinett", in *Bild der Wissenschaft*, Mai 1970

[BILLET 1965] REINHARD BILLET, *Verdampfertechnik*, Bibliographisches Institut Mannheim, B.I.-Wissenschaftsverlag, 1965

[BILSTEIN 2003] ROGER E. BILSTEIN, *Stages to Saturn – A Technological History of the Apollo/Saturn Launch Vehicles*, University Press of Florida, 2003

[BLACK 1937] HAROLD STEPHEN BLACK, *Wave Translation System*, United States Patent 2102671, Dec. 21, 1937

[BLUM et al.] ALFONS BLUM, MANFRED GLESNER, "Macromodeling Procedures for the Computer Simulation of Power Electronics Circuits", F.B. 12.2 Elektrotechnik, Universität des Saarlandes

[BLUM 1982] WERNER BLUM, "Der Integraph im Analysisunterricht – Ein altes Gerät in neuer Verwendung", in *ZDM Zentralblatt für Didaktik der Mathematik*, Jahrgang 14, 1982, Ernst Klett Verlag, pp. 25–30

[BOGHOSIAN et al. 1950] W. H. BOGHOSIAN, S. DARLINGTON, H. G. OCH, *Artillery Director*, United States Patent 2493183, January 3, 1950

[BOHLING et al. 1970] DOROTHEA M. BOHLING, LAWRENCE A. O'NEILL, "An Interactive Computer Approach to Tolerance Analysis", in *IEEE Transactions on Computers*, Vol. C-19, No. 1, January 1970, pp. 10–16

[BORCHARDT] INGE BORCHARDT, *Demonstrationsbeispiel: Elektrisch geladenes Teilchen im Magnetfeld*, AEG-Telefunken, ADB 007

[BORCHARDT 1965] INGE BORCHARDT, *Berechnung von Teilchenbahnen in einem magnetischen Horn am Analogrechner*, DESY – H 11, Hamburg, 12/17/1965

[BORCHARDT et al.] INGE BORCHARDT, P. MAIER, F. HULTSCHIG, *Simulation von Strahlführungssystemen auf dem hybriden Rechnersystem HRS 860*, AEG Telefunken, Datenverarbeitung

[BORCHARDT et al. 1965] INGE BORCHARDT, G. RIPKEN, *Zur Berechnung der Teilchenbahnen in einem Sextupolfeld*, Hamburg, DESY – H 5, Hamburg, 05/24/1965

[BORCHARDT 1966] INGE BORCHARDT, *Strahloptische Gleichungen und ihre Verwendung im Analogrechenprogramm*, DESY-Strahloptik, 04/01/1966

[BORCHARDT et al. 1969] INGE BORCHARDT, P. ZAJÍČEK, *Wechselrichter-Schutzprobleme*, Interner Bericht, DESY K-69/3, November 1969

[BORCHARDT et al. 1976] INGE BORCHARDT, M. LEVY, *Endstufen-Simulation der Quadrupol-Stromversorgung für langsame Ejektion*, Interner Bericht, DESY K-76/01, März 1976

[BORCHARDT et al. 1977] INGE BORCHARDT, M. LEVY, J. MAASS, *Untersuchungen über die Regelung des 200 Hz Wechselrichters für das flat-top-System des Synchrotrons mit der hybriden Rechenanlage HRS 860*, Interner Bericht, DESY R1-77/01, June 1977

[BORDES et al. 1967] H. J. BORDES, K. TE VELDE, "Die Berechnung von Heiz- und Kühllast mit digitalen und analogen Rechenmaschinen", in *Heizung – Lüftung – Haustechnik*, 18, 1967, Nr. 8, pp. 300–307

[BORSEI et al.] A. BORSEI, G. ESTRIN, "An Analog Computer Study of the Dynamic Behaviour of Stressed Thin Ferromagnetic Films", Technical Report No. 62-37, University of California, Los Angeles

[BOSLEY et al. 1956] DONALD B. BOSLEY, ROTH S. LEDDICK, *Simulation of Steam Pressurizing Tank Transients by Analog Computer*, United States Naval Postgraduate School, Monterey, California, 1956

[BRENNAN et al. 1964] R. D. BRENNAN, H. SANO, "PACTOLUS – A digital simulator program for the IBM 1620", in *AFIPS Conference Proceedings, Fall Joint Computer Conference 26*, October 1964, pp. 299–312

[BRENNAN et al. 1967] R. D. BRENNAN, M. Y. SILBERBERG, "Two continuous system modeling programs", in *IBM SYSTEMS JOURNAL*, Vol. 6, No. 4, 1967, S. 242–266

[BREY 1958] R. N. BREY Jr., "Reactor Control", in *Instruments and Automation*, Vol. 31, April 1958, pp. 630–635

[BRISSENDEN et al. 1961] ROY F. BRISSENDEN, BERT B. BURTON, EDWIN C. FOUDRIAT, JAMES B. WHITTEN, *Analog Simulation of a Pilot-Controlled Rendezvous*, Technical Note D-747, National Aeronautics and Space Administration Washington, April 1961

[BRITAIN 2011] JAMES E. BRITAIN, "Electrical Engineering Hall of Fame: Harold S. Black", in *Proceedings of the IEEE*, Vol. 99, No. 2, February 2011, pp. 351–353

[MEYER-BRÖTZ et al. 1966] GÜNTER MEYER-BRÖTZ, E. HEIM, "Ein breitbandiger Operationsverstärker mit Silizium-Transistoren", in *Telefunken Zeitung*, Jahrgang 39, Heft 1, 1966, pp. 16–32

[BROMLEY 1984] ALLAN G. BROMLEY, *British Mechanical Gunnery Computers of World War II*, Technical Report 223, January 1984

[BROUWER 2007] JENS BROUWER, *Das FitzHugh-Nagumo Modell einer Nervenzelle*, Universität Hamburg, Department Mathematik, 20.8.2007

[BROWN 1969] FRANK M. BROWN, "Comment on Canonical Programming of Nonlinear and Time-Varying Differential-Equations", in *IEEE Transactions on Computers*, Vol. C-18, No. 6, June 1969, p. 566

[BRYANT et al. 1962] LAWRENCE T. BRYANT, MARION J. JANICKE, LOUIS C. JUST, ALAN L. WINIECKI, *Engineering Applications of Analog Computers*, Argonne National Laboratory, ANL-6319 Rev., Revised October 1962

[BRYANT et al. 1966] LAWRENCE T. BRYANT, LAWRENCE W. AMIOT, RALPH P. STEIN, "A hybrid computer solution of the co-current flow heat exchanger Sturm-Liouville problem", in *AFIPS '66 (Fall): Proceedings of the November 7–10, 1966, fall joint computer conference*, November 1966, pp. 759–769

[BÜCKNER 1950] HANS BÜCKNER, "Ein neuer Typ einer Integrieranlage zur Behandlung von Differentialgleichungen", in *Archiv der Mathematik*, 1949/50, Volume 7, Issue 6, pp. 424–433

[BÜCKNER 1953] HANS BÜCKNER, "Über die Entwicklung des Integromat", in [CREMER 1953, pp. 1–16]

[Bureau of Ordnance 1940] N. N., *Basic Fire Control Mechanisms*, OP1140, September 1940

[Bureau of Ordnance 1944] N. N., *Torpedo Data Computer, Mark 3, Mods. 5 to 12 inclusive*, June 1944

[BURNS et al. 1961] ARTHUR J. BURNS, RICHARD E. KOPP, "Combined analog-digital simulation", in *AFIPS '61 (Eastern): Proceedings of the December 12–14, 1961, eastern joint computer conference: computers – key to total systems control*, December 1961, pp. 114–123

[BUSH 1912] VANNEVAR BUSH, *Profile Tracer*, United States Patent 1048649, Dec. 31, 1912

[BUSH et al. 1927] VANNEVAR BUSH, F. D. GAGE, H. R. STEWART, "A Continnuous Integraph", in *Journal of the Franklin Institute*, Vol. 203, No. 1, 1927, pp. 63–84

[BUSH et al. 1945] VANNEVAR BUSH, S. H. CALDWELL, "A New Type of Differential Analyzer", in *Journal of The Franklin Institute – Devoted to Science and the Mechanic Arts*, Vol. 240, No. 4, October 1945, pp. 255–326

[BYWATER 1973] R. E. H. BYWATER, *The Systems Organisation of Incremental Computers*, Thesis, Faculty of Engineering, Department of Electronic and Electrical Engineering of the University of Surrey, April 1973

[CAJORI 1994] FLORIAN CAJORI, *A History of the Logarithmic Slide Rule and Allied Instruments*, Astragal Press, 1994

[CALAMORE et al. 1963] D. V. CALAMORE, D. H. CRIMMINS, E. TOBEY, *Army Gas-Cooled Reactor Systems Program – Analog Computer Studies of the ML-1 Power Plant*, Report No. AGN-TM-400, AEC Research and Development Report UC-32, 2nd printing, 1963

[CALDWELL et al. 1955] R. R. CALDWELL, V. C. RIDEOUT, "A Differential-Analyzer Study of Certain Nonlinearly Damped Servomechanisms", in [PAYNTER ed. 1955, pp. 193–198]

[Caltech 1949] N. N., "The Electric Analog Computer", in *Research in Progress*, April 1949, pp. 17–18

[CAMERON et al. 1961] W. D. CAMERON, R. E. TILLER, "Analog Program in Reactor Speed of Control Systems", Technical Report, HW-69940, 1961 Jun. 14

[CAMPEAU 1969] JOSEPH O. CAMPEAU, "The Block-Oriented Computer", in *IEEE Transactions on Computers*, Vol. C-18, No. 8, August 1969, pp. 706–718

[CARE 2006] CHARLES PHILIP CARE, "A Chronology of Analogue Computing", in *The Rutherford Journal*, Volume 2, 2006–2007

[CARE 2008] CHARLES PHILIP CARE, *From analogy-making to modelling: the history of analog computing as a modelling technology*, PhD. thesis, University of Warwick, Department of Computer Science, September 2008

[CARLSON et al. 1967] ALAN CARLSON, GEORGE HANNAUER, THOMAS CAREY, PETER J. HOLSBERG (eds.), *Handbook of Analog Computation*, 2nd edition, Electronic Associates, Inc., Princeton, New Jersey, 1967

[CARLSON et al. 1968] ALAN M. CARLSON, "Hybrid Simulation of an Exchanger/Reactor Control System", in *EAI applications reference library*, 6.4.20h, October 1968

[CARSON 1926] JOHN R. CARSON, "Wave propagation in overhead wires with ground return", in *The Bell System Technical Journal*, Volume 5, Issue 4, Oct. 1926, pp. 539–554

[CAVANA et al. 2021] ROBERT Y. CAVANA, BRIAN C. DANGERFIELD, OLEG V. PAVLOV, MICHAEL J. RADZICKI, I. DAVID WHEAT (eds.), *Feedback Economics – Economic Modelling with System Dynamics*, Springer, 2021

[CELINSKI et al. 1964] O. CELINSKI, I. H. RIMAWI, "Classification of Analog Multipliers", in *Instruments & Control Systems*, June 1964, pp. 149–156

[CELMER et al. 1970] JOHN CELMER, MARY ROULAND, *Automatic Analog Computer Scaling Using Digital Optimization Techniques*, National Aeronautics and Space Administration, Washington, D. C., March 1970

[CERUZZI 1989] PAUL E. CERUZZI, *Beyond the Limits – Flight Enters the Computer Age*, The MIT Press, 1989

[CHAN 1969] SHU-KWAN CHAN, "The Serial Solution of the Diffusion Equation Using Nonstandard Hybrid Techniques", in *IEEE Transactions on Computers*, Vol. C-18, No. 9, September 1969, pp. 786–799

[CHANCE et al. 1947] BRITTON CHANCE, J. N. THURSTON, P. L. RICHMAN, "Some Designs and Applications for Packaged Amplifiers Using Subminiature Tubes", in *The Review of Scientific Instruments*, Volume 18, Number 9, September, 1947, pp. 610–616

[CHANCE et al. 1949] BRITTON CHANCE, VERNON HUGHES, EDWARD F. MACNICHOL, DAVID SAYRE, FREDERIC C. WILLIAMS (ed.), *Waveforms*, McGraw-Hill Book Company, Inc., 1949

[CHANDLER et al. 1961] W. G. CHANDLER, P. L. DANDENO, A. F. GLIMN, "Short-range economic operation of a combined thermal and hydroelectric power system", in *AIEE Transactions*, Part 3, Vol. 80, May 1961, pp. 1219–1228

[CHANNAMADHAVUNI et al. 2021] SHRAVYA CHANNAMADHAVUNI, SVEN THIJSSEN, SUMIT KUMAR JHA, RICKARD EWETZ, "Accelerating AI Applications using Analog In-Memory Coputing: Challenges and Opportunities", in *GLSVLSI '21: Proceedings of the 2021 on Great Lakes Symposium on VLSI*, June 2021, pp. 379–384

[CHARLESWORTH et al. 1974] A. S. CHARLESWORTH, J. R. FLETCHER, *Systematic Analogue Computer Programming*, Pitman Publishing, Second Edition, 1974

[CHENG] SHANG-I CHENG, "Analog Simulation and Polymerization Kinetics", author's archive

[CLARK et al. 1958] DOROTHY K. CLARK, ARCHIE N. COLBY, PAUL IRIBE, NICHOLAS M. SMITH, LLOYD D. YATES, *HUTSPIEL – a Theater War Game*, Operations Research Office, The Johns Hopkins University, 1958

[CLOS 1953] CHARLES CLOS, "A Study of Non-Blocking Switching Networks", in *Bell System Technical Journal*, March 1953, Vol. 32, pp. 406–423

[CLYMER 1993] A. BEN CLYMER, "The Mechanical Analog Computers of Hannibal Ford and William Newell", in *IEEE Annals of the History of Computing*, Vol. 15, No. 2, 1993, pp. 19–34

[CLYNES 1960] MANFRED CLYNES, "Respiratory control of heart rate: laws derived from analog computer simulation", in *IRE Transaction on Medical Electronics*, Jan. 1960, pp. 2–14

[COFFIN 1882] J. COFFIN, *Averageometer or Instrument for Measuring the Average Breadth of Irregular Planes*, United States Patent 258993, June 6, 1882

[COHEN 1971] E. M. COHEN, *Hybrid Simulation-Aided Design of a Pneumatic Relay*, Presented at the Regional Meeting, Eastern Simulation Councils Foxboro, Mass., September 23, 1971

[COOPER 1961] N. R. COOPER, "X-15 analog flight simulation program: systems development and pilot training", in *IRE-AIEE-ACM '61 (Western): Papers presented at the May 9–11, 1961, western joint IRE-AIEE-ACM computer conference*, May 1961, pp. 623–638

[COSTAKIS 1974] WILLIAM G. COSTAKIS, *Analog Computer Implementation of Four Instantaneous Distortion Indices*, National Aeronautics and Space Administration, Washington, D. C., March 1974

[COULTER] DOUG COULTER, *Evolution of a Model Kit of Tools – a Brief History of a Computor-Builder in Terms of Ideas & Instruments*, http://www.philbrickarchive.org/dc032_philbrick_history.htm, retrieved 04/10/2022

[COWAN 2005] GLENN EDWARD RUSSELL COWAN, *A VLSI Analog Computer / Math Co-processor for a Digital Computer*, Columbia University, 2005

[COWAN et al. 2006] GLENN EDWARD RUSSELL COWAN, ROBERT C. MELVILLE, YANNIS P. TSIVIDIS, "A VLSI Analog Computer/Digital Computer Accelerator", in *IEEE Journal of Solid-State Circuits*, Vol. 41, No. 1, January 2006, pp. 42–53

[COWAN et al. 2009] GLENN EDWARD RUSSELL COWAN, Y. TSIVIDIS, *Analog and Digital Continuous-Time Computation and Signal Processing*, CMOSET 2009

[CREMER 1953] HUBERT CREMER ed., *Probleme der Entwicklung programmgesteuerter Rechengeräte und Integrieranlagen*, Rhein.-Westf. Technische Hochschule Aachen, Mathematisches Institut, Lehrstuhl C, 1953

[CROSSLEY 1963] FRANK ERSKINE CROSSLEY, "Die Nachbildung eines mechanischen Kurbelgetriebes mittels eines elektronischen Analogrechners", in *Feinwerktechnik*, Jahrgang 1967, Juni 1963, Heft 6

[CROSSLEY 1965] FRANK ERSKINE CROSSLEY, "Geometric computing: Analogue-simulation of a linkage", in *International Journal of Mechanical Sciences*, Volume 7, Issue 9, September 1965, pp. 595–601

[CUTLER 1965] A. E. CUTLER, "A new colour-matching computer", in *Journal of the Society of Dyers and Colourists*, 81, 1965, pp. 601–608

[DAGBJARTSSON et al. 1976] S. DAGBJARTSSON, D. EMENDÖRFER, "Simulation in der Kernenergietechnik", in [SCHÖNE 1976/1, pp. 298–340]

[DARLINGTON 1965] R. F. DARLINGTON, *Automatic Control of Multistand Cold Rolling Mill*, doctoral thesis, Mechanical Engineering Department, Imperial College of Science and Technology, London, 1965

[DASSOW 2015] ACHIM DASSOW, "Geschichte der nichtlinearen Röhren Raytheon QK-256 und QK-329", https://www.radiomuseum.org/tubes/tube_qk329.html, retrieved 05/22/2022

[DAVIS et al. 1974] FRANK T. DAVIS, ARMANDO B. CORRIPIO, "Dynamic Simulation of Variable Speed Centrifugal Compressors", ISA CPD 74105, 1974

[DAY 1959] RICHARD E. DAY, *Training Considerations During the X-15 Development*, paper presented to the Training Advisory Committee of the National Security Industrial Association, Los Angeles, California, November 17, 1959

[DEKEN 1984] JOSEPH DEKEN, *Computerbilder, Kreativität und Technik*, Birkhäuser Verlag, 1984

[DEHMEL 1949] RICHARD C. DEHMEL, *Aircraft Trainer for Aerial Gunners*, United States Patent 2471315, May 24, 1949

[DEHMEL 1954] RICHARD C. DEHMEL, *Flight Training Apparatus for Computing Flight Conditions and Simulating Reaction of Forces on Pilots*, United States Patent 2687580, Aug. 31, 1954

[DELAROUSSE et al. 1962] P. DELAROUSSE, C. TROUVE, R JACQUES, *Etude Sur Simulateur Des Regimes Transitoires Des Concentrations Dans Une Installation De Diffusion Gazeuse*, in *Service des Etudes sur la Séparation des Isotopes de l'Uranium*, Rapport C. E. A. no. 2010

[DEMLER 2018] MIKE DEMLER, "Mythic Multiplies in a Flash – Analog In-Memory Computing Eliminates DRAM Read/Write Cycles", in *Microprocessor Report*, August 2018, pp. 1–3

[DEMOYER 1980] R. DeMOYER Jr., "Interactive Anti-Aircraft Gun Fire Control Simulation: An Introduction to Hybrid Computation", in *EAI Product Information Bulletin*, February 18, 1980, Bulletin No. 023

[DENNIS 1958] JACK BONNELL DENNIS, *Mathematical Programming and Electrical Networks*, dissertation, Massachusetts Institute of Technology, August, 1958

[Department of the Army 1956] Department of the Army, *NIKE I SYSTEMS, NIKE I COMPUTER (U)*, Department of the Army Technical Manual TM 9-5000-3, April 1956

[DERN et al. 1957] HERBERT DERN, JOHN O'CONNOR, REINHOLD P. VOGEL, "Low Frequency Cumulative Distribution Analyzer", in [White Sands 1957, pp. 171–197]

[DEWDNEY 1988/1] A. K. DEWDNEY, "Rechnen mit Spaghetti – Wie der Spaghetti-Computer und andere kuriose Analoggeräte Probleme im Handumdrehen lösen, an welchen selbst die größten Digitalrechner scheitern.", in *Computer-Kurzweil*, Spektrum der Wissenschaft, 1988, pp. 198–203

[DEWDNEY 1988/2] A. K. DEWDNEY, "Kuriose Analog-Computer – Eine neue Kollektion von Analogrechnern für Heimwerker und eine vertiefte Diskussion ihrer Stärken und Schwächen im Vergleich zu Digitalrechnern.", in *Computer-Kurzweil*, Spektrum der Wissenschaft, 1988, pp. 204–210

[DEZIEL 1966] D. P. DEZIEL, *Applications of Analogue Computers in Operational Research*, Thesis, Imperial College, June 1966

[DHEN 1959] WALTER DHEN, *Entwurf und Aufbau eines repetierenden Analogrechners, unter besonderer Berücksichtigung der Zusammenhänge zwischen den Rechenfehlern und den Regelkreiseigenschaften in elektronischen Rechengeräten*, Fakultät für Mathematik und Physik der Technischen Hochschule Darmstadt, 1959

[DHEN 1960] WALTER DHEN, "Entwurf und Aufbau eines repetierenden Analogrechners unter besonderer Berücksichtigung der Zusammenhänge zwischen den Rechenfehlern und den Regelkreiseigenschaften in elektronischen Rechengeräten", in *Nachrichtentechnische Fachberichte*, Band 17, Vieweg & Sohn, Braunschweig, 1960

[DICK et al. 1967] D. E. DICK, H. J. WERTZ, "Analog and Digital Computation of Fourier Series and Integrals", in *IEEE Transactions on Electronic Computers*, Vol. EC-16, No. 1, February 1967, pp. 8–13

[DODD 1969] K. N. DODD, *Analogue Computers*, The English Universities Press Ltd., 1969

[DONAN 1950] J. DONAN, *MADDIDA (Preliminary Report)*, Northrop Aircraft, Inc., Project MX-775, Report No. GM-545, 26 May 1950

[DOORE et al. 2014] KAREN DOORE, PAUL FISHWICK, "Prototyping an analog computing represetation of predator prey dynamics", in *WSC '14: Proceedings of the 2014 Winter Simulation Conference*, December 2014, pp. 3561–3571a

[Dornier/1] N.N., *Beschleunigung eines PKW mit automatischem Getriebe (vereinfacht)*, Dornier

[Dornier/2] N.N., *Simulation reaktionskinetischer Probleme auf dem Analogrechner*, Dornier

[Dornier/3] N.N., *Der Analogrechner als Hilfsmittel bei der Untersuchung des Schwingungsverhaltens von Mehrmassensystemen*, Dornier

[Dornier/4] N.N., *Feder-Masse-System mit trockener Reibung*, Dornier

[Dornier/5] N.N., *Kurbeltrieb*, Dornier

[Dornier/6] N.N., *Ermittlung der Scherarbeit bei Materialprüfungen mit Hilfe eines Analogrechners*, Dornier

[Dornier/7] N.N., *Rundlaufprüfungen von Rädern mit Hilfe eines Analogrechners*, Dornier

[Dornier/8] N.N., *Anwendung von Analogrechnern bei der Ausbildung im Rahmen der Oberschulen – Kurvendiskussion*, Dornier

[Dornier/9] N.N., *Anwendung von Analogrechnern bei der Ausbildung im Rahmen der Oberschulen – Approximation von trigonometrischen Funktionen durch Reihenentwicklungen*, Dornier

[Dornier/10] N.N., *Anwendung von Analogrechnern bei der Ausbildung im Rahmen der Oberschulen – Parameterdarstellung von geschlossenen Kurven*, Dornier

[DUNGAN 2005] TRACY DWAYNE DUNGAN, *V-2 – A Combat History of the First Ballistic Missile*, Westholme Publishing, 2005

[EAI PACE 231R] EAI, *PACE 231R analog computer*, Electronic Associates, Inc., Long Branch, New Jersey, Bulletin No. AC 6007

[EAI 690] N. N., *690 Hybrid Computing System – Reference Handbook*, Electronic Associates, Inc., Long Branch, New Jersey

[EAI TR-10] N. N., *PACE TR-10 Transistorized Analog Computer – operator's handbook*, Electronic Associates, Inc., Long Branch, New Jersey

[EAI 1964] N. N., *Seminar on Hybrid Computation as applied to the Aerospace Field*, Electronic Associates, Inc., Long Branch, New Jersey, 1964

[EAI 1.3.2 1964] N. N., *Continuous Data Analysis with Analog Computers Using Statistical and Regression Techniques*, EAI Applications Reference Library 1.3.2a, 1964, http://bitsavers.org/pdf/eai/applicationsLibrary/1.3.2a_Continuous_Data_Analysis_with_Analog_Computers_Using_Statistical_and_Regression_Techniques_1964.pdf, retrieved 06/20/2022

[EAI 7.7.4a 1964] N. N., *Solution of Mathieu's Equation on the Analog Computer*, EAI Applications Reference Library, 7.7.4a, 1964

[EAI 7.3.8a 1965] N. N., *Investigation of Heat Transfer by Conduction*, EAI Applications Reference Library, 7.3.8a, 1965

[EAI Primer 1966] N. N., *Primer on Analog Computations and Examples for EAI-180 Series of Computers*, Electronic Associates, Inc., Bulletin No. 957051, 1966

[EAI 1986] N. N., *EAI User's Group Newsletter*, 14th edition, April 1986

[ECK 1954] BRUNO ECK, *Technische Strömungslehre*, Springer-Verlag, Berlin/Göttingen/Heidelberg, 1954, 4. verbesserte Auflage

[ECKDAHL et al. 2003] DONALD E. ECKDAHL, IRVING S. REED, HRANT HAROLD SARKISSIAN, "West Coast Contributions to the Development of the General-Purpose Computer: Building Maddida and the Founding of Computer Research Corporation", in *IEEE Annals*, January-March 2003 (vol. 25 no. 1), pp. 4–33

[EDWARDS 1954] C. M EDWARDS, "Survey of Analog Multiplication Schemes", in *Journal of the ACM*, Vol. 1, Issue 1, Jan. 1954, pp. 27–35

[EGGERS 1954] K. EGGERS, *Über die Integrieranlage 'Integromat' des Instituts für angewandte Mathematik der Universität Hamburg*, Schriftenreihe Schiffbau, 3, Februar 1954

[EHRICKE 1960] KRAFFT A. EHRICKE, *Space Flight – Environment and Celestial Mechanics*, D. Van Nostrand Company, Inc., Princeton, N. J., 1960

[ELFERS 1957] W. A. ELFERS, "Simulation Analysis Technique", in [White Sands 1957, pp. 35–46]

[ELSHOFF et al. 1970] J. L. ELSHOFF, P. T. HULINA, "The binary floating point digital differential analyzer", in AFIPS '70 (Fall) Proceedings of the November 17–19, 1970, fall joint computer conference, pp. 369–376

[Elton et al. 1942] Charles Elton, Mary Nicholson, "The Ten-Year Cycle in Numbers of the Lynx in Canada", in *Journal of Animal Ecology*, Vol. 11, No. 2, Nov. 1942, pp. 215-244

[Emblay] Ronald W. Embley, *The Technology behind SIMSTAR, an all-new Simulation Multiprocessor*, Electronic Associates, Inc., West Long Branch, New Jersey, USA

[Engel et al. 1964] Alfred Engel, John J. Kennedy, "Modelling dynamic economic problems on the analog computer", in *International Journal of Computer Mathematics*, 1(1), DOI: 10.1080/00207167508803020, pp. 289–311

[Enns et al.] Mark Enns, Theo C. Giras, Norman R. Carlson, "Load Flows by Hybrid Computation for Power System Operation", IEEE, Paper No. 71 C 26-PWR-XII-A

[Ernst 1960] Dietrich Ernst, *Elektronische Analogrechner – Wirkungsweise und Anwendung*, R. Oldenbourg Verlag München, 1960

[Esch 1969] John William Esch, *RASCEL – A Programmable Analog Computer Based on a Regular Array of Stochastic Computing Element Logic*, Department of Computer Science, University of Illinois, Urbana, Illinois, Report No. 332, June, 1969

[Eterman 1960] I. I. Eterman, *Analogue Computers*, Pergamon Press, 1960

[Euser 1966] Ir. P. Euser, "Elektrisch analogon voor klimaattechnische berekeningen", in *Dr Raadgevend-Ingenieur*, Jaargang 8, no. 10, 1966, pp. 3–7

[Evans 1959] William T. Evans, *Analog Computer to Determine Seismic Weathering Time Corrections*, United States Patent 2884194, April 28, 1959

[Evans et al. 1966] D. C. Evans et al., *Digital Differential Analyzer in Conjunction with a General Purpose Computer*, United States Patent US 3274376, Sept. 20, 1966

[Everyday Science and Mechanics 1932] N. N., "Mechanical SUPER-BRAINS – Calculations in Higher Mathematics Performed by Complex Machinery", in *Everyday Science and Mechanics*, June 1932, pp. 625, 678

[Eyman et al. 1976] Earl D. Eyman, Yevgeny V. Kolchev, "Universal Analog Computer Model for Three-Phase Controlled Rectifier Bridges", in *IEEE PES Winter Meeting & Tesla Symposium*, New York, N.Y., January 25–30, 1976, pp. 1136–1144

[222] M. D. Fagen (ed.), *A History of Engineering and Science in the Bell System – National Service in War and Peace (1925–1975)*, Bell Telephone Laboratories, Inc., First Printing, 1978

[Fano 1950] Robert Mario Fano, "Short-Time Autocorrelation Functions and Power Spectra", in *The Journal of the Acoustical Society of America*, Volume 22, Number 5, pp. 546–550

[FARRIS 1964] GEORGE JOSEPH FARRIS, *A Digital Differential Analyzer Programming System for the I.B.M. 7074 Computr*, dissertation, Iowa State University of Science and Technology, Ames, Iowa, 1964

[FEILMEIER 1974] MANFRED FEILMEIER, *Hybridrechnen*, International Series of Numerical Mathematics, Vol. 2, Birkhäuser Verlag, 1974

[FENECH et al. 1973] HENRI FENECH, JAY BANE, "A nuclear reactor analog simulation for undergraduate nuclear engineering education", in *Simulation*, Volume 20, Issue 4, April 1974, pp. 127–135

[FIFER 1961] STANLEY FIFER, "Analogue Computation – Theory, Techniques and Applications", Vol. III, McGraw-Hill Book Company, Inc., 1961

[FISCHER 2022] THOMAS FISCHER, *THE ANALOG THING – First Steps*, https://the-analog-thing.org/THAT_First_Steps.pdf, retrieved 08/07/2022

[FOOTE et al. 2007] ROBERT L. FOOTE, ED SANDIFER, "Area Without Integration: Make Your Own Planimeter", in *Hands on History – A Resource for Teaching Mathematics*, Amy Shell-Gellash, ed., The Mathematical Association of America (Incorporated), 2007

[FORBES 1957] GEORGES F. FORBES, *Digital Differential Analyzers*, Fourth Edition, 1957

[FORBES 1972] GEORGE FORBES, "The simulation of partial differential equations on the digital differential analyzer", in ACM '72 Proceedings of the ACM annual conference, Volume 2, pp. 860–866

[FORNEL et al. 1981] B. DE FORNEL, H. C. HAPIOT, J. M. FARINES, J. HECTOR, "Hybrid Simulation of a Current Fed Asynchronous Machine", in *Mathematics and Computers in Simulation*, XXIII (1981), pp. 253–261

[Fortune 1952] N. N., "The Moniac – 'Economics in thirty fascinating minutes'", in *Fortune*, March 1952, pp. 101–102

[Fox et al. 1963] J. C. FOX, T. G. WINDEKNECHT, "Six degree-of-freedom simulation of a manned orbital docking system", in *AFIPS '63 (Spring): Proceedings of the May 21-23, 1963, spring joint computer conference*, May 1963, pp. 91–104

[Fox et al. 1969] HARRY W. FOX, RONALD J. BLAHA, *An Analog Computer Study of the Low-Frequency Dynamics of two Nuclear-Rocket Cold-Flow Engine Systems*, National Aeronautics and Space Administration, Washington, D. C., July 1969

[FRANÇOIS-LAVET et al. 2018] VINCENT FRANÇOIS-LAVET, PETER HENDERSON, RIASHAT ISLAM, MARC G. BELLEMARE, JOELLE PINEAU, "An Introduction to Deep Reinforcement Learning", in *Foundation and Trends in Machine Learning*, Vol. 11, No. 3–4, pp. 219–354

[FREEDMAN 2011] IMMANUEL FREEDMAN, *System for forecasting outcomes of clinical trials*, United States Patent US 2011/0238317 A1, Sep. 29, 2011

[FREETH 2008] TONY FREETH, *The Antikythera Mechanism – Decoding an Ancient Greek Mystery*, Whipple Museum of the History of Science, University of Cambridge, 2008

[FREETH et al. 2021] TONY FREETH, DAVID HIGGON, ARIS DACANALIS, LINDSAY MACDONALD, MYRTO GEORGAKOPOULOU, ADAM WOJCIK, "A Model of the Cosmos in the ancient Greek Antikythera Mechanism", in *Scientific Reports*, nature portfolio, 2021, 11:5821, https://doi.org/10.1038/s41598-021-84310-w

[FREETH 2022] TONY FREETH, "Wonder of the Ancient World", in *Scientific American*, 326, 1, January 2022, pp. 24-33

[FREMEREY] JOHAN K. FREMEREY, *Permanentmagnetische Lager*, Forschungszentrum Jülich, Institut für Grenzflächenforschung und Vakuumphysik, 0B30-A30

[FREMEREY 1978] JOHAN K. FREMEREY, KARL BODEN, "Active permanent magnet suspensions for scientific instruments", in *Journal of Physics E – Scientific Instruments*, February 1978, Vol. 11, No. 2, pp. 106–113

[FRIEDMAN 2008] NORMAN FRIEDMAN, *Naval Firepower – Battleship Guns and Gunnery in the Dreadnought Era*, Seaforth Publishing, 2008

[FRISCH 1968] WILLI FRISCH, *Stabilitätsprobleme bei dampfgekühlten schnellen Reaktoren*, Dissertation, Universität Karlsruhe, 1968

[FRISCH et al. 1969] WILLI FRISCH, G. WILHELMI, *Dynamische Simulatoren in der Reaktorentwicklung – ein Vergleich*, Gesellschaft für Kernforschung mbH, Karlsruhe, Januar 1969, 8/69-1

[FRISCH 1971] WILLI FRISCH, *Analogrechnen in der Kernreaktorrechnik*, G. Braun Karlsruhe, 1971

[GAINES 1967] BRIAN R. GAINES, "Stochastic Computing", in *AFIPS '67 (Spring): Proceedings of the April 18–20, 1967, spring joint computer conference*, April 1967, pp. 149–156

[GAINES 2019] BRIAN R. GAINES, "Origins of Stochastic Computing", in [GROSS et al. 2019, pp. 13–38]

[GALLAGHER et al. 1957] JOHN M. GALLAGHER, WILLIAM W. SEIFERT, "Simulation Techniques for Nuclear Reactor Studies", in [White Sands 1957, pp. 69–89]

[GAP/R EAC] N.N., *GAP/R Electronic Analog Computors*, George A. Philbrick Researches, Inc.

[GAP/R Evolution] N.N., *A Brief History of a Computor-Builder in Terms of Ideas & Instrument – Evolution of a* Model *Kit of Tools*, George A. Philbrick Researches, Inc.

[GAP/R K2W] N.N., *Model K2-W Operational Amplifier*, George A. Philbrick Researches, Inc.

[GAP/R 1959] N. N., *Squaring, Rooting, and the Douglas Quadratron*, GAP/R Application Brief, George A. Philbrick Researches, Inc., No. D1, December 1, 1959

[GAUDET et al. 2019] VINCENT C. GAUDET, WARREN J. GROSS, KENNETH C. SMITH, "Introduction to Stochastic Computing", in [GROSS et al. 2019, pp. 1–11]

[GERWIN 1958] ROBERT GERWIN, "Atom-Strom für deutsche Städte", in *Hobby – Das Magazin der Technik*, Nr. 9, September 1958

[GILBERT 1970] LEONARD J. GILBERT, *Analog Computer Simulation of a Parasitically Loaded Rotating Electrical Power Generating System*, National Aeronautics and Space Administration, Washington, D. C., November 1970

[GILBERT 1982] BARRY GILBERT, "A monolithic microsystem for analog synthesis of trigonometrix functions and their inverses", in *IEEE Journal of Solid-State Circuits*, Volume 17, Issue 6, Dec. 1981, pp. 1179–1191

[GILOI 1960] W. GILOI, "Behandlung von Transformatorproblemen mit dem Analogrechner", in *Telefunken Zeitung*, Vol. 33 (1960), No. 129, p. 50

[GILOI 1961] W. GILOI, "Ein Verfahren zur Berechnung von Optimalfiltern auf dem Analogrechner", in *Elektronische Rechenanlagen*, 3 (1961), No. 2, pp. 61–65

[GILOI 1962] W. GILOI, "Über die Behandlung elektrischer und mechanischer Netzwerke auf dem Analogrechner", in *Elektronische Rechenanlagen*, 4 (1962), No. 1, pp. 27–35

[GILOI 1963] W. GILOI, "Hybride Rechenanlagen – ein neues Konzept", in *Elektronische Rechenanlagen*, 5 (1963), No. 6, pp. 262–269

[GILOI et al. 1963] WOLFGANG GILOI, RUDOLF LAUBER, *Analogrechnen*, Springer-Verlag, 1963

[GILOI 1975] W. K. GILOI, *Principles of Continuous System Simulation*, B. G. Teubner, Stuttgart, 1975

[GILOI et al.] W. GILOI, R. HERSCHEL, *Rechenanleitung für Analogrechner*, Telefunken-Fachbuch, AFB 001

[GILPIN 1973] MICHAEL E. GILPIN, "Do Hares Eat Lynx?", in *The American Naturalist*, Vol. 107, No. 957, pp. 727–730

[GLEISER 1980] MOLLY GLEISER, "Analog Inventor", in *DATAMATION*, October 1980, pp. 141–143

[GLUMINEAU et al. 1982] A. GLUMINEAU, R. MEZENCEV, "Hybrid Simulation of a Tanker Moored at a Single Point Subjected to Effects of Wind, Current and Waves", in 10th *IMACS World Congress on System Simulation and Scientific Computation*, 1982, pp. 98–100

[GOLDBERG 1951] EDWIN A. GOLDBERG, "Step Multiplier in Guided Missile Computer", in *ELECTRONICS*, August, 1951, pp. 120–124

[GOLDBERG et al. 1954] EDWIN A. GOLDBERG, JULES LEHMANN, *Stabilized Direct Current Amplifier*, United States Patent 2684999, July 27, 1954

[GOLDBOHM 1993] E. GOLDBOHM, "Near field measurements and the synthesis of linear arrays with prescribed radiation patterns: a survey of some early work at Christian Huygens Laboratory", in *IEEE Antennas and Propagation Magazine*, Volume 35, Issue 3, June 1993, pp. 26–34

[GOLDMAN 1965] MARK W. GOLDMAN, "Design of a High Speed DDA", in AFIPS '65 (Fall, part I) Proceedings of the November 20–December 1, 1965, fall joint computer conference, part I, pp. 929–949

[GOLDSMITH et al. 1948] THOMAS T. GOLDSMITH, ESTLE RAY MANN, *Cathode-Ray Tube Amusement Device*, United States Patent 2455992, Dec. 14, 1948

[GOLTEN et al. 1967] J. W. GOLTEN, DAVID REES, "The Use of Hybrid Computing in the Analysis of Steel Rolling", in *IEEE Transactions on Electronic Computers*, Vol. EC-16, No. 6, December 1967, pp. 717–722

[GOMPERTS et al. 1957] R. J. GOMPERTS, D. W. RIGHTON, "LACE (The Luton Analogue Computing Engine (Part 1)", in *Electronic Engineering*, Vol. 29, July 1957, pp. 306–312

[GOVINDARAO 1975] V. M. H. GOVINDARAO, "Analog Simulation of an Isothermal Semibatch Bubble-Column Slurry Reactor", in *Annales de l'Association internationale pour le Calcul analogique*, No. 2, Avril 1975, pp. 69–78

[GRACON et al. 1970] T. J. GRACON, J. C. STRAUSS, "Design of automatic patching systems for analog computers", in *AFIPS '70 (Spring): Proceedings of the May 5–7, 1970, spring joint computer conference*, May 1970, pp. 31–38

[GRAEFE et al. 1974] P. W. U. GRAEFE, LEO K. NENONEN, "Simulation of combined discrete and continuous systems on a hybrid computer", in *SIMULATION*, May 1974, pp. 129–137

[GRAY 1948] JOHN W. GRAY, "Direct-Coupled Amplifiers", in *Vacuum Tube Amplifiers*, Massachusetts Institute of Technology, Radiation Laboratory Series, pp. 409–495

[GRAY et al. 1955] JOHN W. GRAY, DUNCAN MACRAE, *Bombing Computer*, United States Patent 2711856, June 28, 1955

[GRAY 1958] H. J. GRAY Jr., "Digital Computer Solution of Differential Equations in Real Time", in IRE-ACM-AIEE '58 (Western) Proceedings of the May 6–8, 1958, western joint computer conference: contrasts in computers, pp. 87–92

[GRIERSON et al.] W. O. GRIERSON, D. B. LIPSKI, N. O. TIFFANY, *Simulation Tools: Where can we go?*, author's archive

[GRISWOLD et al. 1957] JAMES A. GRISWOLD, HERBERT V. LOOMIS, "Ship Motion Simulator", in [White Sands 1957, pp. 11–19]

[GRÖBNER 1961] W. GRÖBNER, "Steuerungsprobleme mit Optimalbedingung", in *mtw – Zeitschrift für moderne Rechentechnik und Automation*, 2/61, pp. 62–64

[GROSS et al. 2019] WARREN J. GROSS, VINCENT C. GAUDET (eds.), *Stochastic Computing: Techniques and Applications*, Springer Nature Switzerland, 2019

[GUIBERT et al. 1949] A. G. GUIBERT, E. JANSSEN, W. M. ROBBINS, *Determination of Rate, Area, and Distribution of Impingement of Waterdrops on Various Airfoils from Trajectories Obtaines on the Differential Analyzer*, National Advisory Committee for Aeronautics, Washington, Research Memorandum No. 9A05, February 16, 1949

[GUL'KO 1961] F. B. GUL'KO, "Thyrite Multiplier with an Increased Passband", translated from *Avtomatika i Telemekhanika*, Vol. 22, No. 12, December, 1961, pp. 1649–1655

[GUMPERT 1972] W. GUMPERT, "Analogberechnung selbsterregter Schwingungen beim Anfahren von Fahrzeugen", in *Kraftfahrzeugtechnik*, 2/72, pp. 46–48

[GUNDLACH 1955] F. W. GUNDLACH, "A new electronic-beam multiplier with an electrostatic hyperbolic field", in *Actes. J. Internat. Calcul Analog Analogique*, Brüssel, 1955, pp. 101–103

[GUO 2017] NING GUO, *Investigation of Energy-Efficient Hybrid Analog/Digital Approximate Computation in Continuous Time*, Columbia University, 2017

[GUZDIAL et al. 2013] MARK GUZDIAL, DANIEL REED, "Securing the Future of Computer Science; Reconsidering Analog Computing", in *Communications of the ACM*, April 2013, Vol. 56, No. 4, pp. 11–12

[HABERMEHL et al. 1969] A. HABERMEHL, E. H. GRAUL, *Analog Computer Investigations with a Mathematical Model for the Regulation Mechanism of Calcium Metabolism*, National Aeronautics and Space Administration, Washington, D. C., April 1969

[HABERMEHL et al. 1969] A. HABERMEHL, E. H. GRAUL, H. WOLTER, *Computer in der Nuklearmedizin – Die Anwendung des Analogrechners zur Untersuchung biologischer Systeme und Prozesse*, Dr. Alfred Hüthig Verlag Heidelberg, 1969

[HAEBERLIN et al.2011] BARBARA HAEBERLIN, STEFAN DRECHSLER, *Wie funktioniert ein Kegel-Reibrad-Planimeter?*, RST 22, Worms, 10/22/2011, http://www.rechenschieber.org/haedre2011.pdf, retrieved 11/22/2012

[HALL et al. 1969] CARROLL R. HALL, STEPHEN J. KAHNE, "Automated Scaling for Hybrid Computers", in *IEEE Transactions on Computers*, Vol. C-18, No. 5, May 1969, pp. 416–423

[HALL 1996] ELDON C. HALL, *Journey to the Moon: The History of the Apollo Guidance Computer*, American Institute of Aeronautics and Astronautics, Inc., 1996

[HAMMACK et al. 2014] BILL HAMMACK, STEVE KRANZ, BRUCE CARPENTER, *Albert Michelson's Harmonic Analyzer – A Visual Tour of a Nineteenth Century Machine that Performs Fourier Analysis*, Articulate Noise Books, 2014

[HAMMER 1956] J. A. HAMMER, "Evaluation of Diffration Problems in Acrial Technology by Means of an Analyser for Computer and Fourier Series", in *Principes fondamentaux: Équations de Maxwell, Principe de Huygens et Théory de la diffraction en hyper-fréquences*, 1956

[HAMORI 1972] EUGENE HAMORI, "Use of the Analog Computer in Teaching Relaxation Kinetics", in *Journal of Chemical Education*, Volume 49, Number 1, January 1972, pp. 39–43

[HANNAUER 1968] GEORGE HANNAUER, *Stored Program Concept for Analog Computers*, final report, EAI project 320009, NASA order NAS8-21228

[HANNIGAN] FRANK J. HANNIGAN, "Hybrid Computer Simulation of Fluidic Devices", Electronic Associates, Inc.

[HANSEN et al. 1959] P. D. HANSEN, J. H. EATON, "Control and Dynamics Performance of a Sodium Cooled Reactor Power System", MICROTECH RESEARCH COMPANY, Massachusetts, Report No. 171, December 28., 1959

[HANSEN 2005] JAMES R. HANSEN, *First Man – The Life of Neil A. Armstrong*, Simon & Schuster UK, 2005

[HARNETT et al. 1963] R. T. Harnett, J. F. Sansom, L. M. Warshawsky, "MIDAS... An analog approach to digital computation", in *SIMULATION*, September 1963, S. 17–43

[HART et al. 1967] CLINT E. HART, DALE J. ARPASI, *Frequency Response and Transfer Functions of a Nuclear Rocket Engine System obtained from Analog Computer Simulation*, National Aeronautics and Space Administration, Washington, D. C., May 1967

[HARTMANN et al. 1961] JOHN HARTMANN, GRANINO A. KORN, RICHARD L. MAYBACH, "Low-Cost Triangle-Integration Multipliers for Analog Computers", in ANNALES DE L'ASSOCIATION INTERNATIONAL POUR LE CALCUL ANALOGIQUE, No. 4, Octobre 1961, pp. 167–172

[HAUG 1960] ALBERT HAUG, "Funktionsgeneratoren und Funktionsspeicher der Formen $y = f(x)$ und $z = f(x,y)$", in *Nachrichtentechnische Fachberichte*, Band 17, Vieweg & Sohn, Braunschweig, 1960

[HAVIL 2019] JULIAN HAVIL, *Curves for the Mathematically Curious – an Anthology of the Unpredictable, Historical, Beautiful, and Romantic*, Princeton University Press, 2019

[HAZEN et al. 1940] H. L. HAZEN, G. S. BROWN, WALTER R. HEDEMAN, "The cinema integraph – A machine for evauluating a parametric product integral", in JOURNAL OF THE FRANKFURT INSTITUTE – ENGINEERING AND APPLIED MATHEMATICS, 1. Aug. 1940, pp. 183–205

[HEADRICK et al. 1969] J. B. HEADRICK, D. P. JORDAN, "Analog Computer Simulation of the Heat Gain Through a Flat Composite Roof Section", in *ASHRAE Transactions*, Vol. 75, 1969, pp. 21–33

[HEDEMAN 1941] WALTER R. HEDEMAN, "The Cinema Integraph in Interreflection Problems", April 1941, https://doi.org/10.1002/sapm1941201402, retrieved 02/22/2022

[HEDIN et al.] RONALD A. HEDIN, KENNETH W. PRIEST, "Progress in Hybrid Simulation of Power Systems", author's archive

[HEIDEPRIM 1976] J. HEIDEPRIM, "Modelle und Simulation von Produktionsprozessen in der Stahlindustrie", in [SCHÖNE 1976/1, pp. 206–254]

[HEIDERSBERGER] HEINRICH HEIDERSBERGER, *Rhythmogramme*, CARGO Verlag

[HEINHOLD 1959] J. HEINHOLD, "Konforme Abbildung mittels elektronischer Analogrechner", in *mtw – Zeitschrift für moderne Rechentechnik und Automation*, 1/59, pp. 44–48

[HEINHOLD et al.] JOSEF HEINHOLD, ULRICH KULISCH, *Analogrechnen – Eine Einführung*, Bibliographisches Institut Mannheim/Wien/Zürich, B.I.-Wissenschaftsverlag, 1969

[HELLER et al. 1976] RAINER HELLER, CARL W. MALSTROM, E. HARRY LAW, "Hybrid Simulation of Rail Vehicle Lateral Dynamics", presented at *The 1976 Summer Computer Simulation Conference*, Washington, D. C., July 12–14, 1976

[HEMMENDINGER 2014] DAVID HEMMENDINGER, "COMIC: An Analog Computer in the Colorant Industry", in *IEEE Annals of the History of Computing*, Vol. 36, Issue 3, July–Sept. 2014, pp. 4–18

[HENRICI 1894] O. HENRICI, "Report on Planimeters", in *Report of the Sixty-Fourth Meeting of the British Association for the Advancement of Science*, London: John Murray, Albemarle Street, 1894

[HERSCHEL 1957] R. HERSCHEL, "Elektronische Rechenmaschinen ohne Einmaleins", in *Funkschau*, 1957, Heft 10, pp. 10–11

[HERSCHEL 1961] R. HERSCHEL, "Automatische Optimisatoren", in *Elektronische Rechenanlagen*, 3 (1961), No. 1, pp. 30–36

[HERSCHEL 1962] R. HERSCHEL, "Analogrechenschaltungen für die Entwicklungskoeffizienten nach Orthogonalfunktionen", in *Elektronische Rechenanlagen*, 3 (1961), No. 5, pp. 212–217

[HERSCHEL 1966] R. HERSCHEL, "Zur Programmierung von hybriden Rechenanlagen in ALGOL", in *Telefunken Zeitung*, Vol. 39 (1966), No. 1, pp. 100–109

[HERZOG et al. 2018] BENEDICT HERZOG, LUIS GERHORST, BERNHARD HEINLOTH, STEFAN REIF, TIMO HÖNIG, WOLFGANG SCHRÖDER-PREIKSCHAT, "INTSPECT: Interrupt Latencies in the Linux Kernel", in *2018 VIII Brazilian Symposium on Computing Systems Engineering (SBESC)*, 2018, pp. 83–90

[HICKS 1968] W. D. HICKS, *An Electronic Analogue Computer Representing Twelve Coupled Linear Differential Equations*, Ministry of Technology, Aeronautical Research Council, London, Her Majesty's Stationary Office, 1968

[HIGGINS et al. 1982] W. H. C. HIGGINS, B. D. HOLBROOK, J. W. EMLING, "Defense Research at Bell Laboratories", in *Annals of the History of Computing*, Volume 4, Number 3, July 1982, pp. 218–236

[HINDMARSH et al. 1982] J. L. HINDMARSH, R. M. ROSE, "A model of the nerve impulse using two first-order differential eequations", in *Nature*, Vol. 296, 11 March 1982, pp. 162–164

[HINDMARSH et al. 1984] J. L. HINDMARSH, R. M. ROSE, "A model of neuronal bursting using three coupled first order differential equations", in *Prov. R. Soc. Lond.*, B 221, 87–102 (1984)

[HINTZE 1957] GUENTHER HINTZE, "Mathematical vs. Physical Simulation", in [White Sands 1957, pp. 3–9]

[Hitachi 200X] N. N., *Hitachi Analog Hybrid Computer – Hitachi-200X*, Hitachi Electronics, Ltd.

[Hitachi 505E] N. N., "HITACHI 505E analog/hybrid computer", Hitachi Electronics

[Hitachi 1967] N. N., "Double Integral – Calculation of Volume of Cone", in *Technical Information Series No. 3*, Hitachi Electronics, Ltd., 1967

[Hitachi 1968] N. N., "Analysis of Rolling Theory by Analog Computer – Karman's Differential Equation", in *Technical Information Series No. 8*, Hitachi Electronics, Ltd., 1968

[Hitachi 1969] N. N., "Introduction to Simulator – Part 1", in *Technical Information Series No. 9*, Hitachi Electronics, Ltd., 1969

[Hobby 1969] N. N., "Psychedelic – Explosion der Farben", in *hobby – Das Magazin der Technik*, Nr. 15/69, p. 36–45

[HOELZER 1946] HELMUT HOELZER, *Anwendung elektrischer Netzwerke zur Lösung von Differentialgleichungen*, Dissertation TH Darmstadt, 1946

[HOELZER 1946/2] HELMUT HOELZER, *Anwendung elektrischer Netzwerke zur Stabilisierung von Regelvorgängen und zur Lösung von Differentialgleichungen, gezeigt an der Stabilisierung des Fluges einer selbst- bzw. ferngesteuerten Großrakete*, Dissertation TH Darmstadt, 1946

[HOELZER 1992] HELMUT HOELZER, "50 Jahre Analogcomputer", Rede anlässlich des fünfzigsten Jubiläums des elektronischen Analogrechners im Senatssaal in Berlin, 05/12/1992, Manuskript aus dem Archiv der Familie Hoelzer-Beck

[HOFFMAN 1979] DALE T. HOFFMAN, "Smart Soap Bubbles Can Do Calculus", in *The Mathematics Teacher*, Vol. 72, No. 5, Mai 1979, pp. 377–385

[HOFFMANN 2006] JUSTIN HOFFMANN / Kunstverein Wolfsburg (Hg.), *Der Traum von der Zeichenmaschine – Heinrich Heidersbergers Rhythmogramme und die Computergrafik ihrer Zeit*, Kunstverein Wolfsburg, Kataloge #1/2006

[HOLST 1982] PER A. HOLST, "George A. Philbrick and Polyphemus – The First Electronic Training Simulator", in *Annals of the History of Computing*, Volume 4, Number 2, April 1982

[HOLST 1996] PER A. HOLST, "Svein Rosseland and the Oslo Analyzer", in *IEEE Annals*, Winter 1996, Vol. 18, No. 4, pp. 16–26

[HOLT et al. 1995] FREDERICK B. HOLT, DZIEM D. NGUYEN, *Pseudorandom Stochastic Data Processing*, United States Patent 9412587, May 2, 1995

[HOLZER et al. 2021] MIRKO HOLZER, BERND ULMANN, "Hybrid computer approach to train a machine learning system", in *Handbook in Unconventional Computing*, ed. ANDREW ADAMATZKY, World Scientific Publishers, 2021

[HORLING et al.] JAMES E. HORLING, ESMAT MAHMOUT, FRANK J. HANNIGAN, *Hardware-in-the-Loop Simulation for Evaluating Turbine Engine Fuel System Components*, author's archive

[HOSENTHIEN et al. 1962] H. H. HOSENTHIEN, J. BOEHM, "Flight Simulation of Rockets and Spacecraft", in *From Peenemünde to Outer Space – Commemorating the Fiftieth Birthday of* WERNHER VON BRAUN, March 23, 1962, pp. 437–469

[HOWARD et al. 1953] R. C. HOWARD, C. J. SAVANT, R. S. NEISWANDER, "Linear-to-Logarithmic Voltage Converter", in *Electronics*, July, 1953, pp. 156–157

[HOWARD 1961] WARREN DEE HOWARD, "The computer simulation of a colonial socio-economic system", in *IRE-AIEE-ACM '61 (Western): Papers presented at the May 9–11, 1961, western joint IRE-AIEE-ACM computer conference*, May 1961, pp. 613–622

[HOWE et al. 1953] ROBERT M. HOWE, V. S. HANEMAN, "The solution of partial differential equations by difference methods using the electronic differential analyzer", in AIEE-IRE '53 (Western) Proceedings of the February 4–6, 1953, western computer conference, pp. 208–226

[HOWE 1957] ROBERT M. HOWE, "Coordinate Systems and Methods of Coordinate Transformations for Dimensional Flight Equations", in [White Sands 1957, pp. 49–66]

[HOWE 1962] ROBERT M. HOWE, "Solution of Partial Differential Equations", in [HUSKEY et al. 1962, pp. 5-110–5-132]

[HOWE et al. 1970] ROBERT M. HOWE, RICHARD A. MORAN, THOMAS D. BERGE, "Time sharing of hybrid computers using electronic patching", in *AFIPS '70 (Fall): Proceedings of the November 17–19, fall joint computer conference*, November 1970, pp. 377–386

[HOWE et al. 1975] ROBERT M. HOWE, ALDRIC SAUCIER, "A new fourth generation of hybrid computer systems", in *AFIPS '75: Proceedings of the May 19–22, 1975, national computer conference and exposition*, May 1975, pp. 861–866

[HSIAO et al. 2019] HSUAN HSIAO, JASON ANDERSON, YUKO HARA-AZUMI, "Generating Stochastic Bitstreams", in [GROSS et al. 2019, pp. 137–152]

[HSU et al.1968] STEPHEN K. T. HSU, ROBERT M. HOWE, "Preliminary investigation of a hybrid method for solving partial differential equations", in *AFIPS '68 (Fall, part 1): Proceedings of the December 9–11, 1968, fall joint computer conference, part 1*, December 1968, pp. 601–609

[HU 1972] RICHARD H. HU, "Analog Computer Simulation of an FM Communication System", American Society for Engineering Education, Annual Conference, June 19–22, 1972

[HUANG et al. 2016] YIPENG HUANG, NING GUO, MINGOO SEOK, YANNIS TSIVIDIS, SIMHA SETHUMADHAVAN, "Evaluation of an Analog Accelerator for Linear Algebra", in *43rd Annual Interational Symposium on Computer Architecture*, IEEE, 2016, pp. 570–582

[GUO et al. 2016] NIGN GUO, YIPENG HUANG, TAO MAI, SHARVIL PATIL, CHI CAO, MINGOO SEOK, SIMHA SETHUMADHAVAN, YANNIS TSIVIDIS, "Energy-Efficient Hybrid Analog/Digital Approximate Computation in Continuous Time", in *IEEE Journal of Solid-State Circuits*, Vol. 51, No. 7, July 2016, pp. 1514–1524

[HUANG et al. 2017] YIPENG HUANG, NING GUO, MINGOO SEOK, YANNIS TSIVIDIS, KYLE MANDLI, "Hybrid Analog-Digital Solutions of Nonlinear Partial Differential Equations", in *MICRO-50 '17: Proceedings of the 50th Annual IEEE/ACM International Symposium on Microarchitecture*, October 2017, pp: 665–678

[HUDSON et al. 1945] C. S. HUDSON, I. A. MOOSSOP, *Flight Control Amplifier of the German Long Range Rocket*, Royal Aircraft Establishment, Farnborough, E. A. No. 228-22, Report No. El. 1365, March 1945

[HÜLSENBERG et al. 1975] FRIEDER HÜLSENBERG, UWE KIESSLING, HARTMUT SCHÖNBORN, *Beziehung zwischen Produktion, Lagerhaltung und Marktrealisation – dargestellt an einem Analogie-Rechenmodell für den Analogrechner MEDA T*, VEB Deutscher Verlag für Grundstoffindustrie, 1975

[Hütte 1926] Akademischer Verein Hütte, e.V. in Berlin (ed.), *Hütte – Des Ingenieurs Taschenbuch*, 25. neubearbeitete Auflage, II. Band, Berlin 1926, Verlag von Wilhelm Ernst & Sohn

[HUME et al. 2005] TED HUME, BOB KOPPANY (ed.), *The Oughtred Society Slide Rule Reference Manual*, Striking Impressions, Los Angeles, California, First Edition, 2005

[HUND 1982] MARTIN HUND, *Der Analogrechner – Beschreibung mit 150 Programmbeispielen aus Mathematik, Regelungstechnik und Physik*, Leybold-Heraeus GmbH, 2. Auflage, 1982

[HURST] CHARLES J. HURST, *Computer Simulation in a Mechanical Engineering Laboratory Program*, author's archive

[HUSKEY et al. 1962] HARRY D. HUSKEY, GRANINO A. KORN, *Computer Handbook*, McGraw-Hill Book Company, Inc., 1962

[HYATT et al. 1968] GILBERT P. HYATT, GENE OHLBERG, "Electrically alterable digital differential analyzer", in *AFIPS '68 (Spring) Proceedings of the April 30–May 2, 1968 spring joint computer conference*, pp. 161–169

[Hydrocarbon Processing 1959] Hydrocarbon Processing (Reprint), *Analog Computers Handbook*, Gulf Publishing Company, 1969

[IRWIN et al. 1960] SAMUEL N. IRWIN, ROBERT R. KLEY, "Analog computer serves as both systems analysis tool and operator training facility for Enrico Fermi Atomic Power Plant", in *IRE-AIEE-ACM '60 (Western): Papers presented at the May 3–5, 1960, western joint IRE-AIEE-ACM computer conference*, May 1960, pp. 301–313

[isec/1] N. N., *Instruction and Operating Manual for your isec 250 Model C and D*, isec, Princeton, New Jersey

[isec/2] N. N., *Programming the ISEC 250*, isec, Princeton New Jersey

[JACKSON 1960] ALBERT S. JACKSON, *Analog Computation*, McGraw-Hill Book Company, Inc., 1960

[JAMES et al. 1971] M. L. JAMES, G. M. SMITH, J. C. WOLFORD, *Analog Computer Simulation of Engineering Systems*, Intext Educational Publishers, 1971, 3rd edition

[JAMSHIDI 1976] M. JAMSHIDI, "Optimization of some Dynamic Industrial Control Processes by Analog Simulation", in *Trans. IMACS*, Vol. XVIII, No. 2, April 1976, pp. 93–100

[JANAC 1976] KAREL JANAC, *Control of Large Power Systems Based on Situation Recognition and High Speed Simulation*, presented at 9th-Hawaii International Conference on System Sciences, January 6–8, 1976, University of Hawaii, author's archive

[JANSSEN et al. 1955] J. M. L. JANSSEN, L. ENSING, "The Electro-Analogue, an Apparatus for Studying Regulating Systems", in [PAYNTER ed. 1955, pp. 147–161]

[JEZIERSKI 2000] DIETER VON JEZIERSKI, *Slide Rules – A Journey Through Three Centuries*, Astragal Press, Mendham, New Jersey, 2000

[JIA et al. 2019] XIAOTAO JIA, YOU WANG, ZHE HUANG, YUE ZHANG, JIANLEI YANG, YUANZHUO QU, BRUCE F. COCKBURN, JIE HAN, WEISHENG ZHAO, "Spintronic Solutions for Stochastic Computing", in [GROSS et al. 2019, pp. 165–183]

[JÖNCK et al. 2003] UWE JÖNCK, FLORIAN PRILL, "Das Lorenz-System", in *Seminar über gewöhnliche Differentialgleichungen*, Universität Hamburg, 2003, https://www.math.uni-hamburg.de/home/lauterbach/scripts/seminar03/prill.pdf, retrieved 05/18/2022

[JOHNSON 1915] R. D. JOHNSON, "The Differential Surge Tank", in *Transactions of the American Society of Civil Engineers*, Vol. LXXVIII, 1915

[JOHNSON et al. 1961] R. S. JOHNSON, F. R. WILLIAMSON, R. D. LOFTIN, *Developent of New Methods and Applications of Analog Computation*, George C. Marshall Space Flight Center, Huntsville, Alabama, GIT/EES Report A588/P1, 12 January 1962

[JOHNSON 1962] E. CALVIN JOHNSON, "Computers and Control", in [HUSKEY et al. 1962, pp. 21-62 ff.]

[JOHNSON 1963] CLARENCE L. JOHNSON, *Analog Computer Techniques*, McGraw-Hill Book Company, Inc., Second Edition, 1963

[JONES et al. 1957] J. C. JONES, D. READSHAW, "LACE (The Luton Analogue Computing Engine (Part 2)", in *Electronic Engineering*, Vol. 29, August 1957, pp. 380–385

[JONES 1961] E. D. JONES, *Power Excursion in a Hanford Reactor Due to a Positive Reactivity Ramp*, HW-71119, Hanford Atomic Products Corporation, Richland, Washington, September 20, 1961

[JUNG 2006] WALT JUNG, *Op Amp History*, in [Analog Devices 2006], pp. 765–829

[JUSLIN 1981] KAJ JUSLIN, *Hybrid Computer Model for Synchronous and Asynchronous Motor Interaction Studies*, Scandinavian Simulation Society annual meeting, 18–20th May, 1981 at Royal Institute of Technology, Stockholm

[JUST et al. 1962] L. C. JUST, C. N. KELBER, N. F. MOREHOUSE, *An Analog Computer Model of a Multiple-Region Reactor*, Argonne National Laboratory, ANL-6482, February 1962

[KAHNE 1968] STEPHEN J. KAHNE, "Sensitivity-Function Calculation in Linear Systems Using Time-Shared Analog Integration", in *IEEE Transactions on Computers*, Vol. C-17, No. 4, April, 1968, pp. 375-279

[KAPLAN] PAUL KAPLAN, *A Mathematical Model for Assault Boat Motions in Waves*, Oceanics Inc., Plainview, New York, author's archive

[KARPLUS 1958] WALTER J. KARPLUS, *Analog Simulation – Solution of Field Problems*, McGraw-Hill Book Company, Inc., 1958

[KARPLUS et al. 1958] WALTER J. KARPLUS, WALTER W. SOROKA, *Analog Methods – Computation and Simulation*, McGraw-Hill Book Company, Inc., 1958

[KARPLUS et al. 1972] WALTER J. KARPLUS, RICHARD A. RUSSELL, "Increasing Digital Computer Efficiency with the Aid of Error-Correcting Analog Subroutines", in *IEEE Transactions on Computers*, Vol. C-20, No. 8, August 1972, pp. 831–837

[KASPER 1955] JOSEPH EMIL KASPER, *Construction and application of a mechanical differential analyzer*, Thesis, State University of Iowa, February 1955

[KASTNER 1968] SIGISMUND KASTNER, "Analyse von natürlichem Modellseegang", in *Schriftenreihe Schiffbau*, Technische Universtität Hamburg-Harburg, 223, 1968

[KASTNER 1968/2] SIGISMUND KASTNER, *Das Kentern von Schiffen in unregelmässiger längs laufender See*, Technische Universtität Hamburg-Harburg, 249, 1968

[KAUFMANN et al. 1955] M. G. KAUFMANN, R. E. GARDNER, "An Electronic Slide Rule", in *Radio & Television News*, December 1955, pp 58 ff.

[KAYRAKLIOGLU 2020] ENGIN KAYRAKLIOGLU, JEFF ANDERSON, HAMID REZA IMANI, "Software Stack for an Analog Mesh Computer: The Case of a Nanophotonic PDE Accelerator", in *Proceedings of the 17th ACM Interational Conference on Computing Frontiers*, May 2020, pp. 241–244

[KELLA 1967] J. KELLA, "A Note on the Accuracy of Digital Differential Analyzers", in *IEEE Transactions on Electronic Computers*, Vol. EC-16, No. 2, April 1967, p. 230

[KELLA et al. 1968] J. KELLA, A. SHANI, "On the Reversibility of Computations in a Digital Differential Analyzer", in *IEEE Transactions on Computers*, Vol. C-17, No. 3, 1968, pp. 283–284

[KENNEALLY et al. 1966] W. J. KENNEALLY, E. E. L. MITCHELL, I. HAY, G. BOLTON, "Hybrid simulation of a helicopter", in *AFIPS '66 (Spring): Proceedings of the April 26–28, 1966, Spring joint computer conference*, April 1966, pp. 347-354

[KENNEDY 1962] JEROME D. KENNEDY Sr., "Representation of Time Delays", in [HUSKEY et al. 1962, pp. 6-3–6-16]

[KETTEL 1960] E. KETTEL, "Die Anwendungsmöglichkeiten der Analogrechentechnik in Meßtechnik und Nachrichtenverarbeitung", in *Telefunken Zeitung*, Vol. 33 (September 1960), No. 129, pp. 164–171

[KETTEL et al. 1967] E. KETTEL, A. KLEY, H. MANGOLD, R. MAUNZ, G. MEYER-BRÖTZ, H. OHNSORGE, J. SCHÜRMANN, "Der Einsatz elektronischer Rechner für Aufgaben der nachrichtentechnischen Systemforschung", in *Telefunken Zeitung*, Vol. 40 (1967), No. 1/2, pp. 3–9

[KERR 1978] C. N. KERR, *Use of the Analog/Hybrid Computer in Boundary Layer and Convection Studies*, American Society for Engineering Education, 86th Annual Conference, University of British Columbia, June 19–22, 1978

[KERR 1980] C. N. KERR, "Analog Solution of Free Convection Mass Transfer From Downward-Facing Horizontal Plates", in *Int. J. Heat Mass Transfer*, Vol. 23, 1980, pp. 247–249

[KIDD et al. 1961] E. A. KIDD, G. BULL, R. P. HARPER Jr., "In-Flight-Simulation – Theory and Application", AGARD Report 368, April 1961

[KINZEL et al. 1962/1] B. Kinzel, L. Sengewitz, "Radizierender Verstärker, insbesondere zur Verarbeitung von Einspritzwerten bei Dieselmotoren", in *Elektronische Rundschau*, Januar 1962, Vol. 16, No. 1, pp. 21–23

[KINZEL et al. 1962/2] B. Kinzel, L. Sengewitz, "Erweiterter radizierender Verstärker", in *Elektronische Rundschau*, Mai 1962, Vol. 16, No. 5, p. 223

[KLEIN et al. 1957] MARTIN L. KLEIN, FRANK K. WILLIAMS, HARRY C. MORGAN, "Digital Differential Analyzers", in *Instruments and Automation*, June 1957, pp. 1105–1109

[KLEIN 1965] MANFRED KLEIN, *Eine Hyperbelfeldröhre für die Multiplikation in elektronischen Analogrechnern*, Dissertation, Technische Universität Berlin, 1965

[KLEINWÄCHTER 1938] HANS KLEINWWÄCHTER, "Anwendung der Braunschen Röhre für die Auflösung von Differentialgleichungen auf elektrischem Wege", in *Archiv für Elektrotechnik*, 1938, pp. 118–120

[KLEY et al. 1966] A. KLEY, E. HEIM, "Ein elektronischer Koordinatenwandler", in *Telefunkenzeitung*, Vol. 39, No. 1, 1966, pp. 60–65

[KLINE 1993] RONALD KLINE, "Harold Black and the Negative-Feedback Amplifier", in *IEEE Control Systems*, August 1993, pp. 82–85

[KLIPSCH 1981] PAUL W. KLIPSCH, "In Memoriam: Paul G. A. H. Voigt", in *J. Audio Eng. Soc.*, Vol. 29, No. 4, 1981 April, p. 308

[KLITTICH 1966] MANFRED KLITTICH, "Über die Nachbildung von Getrieben auf dem Analogrechner", in *Elektronische Rechenanlagen*, 8, Heft 3, 1966, pp. 125–130

[KLITTICH 1974] MANFRED KLITTICH, "Entwurf und Inbetriebnahme des Nachführ- und Positioniersystems für das Radioteleskop Effelsberg", in *IFAC Symposium*, Düsseldorf, 1974, pp. 301–316

[KNORRE 1971] WOLFGANG A. KNORRE, *Analogcomputer in Biologie und Medizin – Einführung in die dynamische Analyse biologischer Systeme*, VEB Gustav Fischer Verlag Jena, 1971

[KOENIG et al. 1955] ELDO C. KOENIG, WILLIAM C. SCHULTZ, "How to Select Governor Parameters with Analog Computers", in [PAYNTER ed. 1955, pp. 237–238]

[KÖPPEL et al. 2021] SVEN KÖPPEL, BERND ULMANN, LARS HEIMANN, DIRK KILLAT, "Using analog computers in today's largest computational challenges", in *Advances in Radio Science*, 19, 2021, pp. 105–116, https://doi.org/10.5194/ars-19-105-2021

[KÖPPEL et al. 2022] SVEN KÖPPEL, ALEXANDRA KRAUSE, BERND ULMANN, "Analog Computing for Molecular Dynamics", in *International Journal of Unconventional Computing*, Vol. 17, 2022, pp. 259–282

[KOHR 1960] ROBERT H. KOHR, "Real-Time Automobile Ride Simulation", in *IRE-AIEE-ACM '60 (Western): Papers presented at the May 3-5, 1960, western joint IRE-AIEE-ACM computer conference*, May 1960, pp. 285–300

[KOPACEK] P. KOPACEK, "Testing Various Identification Algorithms for Control Systems with Stochastically Varying Parameters by a Hybrid Computer", in *10th IMACS World Congress on System Simulation and Scientific Computation*, pp. 69–71

[KOLMS et al. 2020] THORE KOLMS, ANDREAS WALDNER, CHRISTINE LANG, PHILIPP GROTHE, JAN HAASE, "Analog implementation of arithmetic operations on real memristors", in *SCOPES '20: Proceedings of the 23th Interational Workshop on Software and Compilers for Embedded Systems*, ay 2020, pp. 54–57

[KOPPE et al. 1971] IR. R. KOPPE, IR. P. EUSER, "Analoge simulatie van warmte- en waterdamp-transport in klimaatkamers voor planten en in de daarbij toegepaste luchtbehandelingsinstallaties", in *TNO-nieuws*, 1971, 26, pp. 497–513

[KORN et al. 1954] GRANINO A. KORN, THERESA M. KORN, "Relay Time-Division Multiplier", in *The Review of Scientific Instruments*, Vol. 25, No. 10, October 1954, pp. 977–982

[KORN et al. 1956] GRANINO A. KORN, THERESA M. KORN, *Electronic Analog Computers (D-c Analog Computers)*, McGraw-Hill Book Company, Inc., 1956

[KORN 1962] GRANINO A. KORN, "Electronic Function Generators, Switching Circuits and Random-Noise Generators", in [HUSKEY et al. 1962, pp. 3-62–3-84]

[KORN et al. 1964] GRANINO A. KORN, THERESA M. KORN, *Electronic Analog and Hybrid Computers*, McGRAW-HILL BOOK COMPANY, 1964

[KORN 1966] GRANINO A. KORN, *Random-Process Simulation and Measurements*, McGraw-Hill Book Company, 1966

[KORN et al. 1970] GRANINO A. KORN, H. KOSAKO, "A Proposed Hybrid-Computer Method for Functional Optimization", in *IEEE Transactions on Computers*, Vol. C-19, No. 2, February 1970, pp. 149–153

[KORN 2005] GRANINO A. KORN, "Continuous-System Simulation and Analog Computers – From op-amp design to aerospace applications", in *IEEE Control Systems Magazine*, June 2005, pp. 44–51

[KOVACH et al. 1962] L. D. KOVACH, H. F. MEISSINGER, "Solution of Algebraic Equations, Linear Programming, and Parameter Optimization", in [HUSKEY et al. 1962, pp. 5-133–5-154]

[KOVACH 1952] L. D. KOVACH, "Solution of Difference Equations", in [HUSKEY et al. 1962, pp. 6-52–6-56]

[KRAFT et al. 2002] CHRIS KRAFT, JAMES L. SCHEFTER, *Flight – My Life in Mission Control*, First Plume Printing, March 2002

[KRAMER] H. KRAMER, *Parameteroptimierung mit einem hybriden Analogrechner an einem Beispiel aus der chemischen Reaktionskinetik*, AEG-Telefunken, AFA 003 0570

[KRAMER 1968] H. KRAMER, "Optimierung eines Regelkreises mit Tischanalogrechner und Digitalzusatz", in *elektronische datenverarbeitung*, (1968) 6, pp. 293–297

[KRÄMER 1989] K. KRÄMER, "Das Modell. Antoni Gaudis Hängemodell und seine Rekonstruktion. Neue Erkenntnisse zum Entwurf für die Kirche der Colonia Güell", in *Mitteilungen des Instituts für leichte Flächentragwerke*, Universität Stuttgart, Nr. 34, 1989

[KRAUSE et al. 1963] WALTER P. KRAUSE, FRANK L. VINZ, *Study of Large Angle Attitude Maneuvers of Bodies in Space*, Georce C. Marshall Space Flight Center, Huntsville, Alabama, Report IN-M-COMP-S-63-3, April 9, 1963

[KRAUSE 1964] WALTER P. KRAUSE, *Dynamics of a Rotating Space Station*, Research & Development Applications Division, Computation Laboratory, George C. Marshall Space Flight Center, Report R-COMP-RS-64-1, February 1964

[KRAUSE 1970] PAUL C. KRAUSE, "Applications of analog and hybrid computers in electric power research", in *SIMULATION*, August 1970, pp. 73–79

[KRAUSE 1971] PAUL C. KRAUSE, *Hybrid Computation Techniques Applied to Power Systems Simulation*, Purdue University, School of Electrical Engineering, November, 1971

[KRAUSE 1974] PAUL C. KRAUSE, "Applications of Analog and Hybrid Computation in Electric Power System Analysis", in *Proceedings of the IEEE*, Vol. 62, No. 7, July 1974, pp. 994–1009

[KRAUSE et al. 1977] P. C. KRAUSE, W. C. HOLLOPETER, D. M. TRIEZENBERG, "Sharp Torques During Out-Of-Phase Synchronization", in *IEEE Transactions on Power Apparatus and Systems*, Vol. PAS-96, No. 4, July/August 1977, pp. 1318–1323

[KRAUSE 2006] CHRISTINE KRAUSE, *Die Entwicklung der Analogrechentechnik in Thüringen und Sachsen*, 3. Greifswalder Symposium zur Entwicklung der Rechentechnik und 12. Internationales Treffen für Rechenschieber- und Rechenmaschinensammler, 2006

[KREGELOH 1956] H. KREGELOH, "Analogrechner und ihre Anwendung auf ein volkswirtschaftliches Modell", in *Mathematical Methods of Operations Research*, Springer-Verlag, Vol. 1, No. 1, Dezember 1956, pp. 97–106

[KRON 1945/1] GABRIEL KRON, "Electric Circuit Models of the Schrödinger Equation", in *Physical Review*, Vol. 67, No. 1/2, January 1 and 15, 1945, pp. 39–43

[KRON 1945/2] GABRIEL KRON, "Numerical Solution of Ordinary and Partial Differential Equations by Means of Equivalent Circuits", in *Journal of Applied Physics*, 16, 1945, pp. 172–186

[KUBELKA et al. 1931] PAUL KUBELKA, FRANZ MUNK, "Ein Beitrag zur Optik der Farbanstricht", in *Zeitschrift für technische Physik*, 12, 1931, pp. 593–601

[449] CHRISTIAN KUEHN, *Multiple Time Scale Dynamics*, Springer, 2015

[KÜNKEL 1961] H. KÜNKEL, "Beitrag zu einer regeltheoretischen Analyse der Pupillenreflexdynamik", in *Kybernetik*, 1, 1961, pp. 69–75

[KUHN 1996] THOMAS SAMUEL KUHN, *The Structure of Scientific Revolutions*, Universty of Chicago Press, 1996

[LANCHESTER 1908] FREDERICK WILLIAM LANCHESTER, *Aerial Flight: Aerodonetics*, London, Constable, 1908

[LAND] BRUCE LAND, *DDA on FPGA – a modern Analog Computer*, https://instruct1.cit.cornell.edu/Courses/ece576/DDA/index.htm, retrieved 03/03/2013

[LANDAUER 1974] J. PAUL LANDAUER, "Personal Rapid Transit (PRT) System Design by Hybrid Computation", in *EAI Scientific Computation Report*, No. SCR 74-17, November 11, 1974

[LANDAUER 1975] J. PAUL LANDAUER, "Non-Destructive Destructive Testing", in *Industrial Research*, March 1975

[LANDAUER 1983] J. PAUL LANDAUER, "SIMSTAR – an attached multiprocessor for dynamic system engineering", in *Informatik Fachberichte: First European Simulation Congress ESC 83*, Springer-Verlag, 1983

[LANGE 2006] THOMAS H. LANGE, *Peenemünde – Analyse einer Technologieentwicklung im Dritten Reich*, Reihe Technikgeschichte in Einzeldarstellungen, VDI-Verlag, GmbH, Düsseldorf 2006

[LANGE et al. 2020] STEFFEN LANGE, JOHANNA POHL, TILMAN SANTARIUS, "Digitalization and energy consumption. Does ICT reduce energy demand?", in *Ecological Economics*, 176, 2020

[LANGHELD 1979/1] ERWIN LANDHELD, "Praxis der stochastischen Rechentechnik", in *Elektronik*, 25, 13. Dezember 1979, pp. 43–48

[LANGHELD 1979/2] ERWIN LANDHELD, "Praxis der stochastischen Rechentechnik", in *Elektronik*, 26, 27. Dezember 1979, pp. 39–42

[LANING et al. 1956] J. HALCOMBE LANING, RICHARD H. BATTIN, *Random Processes in Automatic Control*, Mcgraw-Hill Book Company, Inc., 1956

[LARROWE 1955] VERNON L. LARROWE, "Direct Simulation – Bypasses Mathematics, Simplifies Analysis", in [PAYNTER ed. 1955, p. 127–133]

[LARROWE 1966] VERNON L. LARROWE, "Band-Pass Quadrature Filters", in *IEEE Transactions on Electronic Computers*, Vol. EC-15, No. 5, October 1966, pp. 726–731

[LEATHERWOOD 1972] JACK D. LEATHERWOOD, "Analog Analysis of a Tracked Air-Cushion Vehicle", in *INSTRUMENTS and CONTROL SYSTEMS*, April 1972, pp. 81–86

[LEISE 2007] TANYA LEISE, "As the Planimeter's Wheel Turns: Planimeter Proofs for Calculus Class", in *College Mathematics Journal*, January 2007

[LEMAITRE et al. 1936] G. LEMAITRE, M. S. VALLARTA, "On the Geomagnetic Analysis of Cosmic Radiation", in *Physical Review*, Volume 49, 1936, pp. 719–726

[LEMAITRE et al. 1936/2] G. LEMAITRE, M. S. VALLARTA, "On the Allowed Cone of Cosmic Radiation", in *Physical Review*, Volume 50, 1936, pp. 493–504

[LEVINE 1964] LEON LEVINE, *Methods for Solving Engineering Problems Using Analog Computers*, McGraw-Hill Book Company, 1964

[LEWIN 1972] JOHN ERNEST LEWIN, *Area Measurement*, United States Patent 3652842, Mar. 28, 1972

[LEWIS 1958] LLOYD G. LEWIS, "Simulation of a Solvent Recovery Process", in *Instruments and Automation*, Vol. 31, April 1958, pp. 644–647

[LIBAN 1962] ERIC LIBAN, "The application of finite fourier transforms to analog computer simulations", in *AIEE-IRE '62 (Spring): Proceedings of the May 1–3, 1962, spring joint computer conference*, May 1962, pp. 255-265

[Librascope 1957] N. N., "Ball/Disc Integrator", in *Instruments and Automation*, April 1957, p. 769

[LIEBER et al. 1969] R. E. LIEBER, T. R. HERNDON, "Analog Simulation Spells Safe Startups", in [Hydrocarbon Processing 1959, pp. 78–82]

[LIENHARD 1969] K. LIENHARD, "Planung und Berechnung von Rohrnetzen", in *Gas – Wasser – Abwasser*, 49. Jahrgang, 1969, Nr. 9

[LIGHT et al. 1966] L. LIGHT, J. BADGER, D. BARNES, "An Automatic Acoustic Ray Tracing Computer", in *IEEE Transactions on Electronic Computers*, Vol. EC-15, No. 5, October, 1966, pp. 719–725

[LIGHT 1975] LEON HENRY LIGHT, *Apparatus for Integration and Averaging*, United States Patent 3906190, Sept. 16, 1975

[LILAMAND 1956] M. LEJET LILAMAND, "A Time-Division Multiplier" in *IRE Transactions on Electronic Computers*, March 1956, pp. 26–34

[Linear Technology] Linear Technology, *LTC6943 – Micropower, Dual Precision Instrumentation Switched Capacitor Building Block*, Linear Technology Corporation

[LITTLE 1966] WARREN D. LITTLE, "Hybrid computer solutions of partial differential equations by Monte Carlo methods", in *AFIPS '66 (Fall): Proceedings of the November 7-10, 1966, fall joint computer conference*, November 1966, pp. 181–190

[Litton 1963] N. N., *Digital Differential Analyzer*, Technical Documentation Report ASD-TDR-63-158, June 1963

[LORENZ 1963] EDWARD NORTON LORENZ, "Deterministic Nonperiodic Flow", in *Journal of the Atmospheric Sciences*, Vol. 20, March 1963, pp. 130–141

[LOTZ 1969] HERMANN LOTZ, "Einsatz des Analogrechners in der Regelungs- und Steuerungstechnik", in *Steuerungstechnik*, 2 (1969) 11, pp. 430–435

[LOTZ 1970/1] HERMANN LOTZ, "Programmiertes Spielen auf einem Hybriden Analogrechner: Billardsimulator", in *Datenverarbeitung*, AEG-TELEFUNKEN, 1970, 2, pp. 64–67

[LOTZ 1970/2] HERMANN LOTZ, "Fertigungskontrolle mit Hilfe analoger Meßwertverarbeitung bei der Rundlaufabweichung bei Kraftfahrzeugrädern", in *Datenverarbeitung*, AEG-TELEFUNKEN, 1970, 2, pp. 68–71

[LOTZ et al.] HERMANN LOTZ, G. RAMSAUER, *Hybride Simulation einer direkten digitalen Regelung auf dem Hybriden Rechnersystem HRS 860 am Beispiel eines dampfgekühlten schnellen Reaktors*, AEG-TELEFUNKEN, Datenverarbeitung

[LOVELL et al. 1946] CLARENCE A. LOVELL, DAVID B. PARKINSON, BRUCE T. WEBER, *Electrical Computing System*, United States Patent 2404387, July 23, 1946

[LOVEMAN 1962] BERNARD D. LOVEMAN, "Computer Servomechanisms and Servo Resolvers", in [HUSKEY et al. 1962, pp. 3-1–3-40]

[LOWE] WILLIAM LOWE, *NUC Simulation Facility*, Naval Undersea Center, San Diego, California, author's archive

[LUDWIG 1966] R. LUDWIG, *Stability Research on Parachutes using Digital and Analog Computers*, National Aeronautics and Space Administration, Washington, D. C., November 1966

[LUDWIG et al. 1974] MANFRED LUDWIG, KLAUS KAPLICK, *Elektronische Analogrechner und Prozeßrechnereinsatz*, Reihe *Programmierung und Nutzung von Rechenanlagen, Teil 6*, Verlag "Die Wirtschaft", Berlin, 1974

[LUKES 1967] JAROSLAV H. LUKES, "Oscillographic Examination of the Operation of Function Generators", in *IEEE Transactions on Electronic Computers*, Vol. EC-16, No. 2, April 1967, pp. 133–139

[LUNDERSTÄDT et al. 1981] R. LUNDERSTÄDT, W. MENSSEN, *Regelmechanismus der menschlichen Pupille – Stabilität und Simulation*, Dornier System GmbH, 1981

[MACDUFF et al. 1958] JOHN N. MACDUFF, JOHN R. CURRERI, *Vibration Control*, McGraw-Hill Book Company, Inc., 1958

[MACKAY 1962] DONALD M. MACKAY, MICHAEL E. FISHER, *Analogue Computing at Ultra-High Speed*, John Wiley & Sons Inc., 1962

[MACNEAL 2002] RICHARD H. MACNEAL, *Richard H. MacNeal (b. 1923) Interviewed by Shirley K. Cohen*, in Archives California Institute of Technology, Pasadena, California, January 23, 2002, https://oralhistories.library.caltech.edu/262/1/MacNeal%20OHO.pdf, retrieved 02/28/2021

[MACNEE 1948] A. B. MACNEE, *An Electronic Differential Analyzer*, Technical Report No. 90, December 16, 1948, Research Laboratory of Electronics, Massachusetts Institute of Technology

[MACNEE 1953] A. B. MACNEE, "A High Speed Product Integrator", in *The Review of Scientific Instruments*, Volume 24, Number 3, March 1953, pp. 207–211

[MAHRENHOLTZ 1968] O. MAHRENHOLTZ, *Analogrechnen in Maschinenbau und Mechanik*, Bibliographisches Institut, Mannheim/Zürich, 1968

[MALSTROM et al. 1977] CARL W. MALSTROM, RAINER HELLER, MOHAMMAD S. KHAN, "Hybrid Computation – an Advanced Computation Tool for Simulating the Nonlinear Dynamic Response of Railroad Vehicles", Pre-Publication Copy of Submission to the Post Conference Proceedings, *Advanced Techniques in Track/Train Dynamics and Design Conference*, Chicago, Illinois, September 27 and 28, 1977

[MANOHAR 2015] RAJIT MANOHAR, "Comparing Stochastic and Deterministic Computing", in *IEEE Computer Architecture Letters*, Vol. 14, No. 2, July-Dec. 2015, pp. 119–122

[MANSKE 1968] R. A. MANSKE, "Computer Simulation of Narrowband Systems", in *IEEE Transactions on Computers*, Vol. C-17, No. 4, April, 1968, pp. 301–308

[MARKSON 1958] A. A. MARKSON, "Introduction to Reactor Physics", in *Instruments and Automation*, Vol. 31, April 1958, pp. 616–623

[MARQUITZ et al. 1968] W. T. MARQUITZ, Y. TOKAD, "On Improving the Analog Computer Solutions of Linear Systems", in *IEEE Transactions on Computers*, Vol. C-17, No. 3, March, 1968, pp. 268–270

[MARTIN 1969] GEORGE J. MARTIN, "Hybrid Computation in the Engineering College", in *Engineering Education*, January 1969, pp. 395–400

[MARTIN 1970] DONALD C. MARTIN, "Development of analog/hybrid terminals for teaching system dynamics", in *AFIPS '70 (Fall): Proceedings of the November 17–19, 1970, fall joint computer conference*, November 1970, pp. 241–249

[MARTIN 1972] GEORGE J. MARTIN, "Analog and Hybrid Simulation in Science Education", in *Educational Technology*, April, 1972, pp. 62–63

[MASLO 1974] RONALD M. MASLO, "Dynamic Response of a Ship in Waves", in EAI Scientific Computation Report, No. 74-14, September 20, 1974

[MASSEN 1977] ROBERT MASSEN, *Stochastische Rechentechnik – Eine Einführung in die Informationsverarbeitung mit zufälligen Pulsfolgen*, Carl Hanser Verlag, 1977

[MASTER et al. 1955] R. C. MASTER, R. L. MERRILL, B. H. LIST, "Analogous Systems in Engineering Design", in [PAYNTER ed. 1955, p. 134–145]

[MBB] N. N., *MBB Simulation*, Firmenschrift Messerschmitt Bölkow Blohm GmbH, Unternehmensbereich Flugzeuge

[MCCALLUM] I. R. MCCALLUM, *Horses for Courses: The Mathematical Modelling Requirements of Maritime Simulators*, author's archive

[MCCANN 1949] G. D. MCCANN, "The California Institute of Technology Electric Analog Computer", in *Mathematical Tables and Other Aids to Computation*, Vol. 3, No. 28, Oct. 1949, pp. 501–513

[MCCANN et al. 1949] G. D. MCCANN, C. H. WILTS, B. N. LOCANTHI, "Application of the California Institute of Technology Electric Analog Computer to Nonlinear Mechanics and Servomechanisms", in *AIEE Transactions*, Volume 68, 1949, pp. 652–660

[MCCARTHY 2009] JERRY MCCARTHY, *Der Mechanismus von Antikythera*, 15. Internationales Treffen der Rechenschiebersammlung und 4. Symposium zur Entwicklung der Rechentechnik, Ernst Moritz Arndt Universität Greifswald, 2009

[MCDONAL 1956] FRANK J. MCDONAL, *Wave Analysis*, United States Patent 2752092, June 26, 1956

[MCFADDEN et al. 1958] NORMAN M. MCFADDEN, FRANK A. PAULI, DONOVAN R. HEINLE, *A Flight Study of Longitudinal-Control-System Dynamic Characteristics by the Use of a Variable-Control-System Airplane*, NACA RM A57L10, 1958

[MCGHEE et al. 1970] ROBERT B. MCGHEE, RAGNAR N. NILSEN, "The Extended Resolution Digital Differential Analyzer: A New Computing Structure for Solving Differential Equations", in *IEEE Transactions on Computers*, Vol. C-19, No. 1, January 1970, pp. 1–9

[MCLACHLAN 1947] N. W. MCLACHLAN, *Theory and Applications of Mathieu Functions*, Oxford at the Clarendon Press, 1947

[MCLEAN et al. 1977] L. J. MCLEAN, E. J. HAHN, "Simulation of the Transient Behaviour of a Rigid Rotor in Squeeze Film Supported Journal Bearings", 2nd AINSE Engineering Conference, 1977

[MCLEOD et al. 1957] JOHN H. MCLEOD, ROBERT M. LEGER, "Combined Analog and Digital Systems – Why, When, and How", in *Instruments and Automation*, June 1957, pp. 1126–1130

[MCLEOD et al. 1958/1] JOHN H. MCLEOD, SUZETTE MCLEOD, "The Simulation Council Newsletter", in *Instruments and Automation*, Vol. 31, January 1958, pp. 119–124

[MCLEOD et al. 1958/2] JOHN H. MCLEOD, SUZETTE MCLEOD, "The Simulation Council Newsletter", in *Instruments and Automation*, Vol. 31, February 1958, pp. 297–300

[MCLEOD et al. 1958/3] JOHN H. MCLEOD, SUZETTE MCLEOD, "The Simulation Council Newsletter", in *Instruments and Automation*, Vol. 31, March 1958, pp. 487–491

[MCLEOD et al. 1958/4] JOHN H. MCLEOD, SUZETTE MCLEOD, "The Simulation Council Newsletter", in *Instruments and Automation*, Vol. 31, July 1958, pp. 1219–1225

[MCLEOD et al. 1958/5] JOHN H. MCLEOD, SUZETTE MCLEOD, "The Simulation Council Newsletter", in *Instruments and Automation*, Vol. 31, August 1958, S. 1385–1390

[MCLEOD et al. 1958/6] JOHN H. MCLEOD, SUZETTE MCLEOD, "The Simulation Council Newsletter", in *Instruments and Automation*, Vol. 31, December 1958, pp. 1991–1997

[MCLEOD 1962] JOHN H. MCLEOD, "Electronic-Analog-Computer Techniques for the Design of Servo Systems", in [HUSKEY et al. 1962, pp. 5-35 ff.]

[Meccano 1934] N.N., "Meccano Aids Scientific Research", in *Meccano Magazine*, Vol. XIX, No. 6, June, 1934, p. 441

[Meccano 1934/2] N.N., "Machine Solves Mathematical Problems – A Wonderful Meccano Mechanism", in *Meccano Magazine*, Vol. XIX, No. 6, June, 1934, pp. 442–444

[MEDKEFF et al. 1955] R. J. MEDKEFF, H. MATTHEWS, "Solving process-control problems by ANALOG COMPUTER", in [PAYNTER ed. 1955, pp. 164–166]

[MEISINGER 1978] REINHOLD MEISINGER, "Analog Simulation of Magnetically Levitated Vehicles on Flexible Guideways", in *Simulation of Control-Systems*, I. Troch (ed.), North-Holland Publishing Company, 1978, pp. 207–214

[MEISSL 1960/1] P. MEISSL, "Behandlung von Wasserschloßaufgaben mit Hilfe eines elektronischen Analogrechners, Teil 1", in *mtw – Zeitschrift für moderne Rechentechnik und Automation*, 1/60, pp. 9–13

[MEISSL 1960/2] P. MEISSL, "Behandlung von Wasserschloßaufgaben mit Hilfe eines elektronischen Analogrechners, Teil 2", in *mtw – Zeitschrift für moderne Rechentechnik und Automation*, 2/60, pp. 74–77

[MENZEL et al. 2021] JOHANNES MENZEL, CHRISTIAN PLESSL, TOBIAS KENTER, "The Strong Scaling Advantage of FPGAs in HPC for N-body Simulations", in *ACM Transactions on Reconfigurable Technology and Systems*, Vol. 15, No. 1, pp. 10:1–10:30

[MEYER-BRÖTZ 1960] G. MEYER-BRÖTZ, "RA 800 – Ein transistorisierter Präzisions-Analogrechner", in *Telefunken Zeitung*, Vol. 33 (September 1960), No. 129, pp. 171–182

[MEYER-BRÖTZ 1962] G. MEYER-BRÖTZ, "Die Messung von Kenngrößen stochastischer Prozesse mit dem elektronischen Analogrechner", in *Elektronische Rechenanlagen*, 4 (1962), No. 3, pp. 103–108

[MEYER-BRÖTZ et al. 1966] G. MEYER-BRÖTZ, E. HEIM, "Ein breitbandiger Operationsverstärker mit Silizium-Transistoren", in *Telefunken Zeitung*, Vol. 39 (1966), No. 1, pp. 16–32

[MEYER ZUR CAPELLEN 1949] W. MEYER ZUR CAPELLEN, *Mathematische Instrumente*, Akademische Verlagsgesellschaft Geest & Portig K.-G., Leipzig 1949

[MEZENCEV et al. 1978] R. MEZENCEV, R. LEPEIX, "Hybrid Simulation of a Non Linear Hydro Pneumatic Damper for Ships", in *Simulation of Control Systems*, I. Troch (ed.), North-Holland Publishing Company, 1978, pp. 135–137

[MICHAELS] LAWRENCE H. MICHAELS, *The AC/Hybrid Power System Simulator and its Role in System Security*, author's archive

[MICHAELS et al.] LAWRENCE H. MICHAELS, WILLIAM TESSMER, JOHN MULLER, *The On-Line Power System Simulator*, Electronic Associates, Inc., author's archive

[MICHAELS et al. 1971] G. C. MICHAELS, V. GOURISHANKAR, "Hybrid Computer Solution of Optimal Control Problems", in *IEEE Transactions on Computers*, Vol. C-20, No. 2, February 1971, pp. 209–211

[MICHELS 1954] LOWELL S. MICHELS, *Description of BENDIX D-12 DIGITAL DIFFERENTIAL ANALYZER*, Bendix Computer Division, Bendix Aviation Corporation, 5630 Arbor Vitae Street, Los Angeles 45, California, March 13, 1954

[MICHELSON et al. 1898] ALBERT ABRAHAM MICHELSON, SAMUEL WESLEY STRATTON, "A New Harmonic Analyzer", in *American Journal of Science*, 25, 1898, pp. 1–13

[MILAN-KAMSKI 1969] W. J. MILAN-KAMSKI, "A High-Accuracy, Real-Time Digital Computer for Use in Continuous Control Systems", in *1959 Proceedings of the Western Joint Computer Conference*, pp. 197–201

[MILLER et al. 1954] JOSEPH A. MILLER, AARON S. SOLTES, RONALD E. SCOTT, *A Wide-Band Function Multiplier*, Computer Laboratory, Electronics Research Directorate, Air Force, Cambridge Research Center, Cambridge Massachusetts, December 1954

[MILLER et al. 1955] JOSEPH A. MILLER, AARON S. SOLTES, RONALD E. SCOTT, "Wide-Band Analog Function Multiplier", in *Electronics*, February, 1955, pp. 160–163

[MILLER 2011] DAVID PHILIP MILLER, "The Mysterious Case of James Watt's '1785 Steam Indicator': Forgery or Folklore in the History of an Instrument?", in *Int. J. for the History of Eng. & Tech.*, Vol. 81, No. 1, January, 2011, pp. 129–150

[MILLS/1] JONATHAN W. MILLS, *The Architecture of an Extended Analog Computer Core*, Computer Science Department, Indiana University

[MILLS/2] JONATHAN W. MILLS, *Polymer Processors*, Computer Science Department, Indiana University, http://www.cs.indiana.edu/pub/techreports/TR580.pdf, retrieved 03/03/2013

[MILLS 1995] JONATHAN W. MILLS, *The continuous retina: Image processing with a single-sensor artificial neural field network*, Computer Science Department, Indiana University, technical report 443, November 13, 1995

[MILLS et al. 2006] JONATHAN W. MILLS, BRYCE HIMEBAUGH, BRIAN KOPECKY, MATT PARKER, CRAIG SHUE, Chris Weilemann, "'Empty Space' Computers: The Evolution of an Unconventional Supercomputer", in *CF06*, May 3–8, 2006, Ischia, Italy

[MINDELL 1995] DAVID A. MINDELL, "Automation's Finest Hour: Bell Labs and Automatic Control in World War II", in *IEEE Control Systems*, December 1995, pp. 72–78

[MINDELL 2000] DAVID A. MINDELL, "Opening Black's Box – Rethinking Feedback's Myth of Origin" in *Technology and Culture*, July 2000, Vol. 41, pp. 405–434

[MITCHELL 1960] G. MITCHELL, "An Analogue Computer for Investigating the Directivity Characteristics of Complex Arrays of Unit Aerials", in *The Post Office Electrical Engineers' Journal*, Vol. 52, Part 4, January 1960, pp. 246–250

[MITCHELL et al. 1966] E. E. L. MITCHELL, J. B. MAWSON, J. BULGER, "A Generalized Hybrid Simulation for an Aerospace Vehicle", in *IEEE Transactions on Electronic Computers*, Vol. EC-15, No. 3, June 1966, pp. 304–313

[MITRA 1955] SAMARENDRA KUMAR MITRA, "Electrical Analog Computing Machine for Solving Linear Eequations and Related Problems", in *The Review of Scientific Instruments*, Volume 26, Number 5, May, 1955, pp. 453–457

[MIURA et al. 1967] TAKEO MIURA, JUNJI TSUDA, JUNZO IWATA, "Hybrid Computer Solution of Optimal Control Problems by the Maximum Principle", in *IEEE Transactions on Electronic Computers*, Vol. EC-16, No. 5, October 1967, pp. 666–670

[MIURA et al. 1967/2] TAKEO MIURA, JUNZO IWATA, JUNJI TSUDA, "An application of hybrid curve generation – cartoon animation by electronic computers", in *AFIPS – spring joint computer conference*, April 1967, pp. 141–148

[Montan-Forschung] N. N., *Rechengeräte für Verbundnetze*, Montan-Forschung, Düsseldorf

[MORENO et al. 2020] DANIEL GARCIA MORENO, ALBERTO A. DEL BARRIO, GUILLERMO BOTELLA JUAN, "Simulating and deploying analog arithmetic circuits on FPAAs", in *SummerSim '20: Proceedings of the 2020 Summer Simulation Conference*, July 2020, Article No.: 19, pp. 1–12

[MORRILL 1962] CHARLES D. MORRILL, "Electronic Multipliers and Related Topics", in [HUSKEY et al. 1962, pp. 3-40–3-62]

[E. MORRISON 1962] E. MORRISON, "Nuclear-Reactor Simulation", in [HUSKEY et al. 1962, pp. 5-87–5-93]

[J. E. MORRISON 2007] JAMES E. MORRISON, *The Astrolabe*, Janus Publishing, 2007

[MORTON 1966] R. R. A. MORTON, "A simple d.c. to 10 Mc/s analogue multiplier", in *Journal of Scientific Instruments*, Vol. 43, 1966, pp. 165–168

[MÜLLER 1986] HERIBERT MÜLLER, "Simulation und Lösung physikalischer Probleme mit dem Analogrechner", in *Praxis der Naturwissenschaften, Physik*, Aulis Verlag, Heft 3/35, 15. April 1986, pp. 21–25

[MUSKHELISHVILI 1953] N. I. MUSKHELISHVILI, *Singular Integral Equations*, P. Noordhoff N. V., Groningen-Holland, 1953

[NALLEY 1969] DONALD NALLEY, *Z Transform and the Use of the Digital Differential Analyzer as a Peripheral Device to a General Purpose Computer*, NASA Technical Memorandum, NASA TM X-53866, August 12, 1969

[NAVA-SEGURA et al.] A. NAVA-SEGURA, L. L. FRERIS, *Hybrid Computer Simulation of DC Transmission Systems*, author's archive

[NEUFELD 2007] MICHAEL J. NEUFELD, *Von Braun – Dreamer of Space, Engineer of War*, Borzoi Book, Alfred A. Knopf, 2007

[NISE] NORMAN S. NISE, *Analog Computer Experiments for Undergraduate Courses in Network Analysis and Automatic Controls*, Vol. II, No. 2, author's archive

[Nix 1965] Siegfried H. Nix, "Problematik und Methodik bei der Analyse von Gasrohrnetzen", in *Rohre – Rohrleitungsbau – Rohrleitungstransport*, Heft 5, Oktober 1965, pp. 255–271

[N. N. 1945] N. N., *Das Gerät A4 Baureihe B, Teil III, Gerätebeschreibung V2*, OKH/Wa A/Wa Prüf, Anlage zu Bb.Nr 19/45 gK, 1.2.1945 4/64, p. 175

[N. N. 1956] N. N., *Nike I Systems – Nike I Computer, SAM Problem Analysis, Servo Loop Elements and Power Distribution*, TM9-5000-13, Department of the Army, May 1956

[N. N. 1957/1] N. N., "Berkeley opens its new computer facility", in *Instruments and Automation*, February 1957, p. 288

[N. N. 1957/2] N. N., "Bonneville Power Administration Solves Swing Equations with EASE", in *Instruments and Automation*, March 1957, p. 498

[N. N. 1957/3] N. N., "Eröffnung des ersten europäischen Analog-Rechenzentrums", in *Elektronische Rundschau*, August 1957, Vol. 11, No. 8, p. 253

[N. N. 1957/4] N. N., "New GEDA Power Dispatch Computer", in *Instruments and Automation*, Vol. 30, February 1957, p. 179

[N. N. 1957/5] N. N., "University Research Instrumentation", in *Instruments and Automation*, June 1957, p. 1120

[N. N. 1957/6] N. N., "New Data Handling Centers", in *Instruments and Automation*, April 1957, p. 608

[N. N. 1958/1] N. N., "Computer Designed Rolling Mill", in *Instruments and Automation*, Vol. 31, February 1958, p. 283

[N. N. 1958/2] N. N., "Distillation-Column Dynamic Characteristics", in *Instruments and Automation*, Vol. 31, August 1958, pp. 1357–1359

[N. N. 1960] N. N., "Funktionsgruppen für die Analogrechentechnik", in *Elektronische Rechenanlagen*, 2 (1960), No. 1, pp. 43–44

[N. N. 1961] N. N., "Messen – Datenverarbeiten – Auswerten", advertisement in *Elektronische Rechenanlagen*, 3 (1961), No. 1, p. 44

[N. N. 1964/1] N. N., "Electronic Associates, Inc., Europäisches Rechenzentrum für Analog- und Hybridrechentechnik", in *Elektronische Rechenanlagen*, 6 (1964), No. 4, p. 214

[N. N. 1964/2] N. N., "EAI awarded contract for Hybrid Computing System", in *mtw – Zeitschrift für moderne Rechentechnik und Automation*, 4/64, p. 175

[N. N. 1964/3] N. N., "Nuclear Power Plant of N. S. Savannah simulated by Analog Computers", in *mtw – Zeitschrift für moderne Rechentechnik und Automation*, 3/64, p. 129

[N. N. 1978] N. N. "Pioneer Computer Goes To Washington", in *University Bulletin*, Volume 26, Number 13, January 23, 1978, p. 1

[N. N. 2006] N. N., "Video Games – Did They Begin at Brookhaven?", http://www.osti.gov/accomplishments/videogame.html, retrieved 11/20/2006

[N. N. 2020] N. N., "Highlighting Women in Operations Research and Their Achievements", in *Phalanx*, Vol. 53, No. 4, 2020, pp. 16–19

[Nolan 1955] John E. Nolan, "Analog Computers and their Application to Heat Transfer and Fluid Flow – Part 1, 2, 3", in [Paynter ed. 1955, pp. 109–126]

[Noronha] Leo G. Noronha, *The Benefits of Analog Computation and Simulation in the Electrical Supply Industry*, author's archive

[Northrop 1950] N. N., *MADDIDA Digital Differential Analyzer*, Northrop Aircraft, Inc., Brochure No. 38, December, 1950

[Norum et al. 1962] Vance D. Norum, Marvin Adelberg, Robert L. Farrenkopf, "Analog simulation of particle trajectories in fluid flow", in *AIEE-IRE '62 (Spring): Proceedings of the May 1–3, 1962, spring joint computer conference*, May 1962, pp. 235–254

[Nosker 1957] Paul Nosker, "Dynamic Systems Synthesizer", in [White Sands 1957, pp. 143–168]

[Ochs et al. 2021] Karlheinz Ochs, Sebastian Jenderny, "An equivalent electrical circuit for the Hindmarsh-Rose model", in *International Journal of Circuit Theory and Applications*, 49, 2021, pp. 3526–3539

[O'Grady 1966] Emmett Pearse O'Grady, "A Hybrid-Code Differential Analyzer", in *Annales de l'Association internationale pour le Calcul analogique*, No. 1, Janvier 1966, pp. 13–21

[O'Grady 1967] Emmett Pearse O'Grady, "Correlation Method for Computing Sensitivity Functions on a High-Speed Iterative Analog Computer", in *IEEE Transactions on Electronic Computers*, Vol. EC-16, No. 2, April 1967, pp. 140–146

[Oliver et al. 1974] W. Kent Oliver, Dale E. Seborg, D. Grant Gisher, "Hybrid Simulation of a Computer-Controlled Evaporator", in *SIMULATION*, September 1974, pp. 77–84

[Olson 1943] Harry F. Olson, *Dynamical Analogies*, D. van Nostrand Company, Inc., 1943

[Okah-Avae 1978] B. E. Okah-Avae, "Analogue computer simulation of a rotor system containing a transverse crack", in *SIMULATION*, December 1978, pp. 193–198

[Onizawa et al. 2019] Naoya Onizawa, Warren J. Gross, Takahiro Hanyu, "Brain-Inspired Computing", in [Gross et al. 2019, pp. 185–199]

[Ott 1964] A. Ott, "Zur Bestimmung des Korrelationskoeffizienten zweier Funktionen mit dem Analogrechner", in *Elektronische Rechenanlagen*, 6 (1964), No. 3, pp. 144–148

[OTTERMAN 1960] JOSEPH OTTERMAN, "The Properties and Methods for Computation of Exponentially-Mapped-Past Statistical Variables", in *IRE Transactions on Automatic Control*, Volume: AC-5, Issue: 1, 1. Jan. 1960, pp. 11–17, DOI: 10.1109/TAC.1960.6429289

[OVSYANKO] V. M. OVSYANKO, "The Theory of Synthesis of Electronic Circuits of Linear and Non-linear Object of Structural Mechanics and Applied Elasticity Theory", in *14. seminář MEDA ANALOGOVÁ A HYBRIDNÍ VÝPOČETNÍ TECHNIKA*, Praha 1977, pp. 25–29

[OWEN et al. 1960] P. L. OWEN, M. F. PARTRIDGE, T. R. H. SIZER, *CORSAIR, a Digital Differential Analyser*, Royal Aircraft Establishment (Farnborough), Technical Note No. I.A.P. 1123, December, 1960

[OWENS 1986] LARRY OWENS, "Vannevar Bush and the Differential Analyzer: The Text and Context of an Early Computer", in *Technology and Culture*, Vol. 27, No. 1, Jan. 1986, pp. 63–95

[Packard Bell] N. N., *The HYCOMP Hybrid Analog/Digital Computing System*, Packard-Bell Computer

[PALEVSKY 1962] MAX PALEVSKY, "The Digital Differential Analyzer", in [HUSKEY et al. 1962, pp. 19-14–19-74]

[PALM 2014] MICHAEL PALM, *Historische Integratoren der Firma A. Ott – Anschauliche Darstellung der Funktionsweise und Animation*, Wissenschaftliche Hausarbeit, Darmstadt, 2014

[PARK et al. 1972] WILLIAM H. PARK, JAMES C. WAMBOLD, "Teaching Digital and Hybrid Simulation of Mechanical Systems at the Graduate Level", Delivered during the Joint ACES/ASEE Session No. 3540 at the 1972 Annual Meeting of the American Society for Engineering Education at Texas Tech, Lubbock, Texas, June 19–22, 1972

[PASCHKIS et al. 1968] VICTOR PASCHKIS, FREDERICK L. RYDER, *Direct Analog Computers*, Interscience Publishers, 1968

[PAYNE 1988] PETER R. PAYNE, "An Analog Computer which Determines Human Tolerance to Acceleration", in 39th *Annual Astronautical Congress of the International Astronautical Federation*, Bangalore, 8-15 Oct. 1988, pp. 271–300

[PAYNTER ed. 1955] HENRY M. PAYNTER (ed.), *A Palimpsest on the Electronic Analog Art*, printed by Geo. A. Philbrick Researches Inc., AD 1955

[PAYNTER et al. 1955] HENRY M. PAYNTER, J. M. ASCE, "Surge and Water Hammer Problems", in [PAYNTER ed. 1955, pp. 217–223]

[PAYNTER 1955/1] HENRY M. PAYNTER, "Methods and Results from M.I.T. Studies in Unsteady Flow", in [PAYNTER ed. 1955, pp. 224–228]

[PAYNTER 1955/2] HENRY M. PAYNTER, "A Discussion by H. M. Paynter of AIEE Paper 53 – 172", in [PAYNTER ed. 1955, pp. 229–232]

[PAYNTER] HENRY M. PAYNTER, "A Retrospective on Early Analysis and Simulation of Freeze and Thaw Dynamics", cf. http://www.me.utexas.edu/~lotario/paynter/hmp/PAYNTER_Permafrost.pdf, retrieved 12/04/2008

[PEASE 2003] BOB PEASE, "What's All This K2-W Stuff, Anyhow?", in *electronic design*, January 2003, http://electronicdesign.com/article/analog-and-mixed-signal/what-s-all-this-k-2-w-stuff-anyhow-2530, retrieved 12/12/2012

[PERERA 1969] K. K. Y. WIJE PERERA, "Optimum generating schedule for a hydro-thermal power system / an analog computer solution to the short-range problem", in *SIMULATION*, April 1969, pp. 191–199

[PETZOLD 1992] HARTMUT PETZOLD, *Moderne Rechenkünstler – Die Industrialisierung der Rechentechnik in Deutschland*, Verlag C. H. Beck, 1992

[PFALTZGRAFF 1969] DAVID J. PFALTZGRAFF, "Analog Simulation of the Bouncing-Ball Problem", in *American Journal of Physics*, Volume 37, Number 10, October 1969, pp. 1008–1013

[PHILBRICK 1948] GEORGE A. PHILBRICK, "Designing Industrial Controllers by Analog", in *Electronics*, June, 1948, pp. 108–111

[PHILLIPS 1950] A. W. PHILLIPS, "Mechanical Models in Economic Dynamics", in *Economica*, New Series, Vol. 17, No. 67, Aug. 1950, pp. 283–305

[PICENI et al. 1975] HANS A. L. PICENI, PIETER EYKHOFF, "The Use of Hybrid Computers for System-Parameter Estimation", in *Annales de l'Association internationale pour le Calcul analogique*, No. 1, Janvier 1975, pp. 9–22

[PIERRE 1986] DONALD A. PIERRE, *Optimization Theory with Applications*, Dover Publications, Inc., New York, 1986

[PIRRELLO et al. 1971] C. J. PIRRELLO, R. D. HARDIN, J. P. CAPELLUPO, W. D. HARRISON, *An Inventory of Aeronautical Ground Research Facilities – Volume IV – Engineering Flight Simulation Facilities*, National Aeronautics and Space Administration, Washington, D. C., November 1971, NASA CR-1877

[POPOVIĆ 1964] D. P. POPOVIĆ, "Die Automatisierung des von Mieses'schen Iterationsverfahrens auf dem Analogrechner", in *mtw – Zeitschrift für moderne Rechentechnik und Automation*, 3/64, pp. 104–110

[POPPELBAUM et al. 1967] W. C. POPPELBAUM, C. AFUSO, J. W. ESCH, "Stochastic computing elements and systems", in *Proceedings of the AFIPS Fall Joint Computer Conference*, pp. 635–644

[POPPELBAUM 1968] W. C. POPPELBAUM, "Annual Report Part B, Pattern Processing and Memory Research", in *Annual Report Computer Systems Research*, September 1, 1967 to August 31, 1968

[POPPELBAUM 1979] W. C. POPPELBAUM, *Burst Processing*, Final Report for the Navy for Contract N000014-75-C-0982, Department of Computer Science, University of Illinois at Urbana-Champaign, Urbana, Illinois, 1979

[Popular Mechanics 1950] N. N., "It's Small But Smart, This 'Suitcase Brain'", in *Popular Mechanics*, 8, 1950

[POWELL] FRED O. POWELL, "Analog simulation of an adaptive two-time-scale control system", Advanced Electronic Systems Research Department, Bell Aerospace Division of Textron, Buffalo, New York 14240

[PRESS et al. 2001] WILLIAM H. PRESS, SAUL A. TEUKOLSKY, WILLIAM T. VETTERLING, BRIAN P. FLANNERY, *Numerical Recipes in Fortran 77 – The Art of Scientific Computing, Volume 1 of Fortran Numerical Recipes*, Cambridge University Press, Second Edition, 2001

[PREUSS 1962] HEINZWERNER PREUSS, *Grundriss der Quantenchemie*, Bibliographisches Institut, Mannheim, 1962

[PREUSS 1965] HEINZWERNER PREUSS, *Quantentheoretische Chemie*, Bibliographisches Institut, Mannheim, 1965

[DE SOLLA PRICE 1974] DEREK DE SOLLA PRICE, "Gears from the Greeks: The Antikythera Mechanism – a Calendar Computer from ca. 80 B.C.", in *Transactions of the American Philosophical Society*, Volume 64, Part 7, 1974

[PRONK 2019] KEES PRONK, "De Rekenmachine van Hammer", in *Jubileum Magazine 50 jaar Studieverzameling*, Delft University of Technology, 2019, pp. 37–39

[PUCHTA 1996] SUSANN PUCHTA, "On the Role of Mathematics and Mathematical Knowledge in the Invention of Vannevar Bush's Early Analog Computers", in *IEEE Annals of the History of Computing*, Vol. 18, No. 4, 1996, pp. 49–59

[RAGAZZINI et al. 1947] JOHN R. RAGAZZINI, ROBERT H. RANDALL, FREDERICK A. RUSSEL, "Analysis of Problems in Dynamics by Electronic Circuits", in *Proceedings of the I.R.E.*, Vol. 35, May 1947, pp. 444 ff.

[RAMIREZ 1976] W. FRED RAMIREZ, *Process Simulation*, D. C. Heath and Company, 1976

[RANDERY 1964] VIJAY K. RANDERY, "Study of a Parametron on an Analog Computer", in *IEEE Transactions on Electronic Computers*, October, 1964, pp. 612–614

[RANFFT et al. 1977] ROLAND RANFFT, HANS-MARTIN REIN, "Analog simulation of bipolar-transistor circuits", in *SIMULATION*, September 1977, pp. 75–78

[RASFELD 1983] PETER RASFELD, "Zur Darstellung und Untersuchung von Funktionen im Mathematikunterricht mittels elektronischer Analogrechner", in *Praxis der Mathematik*, Vol. 25, Issue 11, pp. 325–333

[Rationalisierungskuratorium 1957] Rationalisierungskuratorium der Deutschen Wirtschaft (Hg.), *Automatisierung*, Carl Hanser Verlag, München, 1957

[RATZ 1967] ALFRED G. RATZ, "Analog Computation of Fourier Series and Integrals", in *IEEE Transactions on Electronic Computers*, Vol. EC-16, No. 4, August 1967, p. 515

[RECHBERGER 1959] H. RECHBERGER, "Zweite internationale Tagung für Analogierechentechnik", in *mtw – Zeitschrift für moderne Rechentechnik und Automation*, 1/59, pp. 18–19

[REDHEFFER 1953] RAYMOND M. REDHEFFER, *Computing Machine*, United States Patent 2656102, Oct. 20, 1953

[REIHING 1959] JOHN V. REIHING, "A Time-Sharing Analog Computer", in *IRE-AIEE-ACM '59 (Western): Papers presented at the March 3–5, 1959, western joint computer conference*, March 1959, pp. 341–349

[REINEL 1976] K. REINEL, "Bewegungssimulatoren für Raumfahrt-Lageregelungssysteme", in [SCHÖNE 1976/1, pp. 464–475]

[REISIG 1999] GERHARD H. R. REISIG, *Raketenforschung in Deutschland – Wie die Menschen das All eroberten*, Wissenschaft und Technik Verlag, Berlin, 1999

[Remington 1956] Remington Rand Univac, *Increment Computer Logic And Programming*, Bomber Weapons Defense Computer Study, Final Engineering Report, Volume 4, October 1956

[RIEDEL et al. 2019] MARC RIEDEL, WEIKANG QIAN, "Synthesis of Polynomial Functions", in [GROSS et al. 2019, pp. 103–120]

[RIEDEL 2019] MARC RIEDEL, "Deterministic Approaches to Bitstream Computing", in [GROSS et al. 2019, pp. 121–136]

[RIEGER et al. 1974] NEVILLE F. RIEGER, CHARLES H. THOMAS Jr., "Some Recent Computer Studies on the Stability of Rotors in Fluid-Film Bearings", Rochester Institute of Technology, Mechanical Engineering Department, July, 1974

[RIDEOUT 1962] VINCENT C. RIDEOUT, "Random-Process Studies", in [HUSKEY et al. 1962, pp. 5-94–5-110]

[RIGAS et al.] HARRIETT B. RIGAS, ANDREW M. JURASZEK, *Some Approaches to the Design of a Model for an Aquatic Ecosystem*, author's archive

[ROBINSON] TIM ROBINSON, *Torque amplifiers in Meccano*, http://www.meccano.us/differential_analyzers/robinson_da/torque_amplifiers.pdf, retrieved 07/26/2005

[ROBINSON 2008] TIM ROBINSON, *Oral History of Arthur Porter*, Computer History Museum, Recorded March 8, 2008, Advance, North Carolina, https://archive.computerhistory.org/resources/access/text/2015/06/102658245-05-01-acc.pdf, retrieved 02/19/2021

[ROCKCASTLE et al. 1956] C. H. ROCKCASTLE, A. J. YATES, *Simulation of a Turbojet Engine on a Standard Electronic Analogue Computer for the Purpose of Control Study*, Thesis R664, 1956

[ROEDEL 1955] JERRY ROEDEL, "History and Nature of Analog Computors", in [PAYNTER ed. 1955, pp. 27–47]

[ROEDEL 1955/2] JERRY ROEDEL, "Application of an Analog Computer to Design Problems for Transportation Equipment", in [PAYNTER ed. 1955, pp. 199–215]

[RÖPKE et al. 1969] HORST RÖPKE, JÜRGEN RIEMANN, *Analogcomputer in Chemie und Biologie*, Springer-Verlag, 1969

[RÖSSLER 2005] EBERHARD RÖSSLER, *Die Torpedos der deutschen U-Boote*, Verlag E. S. Mittler & Sohn GmbH, 2005

[ROHDE 1977] WOLFGANG H. ROHDE, *Beurteilung und Optimierung von Maschinensystemen in der Entwurfsphase – Dargestellt am Beispiel eines drehzahlgesteuerten Walzwerksantriebes*, Dissertation an der Technischen Universität Clausthal, 1977

[ROHDE et al. 1981] WOLFGANG H. ROHDE, JÜRGEN STELBRINK, "Auslegung und konstruktive Gestaltung von Antriebssystemen schwerer Walzwerke", in *Stahl und Eisen*, No. 13/14/1981, pp. 164–173

[ROSKO 1968] JOSEPH S. ROSKO, "Comments on 'Hybrid Computer Solution of Optimal Control Problems by the Maximum Principle'", *IEEE Transactions on Computers*, Vol. C-17, No. 9, September 1968, p. 899

[RUBY 1996] LAWRENCE RUBY, "Applications of the Mathieu equation", in AMERICAN JOURNAL OF PHYSICS, 64 (1), January 1996, pp. 39-44

[RUDNICKI] MIECZYSLAW RUDNICKI, "Analogrechner MEDA in der Lasertechnik", in *14. seminář MEDA ANALOGOVÁ A HYBRIDNÍ VÝPOČETNÍ TECHNIKA*, Praha 1977, pp. 53–55

[RUSSELL 1962] PAUL E. RUSSELL, "Repetitive Analog Computers", in [HUSKEY et al. 1962, pp. 6-17–6-25]

[RUSSEL et al. 1971] Jack A. Russel, Bradford J. Baldwin, *Golf Game Computing System*, United States Patent 3598976, August 10, 1971

[RUSSELL 1978] A. RUSSELL, "Moon Landing Game", in *Practical Electronics*, November 1978, pp. 1138–1142

[RYDER 2009] WILLIAM H. RYDER, "A System Dynamics View of the Phillips Machine", The 27th Interational Conference of the Systems Dynamics Society, July 26–30, 2009, https://proceedings.systemdynamics.org/2009/proceed/papers/P1038.pdf, retrieved 08/01/2022

[RYDER et al. 2021] WILLIAM H. RYDER, ROBERT Y. CAVANA, "A System Dynamics Translation of the Phillips Machine", in [CAVANA et al. 2021, pp. 97–134]

[SADEK 1976] K. SADEK, "Nachbildung einer Hochspannungs-Gleichstrom-Übertragung", in [SCHÖNE 1976/1, pp. 360–388]

[SANKAR et al. 1979] SESHADRI SANKAR, DAVID R. HARGREAVES, "Hybrid computer optimization of a class of impact absorbers", in *SIMULATION*, July 1979, pp. 11–18

[SANKAR et al. 1980] S. SANKAR, J. V. SVOBODA, "Hybrid Computer in the Optimal Design of Hydro-Mechanical Systems", in *Mathematics and Computers in Simulation*, XXII (1980), pp. 353–367

[SARPESHKAR 1998] RAHUL SARPESHKAR, "Analog Versus Digital: Extrapolating from Electronics to Neurobiology", in *Neural Computation*, 10, 1998, pp. 1601–1638

[SAUER] ALBRECHT SAUER, *Gezeiten – Ein Ausstellungsführer des Deutschen Schifffahrtsmuseums*, Deutsches Schifffahrtsmuseum

[SAURO 2019] HERBERT M. SAURO, *Enzyme Kinetics for Systems Biology*, Ambrosius Publishing, 2nd edition, 2019

[SAVANT et al. 1954] C. J. SAVANT, R. C. HOWARD, "Multiplier for Analog Computers", in *Electronics*, September, 1954, pp. 144–147

[SAVET 1962] PAUL SAVET, "Heat-Transfer Computing Elements", in [HUSKEY et al. 1962, pp. 8-18–8-22]

[Schloemann-Siemag 1978] N. N., *Simulationstechnik im Schwermaschinenbau – Einsatz einer Analogrechenanlage für die Untersuchung und Berechnung von Maschinensystemen*, Sonderdruck der Schloemann-Siemag AG, 2/12.78

[SCHLOTTMANN et al. 2012] CRAIG R. SCHLOTTMANN, SAMUEL SHAPERO, STEPHEN NEASE, PAUL HASLER, "A Digitally Enhanced Dynamically Reconfigurable Analog Platform for Low-Power Signal Processing", in *IEEE Journal of Solid-State Circuits*, Vol. 37, No. 9, September 2012, pp. 2174–2184

[SCHMIDT 1956] W. SCHMIDT, "Die Hyperbelfeldröhre, eine Elektronenstrahlröhre zum Multiplizieren in Analogie-Rechengeräten", in *Zeitschrift für angewandte Physik*, VIII. Band, Heft 2, 1956, pp. 69–75

[SCHNEIDER 1960] G. SCHNEIDER, "Über die Nachbildung und Untersuchung von Abtastsystemen auf einem elektrischen Analogrechner", in *Elektronische Rechenanlagen*, 2 (1960), No. 1, pp. 31–37

[SCHÖNE 1976/1] ARMIN SCHÖNE, *Simulation Technischer Systeme*, Band 2, Carl Hanser Verlag München Wien, 1976

[SCHÖNE 1976/2] A. SCHÖNE, "Modelle von Wärmetauschern", in [SCHÖNE 1976/1, pp. 7–27]

[SCHÜSSLER 1961] W. SCHÜSSLER, "Messung des Frequenzverhaltens linearer Schaltungen am Analogrechner", in *Elektronische Rundschau*, No. 10, 1961, pp. 471–477

[SCHULTZ et al. 1974] HAROLD M. SCHULTZ, ROGER K. MIYASAKI, THOMAS B. LIEM, RICHARD A. STANLEY, "Analog Simulation of Compressor Systems", ISA CPD 74106, 1974

[SCHWARZ 1971] WOLFGANG SCHWARZ, *Analogprogrammierung – Theorie und Praxis des Programmierens für Analogrechner*, VEB Fachbuchverlag Leipzig, 1. Ed., 1971

[SCHWARZE 1972] K. SCHWARZE, "Automatisches Skalieren und statischer Test mit dem hybriden Interpreter HOI", in *Angewandte Informatik*, 3/72, pp. 127–138

[SCHWEIZER 1976/1] G. SCHWEIZER, "Beispiele für die Simulation von Luftfahrzeugen", in [SCHÖNE 1976/1, pp. 414–426]

[SCHWEIZER 1976/2] G. SCHWEIZER, "Das mathematische Modell für die Echtzeitsimulation von Erdsatelliten", in [SCHÖNE 1976/1, pp. 434–453]

[SCHWEIZER 1976/3] G. SCHWEIZER, "Das Systemglied „Mensch" in der Simulation", in [SCHÖNE 1976/1, pp. 555–580]

[SCHWEIZER 1976/4] G. SCHWEIZER, "Der Aufbau der Hybridsimulation für das Strahlflugzeug", in [SCHÖNE 1976/1, pp. 520–526]

[SCHWEIZER 1976/5] G. SCHWEIZER, "Die Aufbereitung der Gleichungen des mathematischen Modells eines Flugzeugs zur Simulation auf dem Analogrechner", in [SCHÖNE 1976/1, pp. 515–519]

[SCHWEIZER 1976/6] G. SCHWEIZER, "Die Sichtsimulation", in [SCHÖNE 1976/1, pp. 526–533]

[SCHWEIZER 1976/7] G. SCHWEIZER, "Simulationsprobleme aus der Luft- und Raumfahrt", in [SCHÖNE 1976/1, pp. 389–398]

[Scientific Data Systems/2] N.N., *SDS 9300 Computer Reference Manual*, Scientific Data Systems, July 1969

[SCOTT 1958] CLYDE C. SCOTT, "Power Reactor Control", in *Instruments and Automation*, Vol. 31, April 1958, pp. 636–637

[SELFRIDGE 1955] R. G. Selfridge, "Coding a general-purpose digital computer to operate as a differential analyzer", in *Proceedings of the Western Joint Computer Conference*, The Institute of Radio Engineers, New York, 1955, S. 82–84

[SEYFERTH 1960] H. SEYFERTH, "Über die Behandlung partieller Differentialgleichungen auf dem elektronischen Analogrechner", in *Elektronische Rechenanlagen*, 2 (1960), No. 2, p. 85–92

[SHEN 1970] KUEI SHEN, *Analog Computer Simulation in Plasma Physics*, Thesis, University of Iowa, Electrical Engineering, 1970

[SHEN et al. 1970] KUEI SHEN, E. D. ALTON, H C. S. HSUAN, "Analog Computer Simulation in Plasma Physics", in *American Journal of Physics*, 38, 1970, pp. 1133–1135

[SHERMAN et al. 1958] WINDSOR L. SHERMAN, STANLEY FABER, JAMES B. WITTEN, *Study of Exit Phase of Flight of a Very High Altitude Hypersonic Airplane by Means of a Pilot-Controlled Analog Computer*, Langley Aeronautical Laboratory, 1958

[SHILEIKO 1964] A. V. SHILEIKO, *Digital Differential Analysers*, Pergamon Press, The Macmillan Company, New York, 1964

[SHORE 1977] NIGEL LESLIE SHORE, *Hybrid Computer Simulation and On-Line Digital Computer Control of D. C. Link*, Thesis, Faculty of Engineering, University of London, Imperial College of Science and Technology, Department of Electrical Engineering, 1977

[SHRESTHA et al. 2022] AMAR SHRESTHA, HAOWEN FANG, ZAIDAO MEI, DANIEL PATRICK RIDER, QING WU, QINRU QIU, "A Survey on Neuromorphic Computing: Models and Hardware", in *IEEE Circuits and Systems Magazine*, second quarter 2022, pp. 2–35

[SIERCK 1963] JOACHIM SIERCK, *Untersuchung der nach dem Rückmischprinzip aufgebauten Frequenzteiler-Schaltungen unter besonderer Berücksichtigung der anomalen Mischrückkopplung*, Dissertation an der Fakultät für Elektrotechnik der RWTH Aachen

[SIMANCA et al. 2002] SANTIAGO R. SIMANCA, SCOTT SUTHERLAND, *Notes for MAT 331 – Mathematical Problem Solving with Computers*, The University at Stony Brook, https://www.math.stonybrook.edu/~scott/Book331/331book.pdf, retrieved 06/20/2022

[SIMONS et al.] FRED O. SIMONS, RICHARD C. HARDEN, SAM J. MONTE, *Perfected Analog/Hybrid Simulations of all Classes of Sampled-Data Systems*, author's archive

[SKRAMSTAD 1957] HAROLD A. SKRAMSTAD, "Some Simulation Problems Under Study at the National Bureau of Standards", in [White Sands 1957, pp. 91–107]

[SKRAMSTAD 1959] HAROLD K. SKRAMSTAD, "A Combined Analog-Digital Differential Analyzer", in *Proc. EJCC*, Volume 16, December 1959, pp. 94–101

[SMITH et al. 1959] F. SMITH, W. D. T. HICKS, *The R. A. E. Electronic Simulator for Flutter Investigations in Six Degrees of Freedom or Less*, Ministry of Technology, Aeronautical Research Council, London, Her Majesty's Stationary Office, 1959

[SNOW 1930] L. T. SNOW, *Planimeter*, United States Patent 718166, January 13, 1930

[SOMERS 1980] ERIC SOMERS, "Computer graphics for television", in *Video Systems*, June 1980, pp. 11–21

[SOMERVILLE] ALEXANDER SOMERVILLE, "A Beam-Type Tube That Multiplies", in *Proc. Natl. Elec. Conf.*, 6, 1950, pp. 145–154

[SONI et al. 2017] JIMMY SONI, ROB GOODMAN, *A Mind at Play – How Claude Shannon invented the Information Age*, Simon & Schuster Paperbacks, 2017

[SOROKA 1962] WALTER W. SOROKA, "Mechanical Analog Computers", in [HUSKEY et al. 1962, pp. 8-2–8-16]

[SORONDO et al.] VICTOR J. SORONDO, GEORGE D. WILSON, *Hybrid Computer Simulation of a Circulating Water System*, author's archive

[SOUDACK 1968] A. C. SOUDACK, "Canonical Programming of Nonlinear and Time-Varying Differential Equations", in *IEEE Transactions on Computers*, Vol. C-17, No. 4, April, 1968, p. 402

[SPECKHART et al. 1976] FRANK H. SPECKHART, WALTER L. GREEN, *A Guide to Using CSMP – The Continuous System Modeling Program – A Program for Simulating Physical Systems*, Prentice-Hall, Inc., 1976

[SPIESS 1992] RAY SPIESS, "The Comdyna GP-6 Analog Computer – Twenty Five Years... and still counting", in *Simulation*, Volume 59, Issue 5, pp. 323–325

[SPIESS 2005] RAY SPIESS, "The Comdyna GP-6 Analog Computer", in *IEEE Control Systems Magazine*, June 2005, pp. 68–73

[SPROTT 2016] JULIEN CLINTON SPROTT, *Elegant Chaos – Algebraically Simple Chaotic Flows*, World Scientific, 2016

[STARNICK 1976] J. STARNICK, "Simulation chemischer Reaktoren", in [SCHÖNE 1976/1, pp. 51–205]

[STATA 1968] RAY STATA, "Transconductance Analog Multiplier", in *Instruments and Control Systems*, November 1968, p. 115

[STEIN et al. 1959] MARVIN L. STEIN, J. ROSE, D. B. PARKER, "A Compiler with an Analog-Oriented Input Language", in *IRE-AIEE-ACM '59 (Western): Papers presented at the March 3–5, 1959, western joint computer conference*, March 1959, pp. 92–102

[STEIN et al. 1970] MARVIN L. STEIN, E. JAMES MUNDSTOCK, „Sorting Implicit Outputs in Digital Simulation", in *IEEE Transactions on Computers*, Vol. C-19, No. 9, September 1970, S. 844–847

[STEINHOFF et al. 1957] E. A. STEINHOFF, M. C. GREEN, "Real-Time Flight Performance Analysis", in [White Sands 1957, pp. 241-245]

[STEPANOW 1956] W. W. STEPANOW, *Lehrbuch der Differentialgleichungen*, VEB Deutscher Verlag der Wissenschaften Berlin, 1956

[STEWART 1979] PETER A. STEWART, "The Analog Computer as a Physiology Adjunct", in *The Physiologist*, 1979, Issue 1, pp. 43–47

[STILLWELL 1956] WENDELL H. STILLWELL, "Studies of Reaction Controls", in *Control Studies, Part B, Studies of Reaction Controls*, NASA, Document ID 19930092438, 1956

[STILLWELL et al. 1958] WENDELL H. STILLWELL, HUBERT M. DRAKE, "Simulator Studies of Jet Reaction Controls for Use at High Altitude", NACA Research Memorandum, September 26, 1958

[STINE 2014] KYLE STINE, "The Coupling of Cinematics and Kinematics", in *Grey Room*, 56, Summer 2014, pp 34–57

[STONE et al.] JOHN STONE, KA-CHEUNG TAUI, EUGENE PACK, "Computer Control Study for a Manned Centrifuge", NASA Technical Report, F-B2300-1

[STROTZ et al. 1951] R. H. STROTZ, J. F. CALVERT, N. F. MOREHOUSE, "Analogue Computing Techniques Applied to Economics", *Transactions of the American Institute of Electrical Engineers*, 70(1), DOI: 10.1109/T-AIEE.1951.5060443, pp. 557–563

[STROTZ et al. 1953] R. H. STROTZ, J. C. MCANULTY, J. B. NAINES, "Goodwin's Nonlinear Theory of the Business Cycle: An Electro-Analog Solution", in *Econometrica*, 21(3), DOI: 10.2307/1905446, pp. 390–411

[STUBBS et al. 1954] G. S. STUBBS, C. H. SINGLE, *Transport Delay Simulation Circuits*, Westinghouse, Atomic Power Division, 1954

[SURYANARAYANAN et al. 1968] K. L. SURYANARAYANAN, A. C. SOUDACK, "Analog Computer Automatic Parameter Optimization of Nonlinear Control Systems with Specific Inputs", in *IEEE Transactions on Computers*, Vol. C-17, No. 8, August, 1968, pp. 782–788

[SUTTON et al. 1963] GEORGE H. SUTTON, PAUL W. POMEROY, "Analog Analyses of Seismograms Recorded on Magnetic Tape", in *Journal of Geophysical Research*, Vol. 68, No. 9, May 1, 1963, pp. 2791–2815

[SUTTON et al. 2018] RICHARD S. SUTTON, ANDREW G. BARTO, *Reinforcement Learning: An Introduction*, second edition, The MIT Press, 2018

[SVOBODA 1948] ANTONIN SVOBODA, *Computing Mechanisms and Linkages*, McGraw-Hill Book Company, Inc., 1948

[SWADE 1995] DORON SWADE, "The Phillips Economic Computer", in *Resurrection – The Bulletin of the Computer Conservation Society*, Issue Number 12, Summer 1995, pp. 11–18

[SWARTZEL 1946] KARL D. SWARTZEL, *Summing Amplifier*, United States Patent 2401779, June 11, 1946

[SWITHENBANK 1960] J. SWITHENBANK, *Design of an Electrical Analogue Computer to Simulate Regenerative Heat Exchangers*, Mechanical Engineering Research Laboratories, Report No. SCS 12, McGill University, Montreal, February 1960

[SYDOW 1964] ACHIM SYDOW, *Programmierungstechnik für elektronische Analogrechner*, VEB Verlag Technik Berlin, 1964

[SZALAI 1971] KENNETH J. SZALAI, "Validation of a general purpose airborne simulator for simulation of large transport aircraft handling qualities", NASA TN D-6431, October 1971

[SZUCH et al. 1965] JOHN R. SZUCH, LEON M. WENZEL, ROBERT J. BAUMBICK, *Investigation of the Starting Characteristics of the M-1 Rocket Engine Using the Analog Computer*, NASA Technical Note TN D-3136, December 1965

[TABBUTT 1967] FREDERICK D. TABBUTT, "The Use of Analog Computers for Teaching Chemistry", in *Journal of Chemical Education*, Volume 44, Number 2, February 1967, pp. 64–69

[TABBUTT 1969] FREDERICK D. TABBUTT, "A New Concept in Graphic Computer Instruction", in *Reed College Science Journal*, Spring 1969, Volume 1, Number 1, pp. 50–53

[TACKER et al.] EDGAR C. TACKER, THOMAS D. LINTON, *Hybrid Simulation of an Optimal Stochastic Control System*, Louisiana State University, author's archive

[TAKAISHI 1965] YOSHIFUMI TAKAISHI, *Untersuchungen über den Einfluß der Tanklage und nichtlinearer Dämpfungseffekte auf die Wirkung eines Schlingertanks*, in *Schriftenreihe Schiffbau*, Technische Universität Hamburg-Harburg, 155, 1965

[Telefunken 1958] N. N., *Elektronischer Analogrechner RA 463/2*, Telefunken, AH 5.2 Apr. 58

[Telefunken/1] N. N., *Demonstrationsbeispiel Nr. 5, Ball im Kasten*, AEG Telefunken

[Telefunken/2] N. N., *Schwingungsberechnung eines Zwei-Massen-Systems*, AEG-Telefunken

[Telefunken/3] N. N., *Perspektivische Darstellung von Rechenergebnissen mit Hilfe eines Analogrechners*, AEG-TELEFUNKEN Datenverarbeitung

[Telefunken/4] N. N., *Darstellung von Tragflügeln und ihren Stromlinien mit einem Analogrechner*, AEG Telefunken

[Telefunken/5] N. N., *Kepler und die Atomphysik – Der Beschuß eines Atomkerns mit Alphateilchen auf einem Tischanalogrechner*, Demonstrationsbeispiel 3, AEG-Telefunken, ADB 003 0570

[Telefunken/6] N. N., *Steuermanöver eines Satelliten*, AEG-Telefunken, AB 009/10 70

[Telefunken/7] N. N., *Echo I – Simulation der Umlaufbahn auf dem Analogrechner*, AEG-Telefunken, ADB 004 0770

[Telefunken 1958] N. N., *Angebot Nr. 557/0010 der Telefunken GmbH an die Technische Hochschule München*, 1958, author's archive

[Telefunken 1963/1] N. N., *Anwendungsbeispiele für Analogrechner – Wärmeleitung*, Telefunken, 15. Oktober 1963

[Telefunken 1963/2] N. N., *Anwendungsbeispiele für Analogrechner – Transformator*, Telefunken, 15. Oktober 1963

[Telefunken 1966] N. N., *Demonstrationsbeispiele für Analogrechner RAT 740, Beispiel 2, Einfache Darstellung einer Planetenbahn*, Mitteilungen der Fachabteilung Analogrechner, TELEFUNKEN, 10.1.66

[TEUBER 1964] D. L. TEUBER, *Nachbildung der Saturn V-Rakete auf elektronischen Analogrechnern*, Tagungsberichte Hermann Oberth-Gesellschaft, 13. Raketen- und Raumfahrttagung vom 25.–28. Juni 1964 in Darmstadt

[THALER et al. 1980] G. J. THALER, T. S. NELSON III, A. GERBA, "Real Time, Man-Interfaced Motion Analysis of a 3000 Ton Surface Effect Ship", in *Summer Computer Simulation Conference*, 1980, pp. 593–598

[The Times 1954] N. N., "The Royal Aircraft Establishment's analogue computer...", in *The Times*, October 8, 1954

[THOMAS 1968] CHARLES H. THOMAS, "Transport Time-Delay Simulation for Transmission Line Representation", in *IEEE Transactions on Computers*, Vol. C-17, No. 3, March, 1968, pp. 205–214

[THOMAS et al./1] C. H. THOMAS, D. H. WELLE, R. A. HEDIN, R. W. WEISHAUPT, "Switching Surges on Parallel HV and EHV Untransposed Transmission Lines Studied by Analog Simulation", IEEE, Paper No. 71 TP 128-PWR

[THOMAS et al./2] CHARLES H. THOMAS, A. E. KILGOUR, D. H. WELLE, T. A. KARNOWSKI, *Transient Performance Study of a Parallel HV and EHV Transmission System*, author's archive

[THOMAS et al. 1968] CHARLES H. THOMAS, RONALD A. HEDIN, "Switching Surges on Transmission Lines Studied by Differential Analyzer Simulation", IEEE, Paper No. 68 RP 4-PWR, 1968

[THOMPSON] R. V. THOMPSON, *Application of Hybrid Computer Simulation Techniques to Warship Propulsion Machinery Systems Design*, author's archive

[THOMSON 1876] Sir WILLIAM THOMSON, "Mechanical Integration of linear differential equations of the second order with variable coefficients", in *Proceedings of the Royal Society*, Volume 24, No. 167, 1876, pp. 269–270,

[THOMSON 1878] Sir WILLIAM THOMSON, "Harmonic Analyzer", in *Proceedings of the Royal Society of London*, Vol. 27, 1878, pp. 371–373

[THOMSON 1882] Sir WILLIAM THOMSON, "Tides", *Evening Lecture To The British Association At The Southhampton Meeting*, source: http://www.fordham.edu/halsall/mod/1882kelvin-tides.html, retrieved 11/24/2012

[THOMSON 1911] Sir WILLIAM THOMSON, "The tidal gauge, tidal harmonic analyser, and tide predictor", in *Kelvin, Mathematical and Physical Papers*, Volume VI, Cambridge 1911, pp. 272–305

[THOMSON 1912] JAMES THOMSON, *Collected Papers in Physics and Engineering – selected and arragned with unpublished material and annotations by Sir Joseph Larmor and James Thomson*, Cambridge at the University Press, 1912

[THWAITES ed. 1987] BRYAN THWAITES (ed.), *Incompressible Aerodynamics – An Account of the Theory and Observation of the Steady Flow of Incompressible Fluid past Aerofoils, Wings, and Other Bodies*, Dover Publications, Inc., New York, 1987

[TISDALE] HENRY F. TISDALE, *How a Modern Analog Computer Duplicates Real-World Behaviour of a Thyristor-Controlled Rectifier Bridge*, Electronic Associates, Inc., author's archive

[TISDALE 1981] HENRY F. TISDALE, "Hybrid Computers Retaining Favor with Controls & Design Engineers", in *EAI OPENERS*, Product Information Bulletin #048 – May, 1981

[TITCHENER et al. 1983] M. R. TITCHENER, R. M. STIMPFLE, "Digital Emulation of Analog Computer Techniques for the Solution of Kinetic Systems", in *Journal of Computational Chemistry*, Vol. 4, No. 1, 1983, pp. 58–67

[TOMAYKO 1985] JAMES E. TOMAYKO, "Helmut Hoelzer's Fully Electronic Analog Computer", in *Annals of the History of Computing*, Volume 7, Number 3, July 1985, pp. 227–240

[TOMAYKO 2000] JAMES E. TOMAYKO, *Computers Take Flight – a History of NASA's Pioneering Digital Fly-By-Wire Project*, NASA SP-2000-4224, 2000

[TOMOVIC et al. 1961] R. TOMOVIC, W. J. KARPLUS, "Land Locomotion-Simulation and Control", in *Proc. 3rd AICA Conference on Analog Computation*, Opatija, Yugoslavia, 1961, pp. 385–390

[TRENKLE 1982] FRITZ TRENKLE, *Die deutschen Funklenkverfahren bis 1945*, AEG-TELEFUNKEN AKTIENGESELLSCHAFT, 1982, Anlagentechnik, Geschäftsbereich Hochfrequenztechnik

[TROCH 1977] INGE TROCH, "Eine neue Methode der Parameteroptimierung mit Anwendung auf Randwertaufgaben", in *14. seminář MEDA ANALOGOVÁ A HYBRIDNÍ VÝPOČETNÍ TECHNIKA*, Praha 1977, pp. 121–139

[TRUBERT 1968] MARC R. TRUBERT, *Use of Analog Computer for the Equalization of Electromagnetic Shakers in Transient Testing*, Jet Propulsion Laboratory, January 1, 1968

[TRUITT et al. 1960] THOS. D. TRUITT, A. E. ROGERS, *Basics of Analog Computers*, John F. Rider Publisher, Inc., New York, December 1960

[TSE et al. 1964] FRANCIS S. TSE, IVAN E. MORSE, ROLLAND T. HINKLE, *Mechanical Vibrations*, Allyn and Bacon, Inc., Boston, Second Printing, August 1964

[TUCKER 2002] WARWICK TUCKER, "A Rigorous ODE Solver and Smale's 14th Problem", in *Foundations of Computational Mathematics* (2002) 2, pp. 53–117

[TYROR et al. 1970] J. G. TYROR, R. I VAUGHAN, *An Introduction to the Neutron Kinetics of Nuclear Power Reactors*, Pergamon Press, 1970

[ULMANN 2014] BERND ULMANN, *AN/FSQ-7: the computer that shaped the Cold War*, DeGrouyter / Oldenbourg, 2014

[ULMANN 2016] BERND ULMANN, *The Lorenz-attractor*, Analog Computer Applications, https://analogparadigm.com/downloads/alpaca_2.pdf, retrieved 01/14/2021

[ULMANN 2017] BERND ULMANN, *Celestial mechanics: Three-body problem*, Analog Computer Applications, https://analogparadigm.com/downloads/alpaca_11.pdf, retrieved 01/14/2021

[ULMANN 2019] BERND ULMANN, *Solving the Schrödinger equation*, Analog Computer Applications, https://analogparadigm.com/downloads/alpaca_22.pdf, retrieved 01/14/2021

[ULMANN et al. 2019] BERND ULMANN, DIRK KILLAT, "Solving systems of linear equations on analog computers", IEEE Xplore, Kleinheubach Conference, 2019, pp. 1–4

[ULMANN 2020/1] BERND ULMANN, *Analog and Hybrid Computer Programming*, DeGruyter, 2021

[ULMANN 2020/2] BERND ULMANN, *A passive network for solving the two-dimensional heat equation*, Analog Computer Applications, https://analogparadigm.com/downloads/alpaca_25.pdf, retrieved 08/10/2022

[ULMANN 2020/3] BERND ULMANN, *Simulating the flight of a glider*, Analog Computer Applications, https://analogparadigm.com/downloads/alpaca_26.pdf, retrieved 01/14/2021

[ULMANN 2021/1] BERND ULMANN, *The Hindmarsh-Rose model of neuronal bursting*, Analog Computer Applications, https://analogparadigm.com/downloads/alpaca_28.pdf, retrieved 01/14/2021

[ULMANN 2021/2] BERND ULMANN, *Neutron kinetics*, Analog Computer Applications, https://analogparadigm.com/downloads/alpaca_30.pdf, retrieved 01/14/2021

[ULMANN 2021/3] BERND ULMANN, *The exponentially-mapped-past approach*, Analog Computer Applications, https://analogparadigm.com/downloads/alpaca_32.pdf, retrieved 01/14/2021

[USHAKOV 1958/1] V. B. USHAKOV, "Soviet Trends in Computers for Control of Manufacturing Processes", in *Instruments and Automation*, November 1958, pp. 1810–1813

[USHAKOV 1958/2] V. B. USHAKOV, "Soviet Trends in Computers for Control of Manufacturing Processes", in *Instruments and Automation*, December 1958, pp. 1960–1961

[VALENTIN 2003] FRANZ VALENTIN, *Hydraulik II – Angewandte Hydromechanik*, Skript des Lehrstuhles für Hydraulik und Gewässerkunde der Technischen Universität München, Oktober 2003

[VALISALO et al. 1982] P. E. VALISALO, D. BERGQUIST, V. MCGREW, *A Hybrid Computer Algorithm for Temperature Distribution Analysis of Irregular Two Dimensional Shapes*, in 10th *IMACS World Congress on System Simulation and Scientific Computation, 1982*, pp. 17–19

[VALISALO et al.] P. E. VALISALO, J. K. LEGRO, "Hybrid Computer Utilization for System Optimization", author's archive

[VAN DER POL et al. 1928] BALTHASAR VAN DER POL, J. VAN DER MARK, "The Heartbeat considered as a Relaxation Oscillation, and an Electrical Model of the Heart", in *Phil. Mag.* 7, 1928, pp. 763–775

[VAN SANTEN 1966] N. VAN SANTEN, "Enkele bouwkundige toepassingen van elektrische weerstandsmodellen", in *TNO-Nieuws*, 21, 1966, pp. 192–196

[VAN VEEN 1947/1] JOHAN VAN VEEN, "Analogy between Tides and A. C. Electricity (No. I)", in *The Engineer*, Nov. 28, 1947, pp. 498–500

[VAN VEEN 1947/2] JOHAN VAN VEEN, "Analogy between Tides and A. C. Electricity (No. II)", in *The Engineer*, Dec. 5, 1947, pp. 520–521

[VAN VEEN 1947/3] JOHAN VAN VEEN, "Analogy between Tides and A. C. Electricity (No. III)", in *The Engineer*, Dec. 12, 1947, pp. 544–545

[VAN ZYL 1964] L. L. VAN ZYL, *An Electronic Analogue Computer for the Solution of Non-Linear Partial Differential Equations Encountered in the Study of the Self-Heating of Fishmeal*, Thesis, Faculty of Engineering, University of Cape Town, 1964

[VICHIK 2015] SERGEY VICHIK, *Quadratic and linear optimization with analog circuits*, dissertation, University of California, Berkeley, 2015

[VICHNEVETSKY 1969] ROBERT VICHNEVETSKY, "Use of Functional Approximation Methods in the Computer Solution of Initial Value Partial Differential Equation Problems", in *IEEE Transactions on Computers*, Vol. C-18, No. 6, June 1969, pp. 499–512

[VOCOLIDES 1960] J. VOCOLIDES, "Über die Behandlung linearer algebraischer Gleichungssysteme mit Analogrechnern", in *Elektronische Rechenanlagen*, 2 (1960), No. 3, pp. 136–141

[VOGEL 1977] FRITZ VOGEL, *Ein elektronischer Koordinatenwandler ohne Diodennetzwerke und seine Anwendung bei der Meßwertverarbeitung mechanischer Größen*, Dissertation, Technische Universität Wien, September 1977

[VOLDER 1959] JACK E. VOLDER, "The CORDIC Trigonometric Computing Technique", in *IRE Trans. Electron. Comput.*, EC-8, 1959, pp. 330–334

[VOLYNSKII et al. 1965] B. A. VOLYNSKII, V. YE. BUKHMAN, *Analogues for the Solution of Boundary-Value Problems*, Pergamon Press, 1965

[VON NEUMANN 1956] JOHN VON NEUMANN, "Probabilistic Logics and the Synthesis of Reliable Organisms from Unreliable Components", in *Automata Studies*, ed. CLAUDE SHANNON, Princeton University Press, 1956

[VON THUN] H. J. VON THUN, *Simulation einer lagegeregelten Radiostern-Antenne auf dem hybriden Analogrechner*, BBC-Mannheim, Zentrale Entwicklung für Elektronik, Abteilung Systemtechnik

[WADEL 1956] LOUIS B. WADEL, "Simulation of Digital Filters on an Electronic Analog Computer", in *Journal of the ACM*, Volume 3, Issue 1, Jan. 1956, pp. 16–21

[WAGNER 1972] MANFRED WAGNER, *Analogrechner in der Verfahrenstechnik*, VEB Deutscher Verlag für Grundstoffindustrie, Leipzig, 1972

[WAIT 1963] JOHN V. WAIT, "A hybrid analog-digital differential analyzer system", in *AFIPS '63 (Fall): Proceedings of the November 12–14, 1963, fall joint computer conference*, November 1963, pp. 277–298

[WALTHER et al. 1949] ALWIN WALTHER, HANS-JOACHIM DREYER, "Die Integrieranlage IPM-Ott für gewöhnliche Differentialgleichungen", in *Die Naturwissenschaften*, No. 7, 1949, pp. 199–206

[WALTMAN 2000] GENE L. WALTMAN, *Black Magic and Gremlins: Analog Flight Simulation at NASA's Flight Research Center*, NASA History Division, Monographs in Aerospace History, Number 20, 2000

[WARSCHAWSKI 1945] S. E. WARSCHAWSKI, "On Theodorsen's Method of Conformal Mapping of Nearly Circular Regions", in *Quarterly of Applied Mathematics*, April, 1945, Vol. 3, No. 1, pp. 12–28

[WASHBURN 1962] R. P. WASHBURN, "Economic-Dispatch Computers for Power Systems", in [HUSKEY et al. 1962, pp. 5-155–5-160]

[WASS 1955] C. A. A. WASS, *Introduction to Electronic Analogue Computers*, London, Pergamon Press Ltd., 1955

[VAN WAUVE 1962] ARMAND VAN WAUVE, *Automatic Analog Computer Control by Means of Punched Cards – "CRESSIDA I"*, National Aeronautics and Space Administration, Washington, D.C., April 1962

[WEINBERG et al. 1958] ALVIN M. WEINBERG, EUGENE P. WIGNER, *The Physical Theory of Neutron Chain Reactors*, The University of Chicago Press, 1958

[WEITNER 1955] G. WEITNER, "Grundschaltungen elektronischer Regler mit Rückführung", in *Elektronische Rundschau*, September 1955, Vol. 9, No. 9, pp. 320–323

[WERRELL 1985] KENNETH P. WERRELL, *The Evolution of the Cruise Missile*, Air University Press, Maxwell Air Force Base, Alabama, September 1985

[WHITE 1966] M. E. WHITE, "An Analog Computer Technique for Solving a Class of Nonlinear Ordinary Differential Equations", in *IEEE Transactions on Electronic Computers*, Vol. EC-15, No. 2, April 1966, pp. 157–163

[White Sands 1957] White Sands Proving Ground, *Proceedings of First Flight Simulation Symposium, November 1956*, WSPG Special Report 9, September 1957

[WHITESELL et al. 1969] L. G. WHITESELL, E. H. BOWLES, "Train Power Station Operators by Analog", in [Hydrocarbon Processing 1959, pp. 84–86]

[WIERWILLE et al. 1968] WALTER W. WIERWILLE, JAMES R. KNIGHT, "Off-Line Correlation Analysis of Nonstationary Signals", in *IEEE Transactions on Computers*, Vol. C-17, No. 5, May, 1968, pp. 525–536

[WILLIAMS et al. 1958] THEODORE J. WILLIAMS, R. CURTIS JOHNSON, ARTHUR ROSE, "Computers in the Process Industries", in *Instruments and Automation*, Vol. 31, January 1958, pp. 90–94

[WILLERS 1943] FRIEDRICH ADOLF WILLERS, *Mathematische Instrumente*, Verlag von R. Oldenbourg, München und Berlin 1943

[WINARNO 1982] H. WINARNO, J. JALADE, J. P. GOUYON, "Hybrid Simulation of a Microprocessor Controlled Multi-Convertor", 10th IMACS World Congress on System Simulation and Scientific Computation, 1982, pp. 23–25

[WINKLER 1961] HELMUT WINKLER, *Elektronische Analogieanlagen*, Akademie-Verlag Berlin, 1961

[WINSTEAD 2019] CHRIS WINSTEAD, "Tutorial on Stochastic Computing", in [GROSS et al. 2019, pp.39–76]

[WITSENHAUSEN 1962] HANS S. WITSENHAUSEN, "Hybrid techniques applied to optimization problems", in *AIEE-IRE*, '62 (Spring), Proceedings of the May 1–3, 1962, spring joint computer conference, pp. 377–392

[WOODS 2008] W. DAVID WOODS, *How Apollo Flew to the Moon*, Springer, Praxis Publishing Ltd., 2008

[WORLEY 1962] CHARLES W. WORLEY, "Process-Control Applications", in [HUSKEY et al. 1962, pp. 5-71–5-86]

[WOŹNIAKOWSKI 1977] MIROSLAV WOŹNIAKOWSKI, "Hybrid and Digital Simulation and Optimization of Dynamic Systems", in *14. seminář MEDA ANALOGOVÁ A HYBRIDNÍ VÝPOČETNÍ TECHNIKA*, Praha 1977, pp. 159–163

[ZACHARY 1999] G. PASCAL ZACHARY, *Endless Frontier – Vannevar Bush, Engineer of the American Century*, The MIT Press, 1999

[ZHANG 2015] WEI-BIN ZHANG, *Differential Equations, Bifurcations, and Chaos in Economics*, World Scientific Publishing Company, 2015

[ZHAO et al. 2019] ZHOU ZHAO, ASHOK SRIVASTAVA, LU PENG, QING CHEN, "Long Short-Ter Memory Network Design for Analog Computing", in *ACM Journal on Emerging Technologies in Computing Systems*, Volume 15, Issue 1, January 2019, Article No.: 13, pp. 1–27

[ZIMDAHL 1965] W. ZIMDAHL, "Führungsverhalten des vierrädrigen Straßenfahrzeugs bei Regelung des Kurses auf festgelegter Bahn", in *Regelungstechnik – Zeitschrift für Steuern, Regeln und Automatisieren*, 13. Jahrgang 1965, Heft 5, pp. 221–226

[ZINGG 1989] DAVID W. ZINGG, *Low Mach Number Euler Computations*, NASA Technical Memorandum 102205

[ZOBERBIER 1968] W. ZOBERBIER, "Die Funktionsgleichungen des digitalen Integrators", in *Elektronische Rechenanlagen*, 10 (1968), Heft 5, pp. 234–242

[ZORPETTE 1989] GLENN ZORPETTE, "Parkinson's gun director", in *IEEE Spectrum*, April 1989, p. 43

[Zuse Z80 1961] N. N., "ZUSE Z 80 – Ein lochendes und druckendes Transistorzählwerk", in *mtw – Zeitschrift für moderne Rechentechnik und Automation*, 1/61, p. 33

Index

A4, 41, 218, 329
accumulator, 210
acoustic analog computer, 266
AD-10, 344
AD-2-64-PBC, 257
AD/FOUR, 360
adaptive control, 311
ADC, 123, 201
ADDAVERTER, 201
adding component, 89
additive cell, 20
ADIOS, 141
aerodynamical unit, 328
Aerojet, 282
aeronautical engineering, 320
aerospace medicine, 289
AHCS, 360
AI, 1, 362
air
– density, 67
– rudder, 44
airborne simulator, 334
aircraft
– arresting gear system, 323
– position computer, 328
ALDRIN, BUZZ, 13
algebraic loop, 248, 262
ALGOL, 206
All-Union Scientific Research Oil-Gas Institute, 290
ALS-2000, 353
alternating operation, 205
ALU, 210
AMDAHL's law, 357
AMDAHL, GENE MYRON, 357
AMMON, WERNER, 135
amplifier
– negative-feedback, 75
– open, 102
– summing, 85
AMSLER-LAFFON, JACOB, 15
analog, 2
analog computer, 2
– anatomy, 123
– centers, 354

– chances, 357
– electronic, 41
– future, 357
– mechanical, 9
– precision, 123
Analog Paradigm, 206
analog-digital converter, 123, 201
analogy
– direct, 4
– indirect, 4
Analysis Laboratory, 69
AND, 230
angle of attack, 331
Antikythera mechanism, 9
AOA, 331
API, 362
Apollo, 345
Appalachian surge tank, 274
Application Programming Interface, 362
applications, 243
– aeronautical engineering, 320
  – airborne simulator, 334
  – aircraft arresting gear systems, 323
  – flight simulation, 325
  – guidance and control, 336
  – helicopter, 323
  – jet engines, 323
  – landing gears, 321
  – nike, 336
  – parachutes, 338
  – polaris, 337
  – rotor blade, 323
– aerospace engineering
  – Apollo, 345
  – Gemini, 345
  – Mercury, 345
  – rocket motor simulation, 338
  – rocket simulation, 339
  – spacecraft manoeuvres, 343
– arts, 349
– automation, 305
  – closed loop control, 307
  – correlation analysis, 306
  – data processing, 306
  – embedded systems, 308

- sampling system, 307
- servo systems, 307
- biology, medicine, 284
  - aerospace, 289
  - cardiovascular systems, 285
  - CO_2 regulation, 286
  - ecosystems, 284
  - epidemiology, 288
  - locomotor systems, 289
  - metabolism, 285
  - neurophysiology, 286
  - pupil regulation, 286
- chemistry, 264
  - quantum chemistry, 265
  - reaction kinetics, 264
- economics, 292
- education, 348
- electronics, telecommunication, 301
  - circuit simulation, 301
  - demodulator, 304
  - filter design, 304
  - frequency response, 303
  - modulator, 304
- engineering, 266
- entertainment, 352
- geology, 290
  - ray tracing, 292
  - resources, 290
  - seismology, 291
- mathematics, 243
  - differential equations, 243
  - Eigenvalues, Eigenvectors, 249
  - FOURIER synthesis and analysis, 250
  - integral equations, 244
  - linear algebra, 248
  - multidimensional shapes, 253
  - optimisation, 252
  - orthogonal functions, 247
  - random process, 251
  - systems of linear equations, 248
  - zeros of polynomials, 247
- mechanics, 266
  - bearings, 269
  - compressors, 270
  - crank mechanisms, 271
  - ductile deformation, 272
  - earthquake simulation, 268
  - hydraulic systems, 273
  - machine tool control, 276
- non-destructive testing, 272
- pneumatic systems, 273
- rotating system, 269
- servo system, 277
- shock absorber, 268
- vibrations, 267
- military, 347
- music, 354
- nuclear technology, 278
  - control, 282
  - research, 279
  - training, 281
- physics, 253
  - ferromagnetic films, 264
  - heat-transfer, 258
  - optics, 258
  - orbit calculation, 255
  - particle trajectory, 255
  - semiconductor research, 262
- power engineering, 295
  - frequency control, 297
  - generators, 295
  - power grid simulation, 298
  - power inverter, 296
  - power stations, 301
  - rectifier, 296
  - transformers, 296
  - transmission lines, 297
- process engineering, 308
  - adaptive control, 311
  - distillation column, 309
  - evaporator, 309
  - heat exchanger, 309
  - mixing tank, 309
  - optimisation, 311
  - parameter determination, 311
- rocketry, 338
- transport systems, 313
  - automotive engineering, 313
  - dynamic behaviour, 318
  - hovercraft, 317
  - maglev, 317
  - marshaling hump, 317
  - motor coach simulation, 317
  - nautics, 318
  - propulsion system, 318
  - railway vehicles, 317
  - ride simulation systems, 315
  - ship simulation, 319

– steering system, 313
– torpedo simulation, 320
– traffic flow simulation, 316
– transmissions, 314
Applied Dynamics, 73, 257, 344
– AD-10, 344
– AD-2-64-PBC, 257
– AD/FOUR, 360
approximation
– Padé, 119
– Stubbs-Single, 119
Arduino
– MEGA-2650, 206
arithmetic/logic unit, 210
Arma Corporation, 31
ARMSTRONG, NEIL A., 332, 346
artifical neural network, 248
artificial intelligence, 1, 248, 362
arts, 349
astrolabe, 9
– planispheric, 9
attenuator, 129
attractor, 187
Automatic Digital Input Output System, 141
automatic level recorder, 61
automation, 305
automotive engineering, 313
Autonetics, 326
autopatch, 300

BÜCKNER, HANS F., 38
backpropagation, 248
barrel cam, 19
bearings, 269
Beckman
– EASE 2132, 290
– EASE 1032, 298
– EASE 1132, 355
– EASE 2133, 331
Bell Laboratories, 27, 61, 85, 345
Bell X-2, 330
Bendix
– Corporation, 220
– D-12, 220
– G-15D, 221
BESSEL function, 220
bevel-gear differential, 20
binary digit, 2
biology, 284

bipolar stochastic number, 230
bit, 2
BLACK, HAROLD STEPHEN, 75
BLUMLEIN, ALAN, 75
BODE diagram, 277
BODE, HENDRIK WADE, 277
body, 12
– dynamics, 289
Boeing, 320
Bonneville Power Administration, 298
bouncing ball, 176
boundary value problem, 244
BRATT, J. B., 33
BRIGGS, HENRY, 11
broadside, 29
Brookhaven National Laboratories, 353
BROUWER, BERT, 261
BTL, 61
burst processing, 231
bus rod, 32
BUSH, VANNEVAR, 32, 246

California Institute of Technology, 68, 229
Caltech, 68
– computer, 68
cam
– squaring, 18
– three-dimensional, 19
camoid, 19
capacitor wheel, 118
car suspension, 181
cardiovascular system, 285
Cathode Ray Tube, 284
CAW, LARRY, 336
CDC
– 6800, 318
– 7600, 318
cell
– additive, 20
– linear, 20
Center for Analysis, 34
center slide, 12
CHANCE, BRITTON, 284
check
– rate, 206
– static, 206
chemistry, 264
chopper, 81
– stabilisation, 82

Chrysler Corporation Missile Operations, 319
CI-5000, 204
cinema integraph, 246
circle test, 40
circuit simulation, 301
closed loop control, 307
CNC, 228
CO_2 regulation, 286
coefficient potentiometer, 92
cold war, 13, 347
Colónia Güell, 5
Colorant Mixture Computer, 278
colour matching, 278
Comcor, 203
Comdyna, 349
– GP-6, 349
COMIC, 278
command guidance system, 336
comparator, 115
– electronic, 116
– relay, 115
compressors, 270
computer
– analog, 2
– Caltech, 68
– digital, 2
– incremental, 210
Computer History Museum, 219
conformal mapping, 196
continuous steepest ascent/descent, 252
contour map, 104
Control Data, 318
control of machine tools, 276
Convair Astronautics, 201, 321
Coordinate Rotation Digital Computer, 338
coordination equations, 301
CORDIC, 338
correlation analysis, 306
CORSAIR, 224
COWAN, GLENN EDWARD RUSSEL, 205, 363
crank mechanisms, 271
Cray-1, 344
CRINER, HARRY E., 69
crossbar switch, 37
crossed-fields multiplier, 106
CRT, 284
CSMP, 237
CSS Virginia, 30
cursor, 12

curve follower, 98

D-12, 220
DAC, 118, 123, 201
Daimler Benz, 314
data processing, 306
DAVIDSON, HUGH R., 278
DAY, RICHARD E., 330, 332
DDA, 4, 209
– parallel, 209
– sequential, 210
– simultaneous, 209
DDP-24, 330
DE SOLLA PRICE, DEREK, 10
decay constant, 281
defense calculator, 233
delayed neutron, 280
Deltar, 275
Deltawerken, 275
demodulator, 304
DESY, 257, 297
Deutsches Elektronensynchrotron, 257, 297
DEX 100, 256
DIAN, 236
DIDA, 218
differential, 20
differential analyser, 32
– electromechanical, 35
differential equation, 243
– partial, 157
differential gear, 20
differential surge chamber, 273
digit
– binary, 2
digital computer, 2
digital differential analyser, 4, 209
digital signal processor, 361
digital voltmeter, 121, 140
digital-analog converter, 118, 123, 201
diode
– ZENER, 79
– function generator, 100
direct analogy, 4
distillation column, 309
domain specific language, 361
double-ball integrator, 22
Douglas Aircraft Company, 100
drift stabilisation, 80
DSL, 361

DSL-90, 234
DSP, 361
ductile deformation, 272
duplex, 12
DVM, 121, 140
dynamic behaviour, 318
Dynamic System Synthesizer, 327, 360
dynamometer multiplier, 114

E6-B, 13
EAC, 363
EAFCOM, 259, 324
EAI, 92, 306, 309
– 231R, 138, 281, 307, 330, 345
– 231RV, 141, 257, 355
– ADIOS, 141
– DOS-350, 330
– HYDAC 2000, 141
– Pace 96, 354
– PACER 500, 148
– SIMSTAR, 360
– time delay, 118
– TR-10, 146, 292
– TR-48, 257, 282, 309
– Variplotter, 140
– VDFG, 101
– 380, 259
– 580, 96, 123, 124
– 590, 309
– 640, 202
– 680, 360
– 690, 201, 284
– 693, 202
– 2000, 95
– 6200, 122
– 8400, 330
– 8800, 320
– 8900, 206
earthquake simulation, 268
EASE
– 1032, 298
– 1132, 355
– 2132, 290
– 2133, 331
Echo-1, 343
ECKDAHL, DONALD E., 218
economics, 292
ecosystems, 284
EDC, 301

EDMUNDS, MIKE, 11
education, 348
Effelsberg, 277
EHRICKE, KRAFFT ARNOLD, 321
Eigenvalue, 249
Eigenvector, 249
ELARD, 108
Electro-Analogue, 307
electrolytic tank, 5, 104, 363
electromechanical differential analyser, 35
electron-beam multiplier, 106
Electronic Analog Frost Computor, 259
Electronic Associates Inc., 92
electronic comparator, 116
Electronic Dispatch Computer, 301
Electronic Graph Paper, 138
electronic slide rule, 13
electronic structure, 265
electronics, 301
embedded system, 308
end
– brace, 12
– bracket, 12
engineering, 266
English Electric Company, 329
enrichment, 283
entertainment, 352
epidemiology, 288
ESAKI diode, 262
ESAKI, LEO, 262
ESCH, JOHN WILLIAM, 231
evaporator, 309
examples
– bouncing ball, 176
– car suspension, 181
– Lotka-Volterra equations, 172
– mass-spring-damper system, 168
– predator and prey, 172
– projection of rotating body, 194
– sin(), 165
Examples, 165
Excess-three, 221
exhaust rudder, 44
Explorer I, 57, 341
exponentially mapped past, 251
extended analog computer, 363

Faber Castell
– 2/83, 12

false, 230
FANO, ROBERT MARIO, 251
FARRIS, GEORGE JOSEPH, 234
Fast FOURIER Transformation, 250
feedback
– negative, 75
– technique, 153
ferromagnetic films, 264
FFT, 250
Field Programmable Analog Arrays, 362
Field Programmable Gate Array, 359
filter design, 304
financephalograph, 292
fire control system
– electronic, 61
– mechanical, 29
FISCHER, 309
FITZHUGH, RICHARD, 287
flight
– computer, 13
– simulation, 325
– table, 329
flip-flop, 126, 132, 141, 231
fluid dynamics, 339
fly-by-wire, 335
flying bedstead, 346
FORTRAN, 206
four-quadrant operation, 105
FOURIER analysis, 250
FOURIER synthesis, 250
FOURIER, JEAN-BAPTISTE-JOSEPH, 250
Foxboro, 57
FPAA, 362
FPGA, 359
FRANKE, HERBERT W., 350
FRASER, JAMES EARL, 364
FREDHOLM, IVAR, 245
frequency
– control, 297
– response, 303
friction-wheel, 15
– integrator, 21
frontlash unit, 22
FULTON, FITZ, 336
function
– MATHIEU, 190
– BESSEL, 220
– generator, 97
  – curve follower, 98

– diode, 100
– photoformer, 99
– polygon, 100
– orthogonal, 247

G-15D, 221
GAGE, F. D., 246
gain
– open loop, 82
GAINES, BRIAN R., 229
GAP/R, 60, 136
gate
– logic, 230
GAUß, Carl Friedrich, 248
gearbox
– helical, 32
GEDA, 301, 347
Gemini, 345
General Dynamics, 141
General Electric, 272
General Motors, 182, 313, 315
General Purpose Airborne Simulator, 335
General Purpose Simulator, 341
generators, 295
genetic programming, 312
geology, 290
George A. Philbrick Researches, 60, 136
Gilbert cell multiplier, 114
GILBERT, BARRIE, 114
GODDARD, ROBERT, 44
GOLDBERG, EDWIN A., 50, 74, 81
GOLDMANN, HANS OTTO, 83, 142
Golf Game Computing System, 353
GONNELLA, TITO, 15
Goodyear, 320
Goodyear Electronic Differential Analyser, 301
GP-6, 349
GPAS, 335
GPS, 341
GRAVES, TOM, 227
GREEN theorem, 15
guidance and control, 336
Guided Weapons Division, 329
GUNDLACH, FRIEDRICH-WILHELM, 107
GUO, NING, 205

HALL effect multiplier, 114
halt, 90, 92
HAMMER, J. A., 305

HANNAUER, GEORGE, 300
hardware in the loop, 270
harmonic
– analyser, 28
– synthesizer, 25
harmonics, 28
HARTREE, DOUGLAS, 33
hatchet planimeter, 15
HAZEN, H. L., 246
HCS, 300
Heat Exchange Transient Analog Computer, 259
heat exchanger, 309
heat-transfer, 258
heat-transfer multiplier, 114
HEIDERSBERGER, BENJAMIN, 350
HEIDERSBERGER, HEINRICH, 350
helical gearbox, 32
helicopter, 323
HEMMENDINGER, HENRY, 278
HERMANN, JOHANN MARTIN, 15, 21
Hermes, 56
HETAC, 259, 324
Hewlett-Packard, 218, 338
high
– performance computing, 361
– speed differential analyser, 338
– speed flight station, 330
HIGINBOTHAM, WILLIAM, 352
HINDMARSH, JAMES L., 287
Hitachi-240, 354
HKW 860, 147
HODGKIN, ALAN LLOYD, 286
HOELZER, HELMUT, 41
hovercraft, 317
HP-35, 12, 338
HPC, 361
HRS 860, 297
HSDDA, 338
Hudson Bay Company, 172
hump marshaling, 317
HUTSPIEL, 347
HUXLEY, ANDREW FIELDING, 286
hybrid
– computer, 201
– computer power simulator, 300
– PACTOLUS, 303
hybrides Koppelwerk, 147
HYDAC 2000, 141

hydraulic system, 273
HYPAC, 303
hyperbolic field tube, 107
HYTRAN, 206

IBM, 346
– 360/67, 284
– 7070/7074, 234
– 701, 233
– 704, 202
– 1620, 234
IC, 90
ICBM, 202
ICT, 357
in memory computing, 362
in-flight simulation, 334
incremental
– computer, 210, 224
– value, 210
indicator
– diagram, 14
– steam engine, 14
indirect analogy, 4
initial condition, 90
input table, 34
Instrumentation Laboratory, 337
integral equations, 244
integraph
– cinema, 246
– product, 245
integrator, 21, 89, 210
– double-ball, 22
– friction-wheel, 15, 21
– mechanical, 21
Integrieranlage IPM-Ott, 34
integromat, 39
intercontinental ballistic missile, 202
inverse functions, 102
ISEC 250, 295
Isograph, 27

JACOBI, CARL GUSTAV JAKOB, 248
jerk, 289
jet engines, 323
Jet Propulsion Laboratory, 69
JetStar, 335
jolt, 289
JOUKOWSKY, 197
JPL, 69

JULIE, LOEBE, 74
Jupiter, 57
– C, 341

K2-W, 78
K2-X, 80
KACO, 82
KASTNER, SIGISMUND, 319
KEAR, F. G., 246
KELVIN, LORD, 23
KEMPEL, BOB, 335
kernel, 245
KETTEL, ERNST, 133, 243
KLEINWÄCHTER, HANS, 41
Koppelwerk
– hybrides, 147
Kreiselgeräte GmbH, 44
KRON, GABRIEL, 265
KUBELKA, PAUL, 278
KUBELKA-MUNK theory, 278
KUHN, HANS, 266
KUHN, THOMAS SAMUEL, 357
KULK, HANS, 354

LÄMMLE, 15
LACE, 329
Landing Craft Mechanized, 319
landing gears, 321
LAPOSKY, BEN, 349
Law
– AMDAHL's, 357
LCM-6, 319
LEM, 346
Lewis Flight Propulsion Laboratory, 276
LGP-30, 187
limiter, 116
linear
– algebra, 248
– cell, 20
– ordinary differential equation, 243
– planimeter, 17
– servo system, 277
LIOUVILL, JOSEPH, 261
LISSAJOUS, JULES ANTOINE, 350
Litton Systems, 338
Lockheed JetStar, 335
locomotor system, 289
locus curve, 277
LODE, 243

logarithmic multiplier, 110
logic gate, 230
long-tailed pair, 79
loop
– algebraic, 248, 262
– servo, 213
Lord KELVIN, 22
LORENTZ force, 257
LORENZ attractor, 187
LORENZ, EDWARD N., 187
LOTKA, ALFRED JAMES, 172
LOTKA-VOLTERRA equations, 172
LOVELL, CLARENCE A., 61
Lunar Excursion Module, 346
Luton Analogue Computing Engine, 329

M-1, 339
M-33, 77
M-5, 350
M-9, 66
machine
– time, 153
– tool control, 276
– unit, 73
– variable, 162
machine learning, 362
MACNEE, A. B., 67
MADDIDA, 217, 218
Mader-Ott, 29
maglev, 317
magnetic bearing, 270
majority voting, 229
mark-space multiplier, 109
marshaling hump, 317
mass-spring-damper system, 168
Massachusetts Institute of Technology, 32, 246
mathematics, 243
MATHIEU equation, 68, 190
MATHIEU functions, 190
MATHIEU, ÉMILE LÉONARD, 190
MATLAB, 240
MBB Aircraft Division, 331
MCCANN, GILBERT D., 69
McDonnel Aircraft Corporation, 346
Mean Time between Failure, 85
mechanical
– analog computer, 9
– integrator, 21
– multiplier, 24

mechanics, 266
mechanism
– crank, 271
medicine, 284
MEL APT, 331, 332
memristor, 361
Mercury, 345
metabolism, 285
Metal Oxide Semiconductor, 262
MEYER-BRÖTZ, GÜNTER, 83, 142
MICHELSON, ALBERT ABRAHAM, 28
MIDAS, 233
Mike Boat, 319
military applications, 347
MILLS, JONATHAN W., 358
MIMIC, 234
Minden, 37
Miniscpace, 83
Mischgerät, 42, 45
MISES, RICHARD VON, 249
missile
– Jupiter, 57
– Redstone, 48, 57, 330
MIT, 32, 67, 246
– Instrumentation Laboratory, 337
mixing tank, 309
ML-1, 282
model, 2
Model-1, 206
modulation multiplier, 109
modulator, 304
Monetary National Income Automatic Computer, 292
MONIAC, 292
Monte-Carlo, 251, 252, 259
MORELAND, W. J., 321
MOSFET, 262
motor coach simulation, 317
MTBF, 85
multi-compartment model, 285
multidimensional shapes, 253
multiplication, 105
multiplier, 24, 105
– HALL effect, 114
– crossed-fields, 106
– dynamometer, 114
– electron-beam, 106
– Gilbert cell, 114
– heat-transfer, 114

– hyperbolic field, 107
– logarithmic, 110
– mark-space, 109
– mechanical, 24
– modulation, 109
– pulsed-attenuator, 109
– quarter square, 111
– servo, 105
– strain-gauge, 114
– time division, 109
MUNK, FRANZ, 278
Museum of Electronic Games & Art, 353
music, 354
muzzle velocity, 67
MX-775A, 217

N. S. Savannah, 281
NACA, 330
NAGUMO, JIN-ICHI, 287
NAPIER, JOHN, 11
NASA, 321
National Physical Laboratory, 292
nautics, 318
Naval Undersea Center, 320
Nederlands Radar Proefstation, 305
negative
– feedback amplifier, 75
negative feedback, 75
network analyser, 298
neuromorphic computing, 362
neurophysiology, 286
neutron
– delayed, 280
– density, 281
– flux density, 279
– kinetics, 280
– lifetime, 281
– prompt, 280
Nike, 77, 336
noise generator, 120
non-destructive testing, 272
non-linear servo system, 277
normalization, 151
North American Aviation, 326
Northrop, 209, 217
NOVA, 339
NPL, 292
nuclear
– technology, 278

– weapon effect computer, 13
Nyquist diagram, 277
Nyquist, Harry, 277

OMS 811, 256
OP, 90
open
– amplifier, 102
– loop gain, 82
operate, 90, 91
operational amplifier, 63, 73, 74
– drift stabilisation, 80
– Nike, 77
operational flight trainer, 224
Oppikofer, Johannes, 15
optics, 258
optimisation, 252, 311
OR, 230
orbit calculation, 255
orthogonal function, 247
oscilloscope, 121
Oslo analyser, 21, 33
Ottermann, Joseph, 251
Oughtred, William, 11
output
– device, 121
– multiplier, 214

PACE, 139
PACER 500, 148
Packard Bell, 225
PACTOLUS, 234, 303
Padé approximation, 119
Padé, Henri, 119
parachutes, 338
parallel DDA, 209
parameter determination, 311
parametron, 190
Parkinson, David B., 61
partial
– differential equation, 157
– feedback technique, 155
– tide, 25
particle trajectory, 255
patch
– cable, 123
– panel, 123
PB-250, 226
PDE, 158

Peenemünde, 42, 133
phase space, 189
PHI-4, 336
Philbrick, George A., 57, 136
Phillips, Alban Williams, 292
photoformer, 99
physics, 253
– plasma, 255
Phytotron, 260
planimeter, 13
– hatchet, 15
– linear, 17
– polar, 15
– Prytz, 15
– wheel-and-cone, 15
planispheric astrolabe, 9
plasma physics, 255
pneumatic system, 273
Pogo effect, 267
polar planimeter, 15
Polaris, 337
pole, 16
– arm, 16
– weight, 16
polygon generator, 100
polynomial
– zeros of, 247
Polyphemus, 58
Porter, Arthur, 33
Pot Set, 94
potentiometer
– coefficient, 92
– digital, 206
– set, 94
– tapped, 98
power
– engineering, 295
– grid simulation, 298
– inverter, 296
– stations, 301
Pratchett, Terry, 292
precision analog computer, 123
Precision Analog Computing Equipment, 139
predator and prey, 172
pressurized-water reactor, 279
problem time, 163
process engineering, 308
product integraph, 245
profile tracer, 32

program, 151
programming, 151
– examples, 165
progressive precision, 231
Project Cyclone, 340
projection of rotating body, 194
prompt neutron, 280
propulsion system, 318
PRYTZ, HOLGER, 15
Pullman, 183
pulsed-attenuator multiplier, 109
pupil regulation, 286
PyAnalog, 240
Python, 364

Q-learning, 207
QK-256, 113
QK-329, 113
quadratron, 100
quadrature
– band-pass, 304
– signal pair, 165
quantum chemistry, 265
quarter square multiplier, 111
quotient of differences, 158

RA 1, 133
RA 463/2, 135, 274
RA 741, 169, 173, 256
RA 770, 85, 121, 127, 147, 257, 279
RA 800, 144, 145
RA 800H, 147
radar, 305
radio telescope, 277
RAGAZZINI, JOHN, 74
railway vehicle, 317
RAIMANN, FRANZ, 350
RAM, 118, 264
Ramo-Wooldridge Corporation, 201, 321
random
– access memory, 118, 264
– noise generator, 120
– process, 251
RANFFT, ROLAND, 303
RASCEL, 231
RAT 700, 142, 271
rate check, 206
ray tracing, 292
Raytheon, 113

REAC, 321, 339
reaction
– kinetics, 264
– wheel, 344
real-time, 163
rectifier, 296
Redstone, 48, 57, 330
– Arsenal, 48, 329
Reeves, 320
– Instrument Corporation, 321, 340
regulation
– CO_2, 286
– pupil, 286
REIN, HANS-MARTIN, 303
relay comparator, 115
repetitive operation, 58, 131
resolver, 117
resources, 290
respiratory arrhythmia, 285
Rhythmograph, 350
ride simulation system, 315
ring modulator, 43
ROBINSON, TIM, 35
ROCKEFELLER differential analyser, 37, 360
Rockefeller Foundation, 34
rocket
– A4, 41, 329
– motor simulation, 338
– NOVA, 339
– Saturn, 339, 341
– simulation, 339
rocketry, 338
Rohde & Schwarz, 306
ROSE, R. M., 287
ROSSELAND, SVEIN, 33
rotating system, 269
rotor blade, 323
Royal Aircraft Establishment, 224
RUBEL, LEE ALBERT, 358, 363
rudder
– air, 44
– exhaust, 44
RUTHERFORD, ERNEST, 255

sampling system, 307
Saturn, 339, 341
scaling, 151, 162
Schoppe & Faeser, 38
SCHOTTKY diode, 303

SCHOTTKY, WALTER HANS, 303
SCHRÖDINGER equation, 265
SCHRÖDINGER, ERWIN, 265
scotch yoke, 26, 271
SCR-584, 66
SEDLACEK, JOSEF ADALBERT, 12
SEIDEL, PHILIPP LUDWIG VON, 248
SEIR model, 174, 288
seismology, 291
semiconductor research, 262
separation of variables, 160
sequential DDA, 210
SERAC, 269
servo, 213
– loop, 213
– multiplier, 105
– system, 277, 307
  – linear, 277
  – non-linear, 277
setback leaf system, 347
shape
– multidimensional, 253
Sharon Steel Corporation, 272
ship simulation, 319
shock absorber, 268, 313
Short Brothers, 320
sidelobe, 305
Sigma-5, 334
silicon carbide, 79
simplex, 12
SIMSTAR, 360
simulation, 2
– airborne, 334
– circuit, 301
– earthquake, 268
– epidemic, 174
– flight, 325
– in-flight, 334
– motor coach, 317
– power grid, 298
– rocket, 339
– rocket motor, 338
– ship, 319
– torpedo, 320
– traffic flow, 316
simulator, 57
Simulink, 240
simultaneous
– DDA, 209

– operation, 205
sin(), 165
SINGLE, C. H., 119
Sirutor, 43
slide rule, 11
– body, 12
– center slide, 12
– cursor, 12
– duplex, 12
– electronic, 13
– end
  – brace, 12
  – bracket, 12
– simplex, 12
– stator, 12
– stock, 12
slipstick, 11
slot, 231
SNR, 209
Solartron, 83
– Minispace, 83
SOUTHERN, JOHN, 14
Space Technology Laboratories, 201
spacecraft manoeuvres, 343
spider block, 21
SPIESS, RAY, 349
SPRAGUE, RICHARD, 218
squaring cam, 18
SSM-A-3 Snark, 218
STARDAC, 224
Starfighter, 336
static
– check, 172, 206
– test, 130
stator, 12
steam engine indicator, 14
STEELE, FLOYD, 218
steering system, 313
STEINHOFF, ERNST, 42
step integrator, 39
STEUDING, HERMANN, 42
STIELTJES integral, 158
STL, 333
stochastic
– computing, 229
– number
  – bipolar, 230
  – unipolar, 229
stock, 12

strain-gauge multiplier, 114
STRATTON, SAMUEL WESLEY, 28
strip-chart recorder, 121
Strong Earthquake Response Analog
    Computer, 269
STUBBS, G. S., 119
Stubbs-Single approximation, 119
Stufenintegrator, 39
Sturm, Charles-François, 261
substitution method, 155
summer, 85, 214
summing amplifier, 85
surge
– chamber, 273
    – differential, 273
– tank
    – Appalachian, 274
SWARTZEL, KARL D., 85
switch
– crossbar, 37
Symphony, 343
system
– cardiovascular, 285
– embedded, 308
– hydraulic, 273
– locomotor, 289
– pneumatic, 273
– propulsion, 318
– rotating, 269
– sampling, 307
– servo, 307
– steering, 313
– transport, 313
Systems Engineering Conference, 309
systems of linear equations, 248
Systron Donner, 353

T-10, 66
T-15, 67
Tactical Avionics System Simulator, 330
tank
– electrolytic, 363
tapped potentiometer, 98
target scintillation, 305
TASS, 330
TEICHMANN, ALFRED, 324
telecommunication, 301
Teledyne, 61
Teledyne-Philbrick, 61

Telefunken, 42, 83, 125
– DEX 100, 256
– HKW 860, 147
– HRS 860, 297
– OMS 811, 256
– RA 1, 133
– RA 463/2, 135, 274, 281
– RA 741, 169, 173, 256
– RA 770, 85, 121, 127, 147, 257, 279
– RA 800, 144, 145
– RA 800H, 147
– RAT 700, 142, 271
TELLEGEN, BERNARDUS DOMINICUS
    HUBERTUS, 75
Tennessee Valley Authority, 274
Tennis for Two, 352
THAT, 150
THE ANALOG THING, 150, 349
THEODORSEN, 197
THOMSON, JAMES, 22
THOMSON, MARGARET, 25
THOMSON, WILLIAM, 22
three-dimensional
– analogue computer, 327
– cam, 19
thyrite, 79
tide, 25
– partial, 25
time
– delay, 117
– division multiplier, 109
– problem, 163
– real, 163
time-scale factor, 92
torpedo
– data computer mark 3, 31
– simulation, 320
torque amplifier, 22
TR-10, 146
TR-48, 309
tracer arm, 16
traffic flow simulation, 316
transformer, 296
transmission lines, 297
transmissions, 314
transport systems, 313
TREFFTZ, ERICH, 197
TRICE, 225
TRIDAC, 327

TROPSCH, 309
TRS-2, 306
true, 230
trunk line, 124
TRW Systems Group, 346
tune-up time, 272
tunnel diode, 262

unipolar stochastic number, 229
UNIVAC
– 1103A, 343
– incremental computer, 224
– 1110, 318, 320
– 1230, 320
University of
– Chicago hospital, 290
– Virginia, 259
US-Air Force, 326
USS Monitor, 30

V1, 66
value coefficient, 252
VAN VARK, TATJANA JOËLLE, 28
VAN VEEN, JOHANN, 276
variable-stability aircraft, 335
Variplotter, 140
varistor, 79, 100
VDFG, 101
vibrations, 267
vibrator, 81
VJ 101C-X2, 331

VLSI, 363
VOIGT, PAUL, 75
VOLTERRA, VITO, 172
VON BRAUN, WERNHER, 42
VON NEUMANN, JOHN, 220, 229

WAINWRIGHT, LAWRENCE, 244
war
– cold, 13, 347
– second world-, 30
water hammer, 275
WATT, JAMES, 14
wavelet transform, 250
weathering layer, 291
WEINEL, ERNST, 34
wheel-and-cone planimeter, 15
WIENER, NORBERT, 245, 304
windage, 67
winds unit, 262
Wright Air Development Center, 321

X-15, 201, 332
Xenon-poisoning, 279

Z3, 324
ZENER diode, 79
zeros of polynomials, 247
ZHUKOVSKY, NIKOLAY YEGOROVICH, 197
ZI-S, 290
ZUSE, KONRAD, 324

Printed in the USA
CPSIA information can be obtained
at www.ICGtesting.com
JSHW061416010524
62308JS00008B/231